Particle Analysis in Oceanography

NATO ASI Series

Advanced Science Institutes Series

A series presenting the results of activities sponsored by the NATO Science Committee, which aims at the dissemination of advanced scientific and technological knowledge, with a view to strengthening links between scientific communities.

The Series is published by an international board of publishers in conjunction with the NATO Scientific Affairs Division

A Life Sciences B Physics	Plenum Publishing Corporation London and New York
C Mathematical and Physical Sciences D Behavioural and Social Sciences E Applied Sciences	Kluwer Academic Publishers Dordrecht, Boston and London
F Computer and Systems Sciences G Ecological Sciences H Cell Biology I Global Environmental Change	Springer-Verlag Berlin Heidelberg New York London Paris Tokyo Hong Kong Barcelona Budapest

NATO-PCO DATABASE

The electronic index to the NATO ASI Series provides full bibliographical references (with keywords and/or abstracts) to more than 30000 contributions from international scientists published in all sections of the NATO ASI Series. Access to the NATO-PCO DATABASE compiled by the NATO Publication Coordination Office is possible in two ways:

- via online FILE 128 (NATO-PCO DATABASE) hosted by ESRIN, Via Galileo Galilei, I-00044 Frascati, Italy.
- via CD-ROM "NATO-PCO DATABASE" with user-friendly retrieval software in English, French and German (© WTV GmbH and DATAWARE Technologies Inc. 1989).

The CD-ROM can be ordered through any member of the Board of Publishers or through NATO-PCO, Overijse, Belgium.

Series G: Ecological Sciences Vol. 27

Particle Analysis in Oceanography

Edited by

Serge Demers

Institut Maurice-Lamontagne
Ministère des Pêches et des Océans Canada
850 Route de la Mer
Mont-Joli, Québec G5H 3Z4
Canada

Springer-Verlag
Berlin Heidelberg New York London Paris Tokyo
Hong Kong Barcelona Budapest
Published in cooperation with NATO Scientific Affairs Division

Proceedings of the NATO Advanced Study Institute on Individual Cell and Particle Analysis in Oceanography held at Acquafredda di Maratea, Italy, October 21–30, 1990.

Library of Congress Cataloging-in-Publication Data
NATO Advanced Research Workshop on Individual Cell and Particle Analysis in Oceanography (1990 : Acquafredda di Maratea, Italy) Particle analysis in oceanography / edited by Serge Demers.—(NATO ASI series. Series G, Ecological sciences ; vol. 27) "Proceedings of the NATO Advanced Research Workshop on Individual Cell and Particle Analysis in Oceanography held at Acquafredda di Maratea, Italy, October 21–30, 1990"—T.p. verso. "Published in cooperation with NATO Scientific Affairs Division." Includes index.

ISBN-13: 978-3-642-75123-3 e-ISBN-13: 978-3-642-75121-9
DOI: 10.1007/978-3-642-75121-9

1. Optical oceanography—Congresses. 2. Flow cytometry—Congresses. I. Demers, Serge, 1951– . II. North Atlantic Treaty Organization. Scientific Affairs Division. III. Title. IV. Series: NATO ASI series. Series G, Ecological sciences ; no. 27. GC177.6.N38 1990 551.46'0028—dc20 91-27492

Softcover reprint of the hardcover 1st edition 1991

Typesetting: Camera-ready by authors
31/3140-543210 – Printed on acid-free paper

Table of Contents

From cell to Oceans

Flow cytometry in oceanography

Conclusion

NATO Advanced Study Institute
on Individual cell and particle analysis in oceanography
Hotel Villa del Mare, Acquafredda di Maratea, Italy
October 21-30, 1990

Acosta Pomar, Lucila; Agusti, Susana; Amann Rudolf; Anderson, Donald; Ault, Kenneth; Azam, Farooq; Babin, Marcel; Balfoort, Hans; Bannister, Thomas T.; Benedetti, Alberto; Berman, Thomas; Binder, Brian; Blaise, Christian; Blanco, Juan; Boyd, Carl; Bruce, Barry D.; Burkill, Peter; Button, D. K.; Campbell, Lisa; Casotti, Raffaella; Chisholm, Sally; Cucci, Terry; Cunningham, Alexander; De Long, Edward; Decembrini, Franco; Demers, Serge; Denis, Michel; Denizeau, Francine; Dinet, Marie-Josèphe; Dubelaar, George B.; Dubinsky, Zvi; Dusenberry, Jeffrey; Estrada, Marta; Falkowski, Paul; Frankel, Sheila; Frenette, Jean-Jacques; Gagnon, Réal; Gieskes, Winfried; Gilstad, Mona; Goericke, Ralf; Gudmundsson, Kristinn; Happey-Wood, Christine M.; Harrison, Paul J.; Heldal, Mikal; Hofstraat, J.W.; Holligan, Patrick; Iturriaga, Rodolfo; Jochem, Frank; Johnsen, Geir; Kachel, Volker; Kana, Todd M.; Koray, Tufan; Kraay, Gysbert; Krigsman, Johan; Laflamme, Manon; Lazzara, Luigi; Lefevre, Dominique; Legendre, Louis; Leppanen, Juha-Markku; Levasseur, Maurice; Lewis, Marlon; Li, William; Lipschultz, Fred; Maestrini, Serge Y.; Marino, Donato; Marrase, Celia; Martin, David; Mazel, Charles; Mazumder, Asit; Medlin, Linda; Monfort, Patrick; Montresor, Marina; Morel, André; Olson, Robert; Orellana, Monica; Owens, Thomas; Paran, Nava; Partensky, Frederic; Peeters, Kees; Perry, Mary-Jane; Rassoulzadegan, Fereidoun; Roesler, Collin; Roy, Suzanne; Sakshaug, Egil; Salzman, Gary; Sandvick, Roar; Shumway, Sandra E.; Smith, David C.; Sosik, Heidi; Steen, Harald; Stramsky, Darius; Taguchi, Satoru; Taniguchi, Akira; Taran, Glen; Therriault, Jean-Claude; Valet, Gunter; Vaulot, Daniel; Veldhuis, Marcel; Wehrmeyer, Werner; Wood, Michelle A.; Wyatt, Timothy; Yentsch, Clarice; Yentsch, Charles; Zingone, Adriana; Zohary, Tamar

Foreword

Investing in new technology is part of normal business in marine science. However, getting involved in flow cytometry at a time of fiscal restraint and reduced capital expenditures took a lot of talking and convincing by Serge Demers. He did such a good job of it that the Maurice Lamontagne Institute (MLI), a new research center of the Canadian Ministry of Fisheries and Oceans, rapidly became a strong supporter of this promising tool that is opening new frontiers in the study of individual cells and particles.

When Serge came to see me a few years ago to examine the possibility of holding a major workshop on this topic, he received a warm response and encouragement to proceed with the project. In retrospect, I must admit that allowing Serge to assume the responsability of organizing an international meeting in Italy, while he had no such prior experience was worrying me a bit. I am delighted that the meeting was a great success, owing to the dynamism of the participants and the energy of those who organized it (see Preface).

The Maurice Lamontagne Institute is proud to have been associated with such a success. MLI is a young organization, having opened its doors in December 1986. We are located in Mont-Joli, 350 km to the North-East of Quebec City, on the south shore of the St. Lawrence Estuary. Our activities cover marine biology, physics, and chemistry, and hydrography, with a major emphasis on inland seas under strong estuarine influence. These encompass the Estuary and Gulf of St. Lawrence in eastern Canada as well as James, Hudson and Ungava Bays, and Foxe Basin in the Canadian Arctic.

Our major thrusts of research include the study of marine food chains, fish stock assessments, marine mammals, aquaculture, aquatic pollution, remote sensing and movement of water masses. We are also responsible for the production of navigational charts in the Province of Québec. These activities are conducted in estuarine and marine waters.

With a staff of 250, MLI is rapidly becoming a major contributor to ocean research in Canada. This occurs at a time when these vast expanses are the object of increased interest from the scientific community with respect to their role in the global warming of the planet, and when coastal areas are subject to increased focus due to the negative consequences of human activity on the most productive part of the oceans.

It is my hope that the proceedings from this important meeting will remain for many years a landmark for those wishing to use flow cytometry as a powerful tool in biological oceanography. Also, that scientists like Dr. Demers in laboratories around the world will continue to press sometime hesitant Institute directors to invest part of their resources in new ideas or unproven techniques and tools.

Jean Boulva
Director
Maurice Lamontagne Institute

Preface

Individual cell and particle analysis in aquatic sciences is involved in many aspects of oceanography and limnology, including optical physics of particles, phytoplankton physiology and ecology, marine and aquatic microbiology and food web interactions. This book concentrates on the optimal utilization of flow cytometry and image analysis and the ways in which oceanographic and limnological problems can be uniquely or better addressed using these techniques.

The book stresses the need to work on merging scales in order to increase our understanding of the productivity of aquatic systems and emphasized the use of flow cytometry as a powerful tool for rapidly integrating large amounts of information over a wide range of particle scales. This new approach should help in finding adequate mechanistic solutions to problems of measurement of primary production at small (individual cells) as well as at global scales. Advancement in understanding of the aquatic system appears to be linked to such development. For example, fluorescent oligonucleotide probes recognize specific sequences of ribosomal RNA. Depending of their base composition they can differentiate cells at the kingdom (e.g. eukaryotes vs prokaryotes), group (e.g. sulfate reducing bacteria) or even finer taxonomic levels. This approach appears extremely powerful since probes can be devised to be as specific or as general as needed. Its coupling with flow cytometry will allow to assess in very fine detail the structure of microbial food webs in aquatic environments. These new probes also open up the possibility to better understand how oceanic communities adapt to a changing and unpredictable environment at the genetic level.

In the future, it is recommended that some members of the international community of ocean scientists continue the transfer of techniques and methodologies developed in other fields, taking into account the challenges of analyzing marine samples. These techniques include for example, immunological probes, oligonucleotide probes, fluorescent probes to stain specific components or activities of the cell and multivariate numerical analysis. However, we should not lose sight of the overall scientific goal, which is to resolve major issues in oceanography by understanding them at the level of the individual organism. These issues include global fluxes of particles in the ocean, food web dynamics and mechanisms, effects of pollution and more generally, anthropogenic impacts on the marine environment. In this regard, techniques for individual cell and particle analyses will produce maximum results and understanding if used in connection with the full complement of traditional oceanographic methods as well as recently developed techniques such as HPLC, remote sensing and so on.

The meeting, which was held in Acquafredda di Maratea in Italy in October 1990, was made possible by a grant from NATO Scientific Affairs Division which covered the major part of the expenses. Other organizations also contributed financially to the success of the meeting. These organization are: The Maurice Lamontagne Institute (a governmental laboratory of the Department Fisheries and Oceans Canada), the Centre Saint-Laurent (a governmental laboratory Department of Environment Canada), NASA (USA), ONR (USA), NSF (USA), CNRS (France), European Economic Community, the tourist agency of Basilicata area (Italy), Becton Dickinson, Coulter and Zeiss. I would like to thanks all these organizations for their generous contributions.

The organization of a large workshop is a huge task requiring a lot of time of many people. I would like to thank the organizing committee, Peter Burkill (United Kingdom), Sally Chisholm (USA), Marlon Lewis (Canada), Donato Marino (Italy), Daniel Vaulot (France) and Clarice Yentsch (USA). I would like to address a special thanks to Manon Laflamme (Canada) who made a considerable contribution to the organization of this meeting. The staff of the hotel Villa del Mare and particularly Mister Guzzardi and Sussi Travisano are gratefully acknowledged for their effort in organizing all the social events and in making the stay of the participants as pleasant as possible. I would like to thank Laure Devine for her help during the editing process and Lynn Cleary from the Centre Saint-Laurent for her support.

Serge Demers
Maurice Lamontagne Institute
Fisheries and Oceans Department
850, Route de la Mer
Mont-Joli, Québec
Canada, G5H 3Z4

Instrumentation and calibration

FLOW CYTOMETRY INSTRUMENTATION

H. B. Steen

Dept. of Biophysics
Cancer Research Institute
The Norwegian Radium Hospital
Oslo, Norway

INTRODUCTION

The uniqueness of flow cytometry lies in the ability to perform quantitative measurements of individual cells in large numbers. By recording the fluorescence and light scattering of each cell as it is carried through the excitation focus of the flow cytometer in the course of a few microseconds, the cellular content of specific fluorescent constituents, or constituents that have been fluorescently labelled, can be determined together with information relating to cell size and structure. This ability to quantitatively characterize individual cells obviously has many important applications in marine microbiology. Although flow cytometers have been used to study marine microorganisms for several years, this field of applications is still in its infancy. That may in part be due to the technical problems encountered when trying to put the flow cytometer to work under the field conditions of a seagoing research vessel and on a type of specimens which in some respects are quite different from what current instruments were designed to handle. Flow cytometry was originally developed for measurement of cells in immunology and research in the biology of mammalian cells. Although it is being applied in other fields of biology as well, the bulk of applications is still limited to mammalian cells, a fact which is reflected in the design of the instruments that are now on the market. The majority of mammalian cells lie within a relatively narrow size range, i.e. 5 to 20 μm. With few exceptions they take on a spherical shape in suspension. They all have the same general chemical composition, and any endogenous fluorescence they may have is typically only a weak background, which is usually not a point of investigation in itself. Current flow cytometers were designed to measure this kind of material within the benevolent conditions of the typical research laboratory.

NATO ASI Series, Vol. G 27
Particle Analysis in Oceanography
Edited by S. Demers

In contrast, marine microorganisms span the size range from sub-micron to hundreds of microns, they have all kinds of intricate shapes, they vary greatly with regard to the chemical composition, and they may exhibit an endogenous fluorescence which can be much stronger than that of dyes used to stain specific components of the cell. Many applications of marine microbiology require that the cells are measured as soon as possible after being harvested, that is in the often harsh field conditions of a vessel at sea. Hence, this science makes significant new demands on the technology of flow cytometry. We shall have to develop instruments which can measure sub-micron organisms which are outside the detection limit of most current flow cytometers. At the same time these instruments should be able to handle cells several hundreds of microns in size. That will require new flow chambers and optics which can illuminate and collect the light from an object of this size. Non-spherical, asymmetrical cells create measuring artifacts in current instruments, especially the laser-based ones. To be able to measure organisms reproducibly and reliably, we have to solve this problem. In order to measure some of the endogenous fluorophores of many marine organisms, such as chlorophyll and phycocyanins, we have to employ other light sources and develop new optical filters. And all of this has to be packed into a device which is sufficiently compact and rugged to be practical for field use. All together, it appears that marine microbiology is a greater technical challenge to flow cytometry than the mammalian cell biology and immunology for which this technology was originally developed. But to the extent that this challenge can be met, flow cytometry represents a unique opportunity to harvest large amounts of scientific information from a very important and relatively unknown part of the biosphere.

In this presentation I shall discuss some of the principal aspects of flow cytometers and some of the instrument parameters which determine their performance, especially with regard to sensitivity and resolution. Various basic designs of flow cytometers are described and design specifications which are particular to applications in marine microbiology are discussed.

BASIC CONSIDERATIONS

Common to all flow cytometers is a narrow, laminar flow of water which carries the cells, one by one, through an intense focus of excitation light. The resulting fluorescence, which can stem either from natural pigments or from dyes used to label specific cell constituents, is collected by suitable optics and directed onto sensitive detectors which transform the light pulses into equivalent electrical pulses. The scattered light, which provides information on cell size and structure, is measured at various scattering angles by separate detectors. The electrical pulses from the detectors are measured and digitized by appropriate electronics and the information stored in a computer, which typically presents it in the form of histograms showing the number of cells as a function of cellular fluorescence and/or the light scattering.

Flow cytometers have a remarkable performance. The measuring precision per cell can be about 1%, the detection limit around 1000 molecules per cell, and the measuring rate up to 10 000 cells s^{-1}. However, the performance depends on the instrument characteristics as well as the application. The following discussion will show how the performance depends on some of the basic instrument characteristics.

Sensitivity and resolution: The sensitivity of the flow cytometer is determined primarily by two factors: the magnitude of the signal produced by a given cell or particle and the background noise on which this signal is to be measured. Thus, the sensitivity is essentially a measure of the signal to noise ratio, S/N, under given conditions; in order to increase sensitivity, it is as important to reduce the noise as to increase the signal. As expressed by Eq.1, the number of fluorescence photons emitted by a cell, N_f, is proportional to the intensity of the excitation light, I, as well as to the time spent by the cell in the excitation focus, that is inversely proportional to the flow velocity. This holds true only to the extent that there is no photodegradation, i.e. bleaching, of the fluorophore during the exposure. Furthermore, N_f is proportional to the overlap between the spectrum of the excitation light, $\nu(\lambda)$, and the absorption spectrum of the fluorophore, $\epsilon(\lambda)$. Hence, fluorescence sensitivity may vary significantly between applications which employ different dyes.

$$N_f = \text{const}_1 \cdot I\, v^{-1}\, \nu(\lambda)\, \epsilon(\lambda) d \qquad (1)$$

The intensity of the scattered light, N_l, is also proportional to the excitation intensity and inversely proportional to the flow velocity. In a first approximation, the light scattering of biological cells is independent of their light absorption. In contrast to the fluorescence, which is largely isotropic (although this depends on the degree of polarization of the excitation light as well as that of the fluorescence), the light scattering is strongly dependent on scattering angle (See Fig.6). N_l is a complex function of the size and shape of the cell, its composition, and the refractive index of its various constituents, as described by the theory of light scattering. It also depends on the excitation wavelength.

Each cell emits typically many thousands of photons of fluorescence and scattered light as it passes through the excitation focus. Only a relatively small fraction of these photons fall within the aperture of the fluorescence collecting optics. The number of fluorescence photons, n_f, that reach the light detector depends on the numerical aperture, NA, of this optics and the optical transmission, T, of the lenses and filters in the optical path (Eq. 2):

$$n_f = \text{const}_2 \text{ NA T } N_f \qquad (2)$$

The light detector, which is usually a photomultiplier tube (PMT), converts the light pulses, each of which usually consists of many photons, into equivalent electrical pulses. The sensitivity of the PMT depends primarily on the photoelectron quantum yield, ϕ_e, of its photocathode, that is the probability that a photon shall cause emission of a photoelectron. ϕ_e, which varies with wavelength, usually does not exceed 0.2. Thus the average number of photoelectrons, n_e, produced by n_f photons is:

$$n_e = \phi_e n_f \qquad (3)$$

The electrons released from the photocathode are multiplied as they pass through the PMT, the multiplying factor increasing with the high voltage across the tube, so that the gain of the PMT may be varied typically from 10^2 to 10^7 depending on high voltage.

The noise of the signal is of two principally different kinds: optical noise and electronic noise. At least in principle, the latter can be reduced to insignificant levels by the use of appropriate electronics. In contrast, optical noise is inherent in all kinds of emission of light and cannot be reduced below a certain level. This noise stems from the fact that the emission of light is a stochastic process, which implies that photons are emitted randomly in time. The result is that even light emitted from a perfectly constant light source exhibit a certain fluctuation in the intensity given by Eq.4:

$$\sigma = N^{1/2} = (t \cdot I)^{1/2} \qquad (4)$$

where σ is the standard variation of a repetitive measurement of the number of photons, N, emitted within the time, t, from a perfectly constant light source having an intensity I photons/s. In flow cytometry, t will be approximately equal to the duration of the pulse of the fluorescence or light scattering produced by each cell. Hence, the relative standard variation, or coefficient of variation, cv, is given by Eq.5:

$$cv = 100\ \% \cdot N^{-1/2} \qquad (5)$$

Hence, there is a principal limit to the precision by which a given amount of light can be measured. For example, a light pulse containing 100 photons can be measured with a precision, i.e. cv, no better than 10 %. In practice the precision is even worse than this because the emission of photoelectrons in the detector is also a stochastic process. Assuming $\phi_e = 0.2$ (see above), 100 photons will release no more than 20 electrons from the PMT photocathode, and the relative standard variation in this number is $20^{-1/2}$, or cv = 22 %.

In addition to the uncertainty in the light pulse itself comes the intensity fluctuation of the (constant) background of light caused by fluorescence from optical components, imperfect filters, etc. Thus, if the number of photons reaching the detector from a cell is n_f and the number of photons from the background during the same time period is n_b, the relative standard variation in the number of photons reaching the detector is:

$$cv_p = 100\ \% \cdot (n_f + n_b)^{1/2}/n_f \qquad (6)$$

This adds to the variation associated with the emission of photoelectrons in the photomultiplier detector:

$$cv_d = 100\ \% \cdot [(n_f + n_b)\phi_e]^{1/2}/n_f \cdot \phi_e \qquad (7)$$

so that the total variation, cv_t, becomes:

$$cv_t = 100\ \% \cdot (cv_p^2 + cv_d^2)^{1/2} = 100\ \% \cdot [(1 + 1/\phi)\cdot(n_f + n_b)]^{1/2} / n_f \qquad (8)$$

The equations 2 - 8 apply to light scattering as well as fluorescence. Eq. 8 shows how the magnitude of the signal as well as that of the optical background limit the precision, or resolution, and thereby the sensitivity of flow cytometric measurements.

A practical definition of the detection limit is when $cv_t = 100\ \%$. It is seen from Eq. 8 then that even in the absence of background the limiting value of n_f is 7.7 (assuming $\phi_e = 0.15$). If the detection optics has NA = 0.6, which is typical for instruments with jet-in-air flow chamber, and T = 0.5, only about 2 % of the photons emitted from the cell reach the detector. So that with such an instrument a cell must emit at least 385 photons of fluorescence in order to produce a measurable signal, even in the unrealistic case of no background. In an instrument having oil immersion detection optics with NA = 1.3 the corresponding numbers are lower by a factor of 4.7.

Obviously, it is essential to reduce the optical background as much as possible. This background is mainly due to excitation light which is scattered from the jet or flow chamber into the detection optics where it may leak through imperfect filters as well as induce fluorescence from lenses and other components in the optical path. There is also a certain level of raman scattered light, that is, an inelastic scattering primarily from the molecules of the water. The raman scattering typically has a wavelength which overlaps the fluorescence and can therefore not be easily eliminated by filters. There are primarily two ways to reduce the background: 1) designing the flow chamber and the optics so as to scatter as little light as possible in the direction of the detection optics, and 2) reducing the measuring volume, that is the volume of water and solid materials which is exposed to excitation light and *seen* by the detector(s). Ways to achieve this in practice are discussed below.

FLOW CYTOMETRIC INSTRUMENTATION

Laser-based flow cytometers: The laser-based flow cytometers that are currently commercially available all have the general optical configuration shown in Fig. 1. The laser beam is usually collimated by means of two lenses, one of which is cylindrical, to form a flat focus of nearly parallel light. The focus typically has a width of 100 to 200 μm and a length in the direction of the flow of 10 - 20 μm. The fluorescence as well as the large angle, i.e. 90°, light scattering are collected by a lens in the direction perpendicular to the flow and the laser beam.

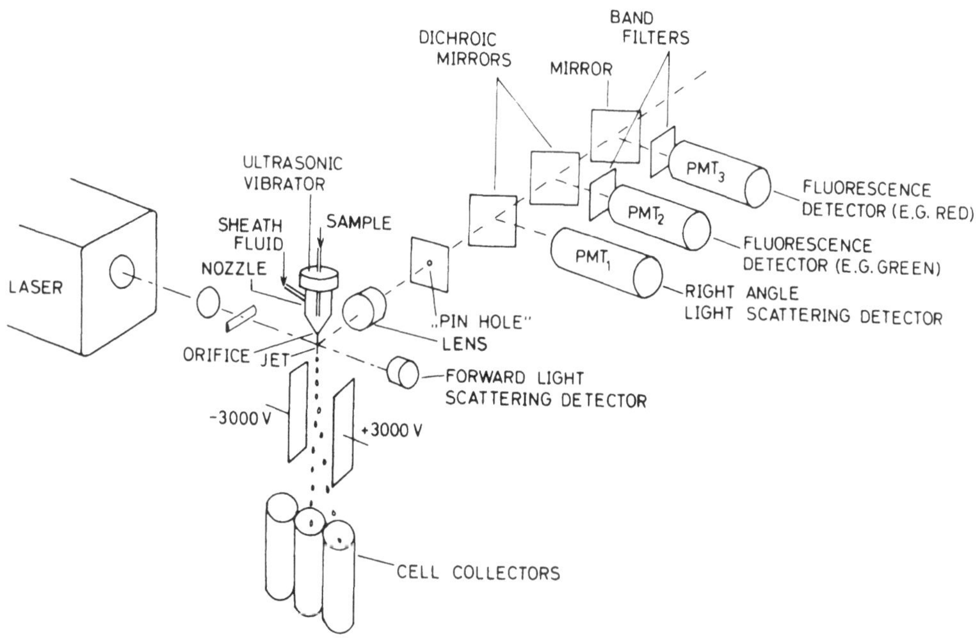

Figure 1. Laser-based flow cytometer with sorting facilities. Dichroic mirrors separate the large angle light scattering, which is detected at right angles to the laser beam, and the various spectral components of the fluorescence. Low angle light scattering is measured by a detector situated close to the forward direction of the laser beam. Sorting of cells according to the signals they give off as they pass through the excitation focus is carried out as follows: a piezoelectric crystal on top of the flow chamber oscillates with a frequency which is typically 40 kHz and thereby causes the jet of water to break into 40 000 droplets per sec just below the focus. The droplets pass through an electric field of a few thousand V/cm, which is maintained between two plates. As the droplets are at the point of breaking away from the jet, they are electrically charged either positively or negatively according to the measuring signal of the cell they contain, by an electrode inside the nozzle. Thus, they are pulled either right or left in the electrical field according to the fluorescence and/or light scattering properties of the cell.

Dichroic mirrors and band filters are employed to separate scattered light from the fluorescence as well as to separate different spectral components of the fluorescence. The low angle scattering is measured by a detector situated close to the forward direction of the laser beam. Hence, such an instrument has three independent, perpendicular axis which must be aligned so as to intersect in a common point to within a few micrometers. This makes the laser-based instruments susceptible to mechanical shock. The alignment is often a tricky and time consuming task, especially since none of the axis can be visualized directly with appropriate magnification.

A main advantage of the laser-based instruments is the very high excitation intensities that can be achieved, partly because of the very powerful lasers which are available, but also because laser light can be focused into a very small and correspondingly intense focus. This facilitates a small measuring volume and low background provided appropriate spatial filtering is employed (see below). Some laser-based flow cytometers have two lasers which are focused to separate foci so that cells can be excited sequentially by two different wavelengths. This device makes it possible to measure two dyes independently in the same cell without interference from spectral overlap and excitation energy transfer.

The near-parallel excitation light beam of laser-based flow cytometers facilitates light scattering measurement at almost any scattering angle. Several workers have developed instruments with light scattering detectors at several different scattering angles and have demonstrated an improved ability to distinguish different types of cells (for review see Salzman *et al.*, 1990). However, commercial instruments only have detectors for low angle and 90° scattering. The disadvantage of the near parallel excitation light of laser instruments is that it produces cell orientation dependent artifacts in the measurement of non-spherical cells (Pinkel *et al.*, 1985).

Laser light is usually polarized, and the orthogonal configuration of laser-based instruments facilitates polarization measurements (Arndt-Jovin *et al.*, 1976; Lindmo and Steen, 1977) which may be useful, for example, to study the binding of fluorochromes. Flow cytometric measurement of the polarization of the light scattering of biological cells has interesting possibilities which have not been exploited.

Arc lamp-based flow cytometers: The optical configuration of arc lamp-based flow cytometers is different from that of laser instruments in several respects (Fig. 2). The light from the high pressure mercury or xenon arc lamp is focused in the sample flow by an oil immersion microscope objective with high numerical aperture, i.e. NA = 1.3. That implies that the cone of excitation light is highly convergent. The arc lamp-based instruments employ epi-illumination, which is to say that the fluorescence is collected by the same lens as that

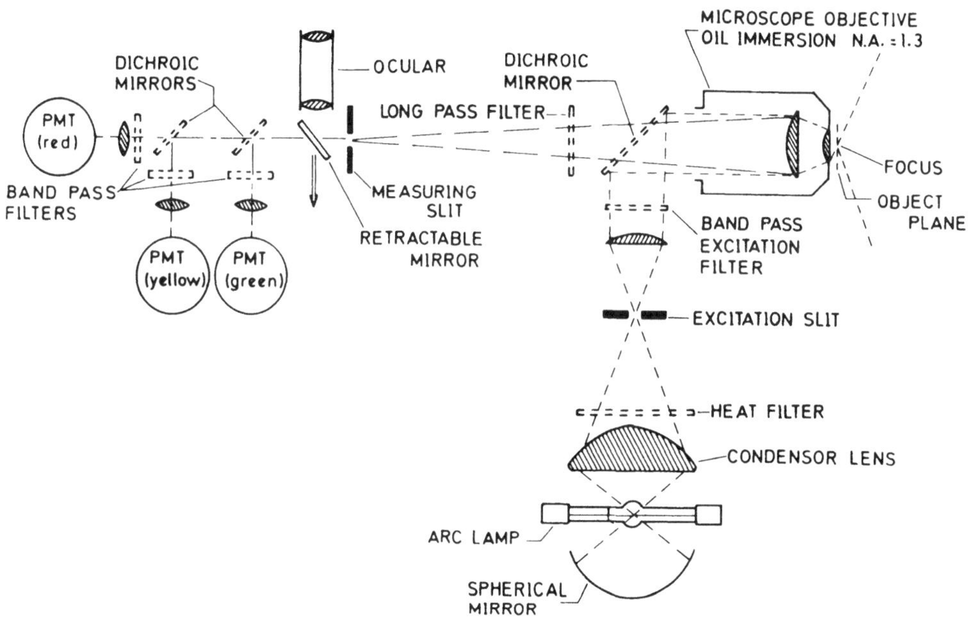

Figure 2. Epi-illumination optical configuration for arc lamp-based flow cytometers. The excitation filter isolates a relatively narrow band of excitation wavelengths from the essentially white light of the light source. The dichroic mirror reflects the excitation light into the microscope objective, while it transmits the fluorescence which has higher wavelength than the excitation light. The emission filter eliminates the traces of excitation light that leak through the dichroic mirror. The excitation slit and the measuring slit are used to eliminate background light from other parts of the focus than the sample flow itself (spatial filtering) and thereby increase the signal to noise ratio. Up to three different dyes or color components can be detected by separate detectors (photomultiplier tubes). Thus, the fluorescence is split by means of dichroic mirrors into the various wavelength components, which are further purified by band filters. To reduce filter fluorescence as much as possible the optical filters, especially the filter in the excitation light path, are usually interference filters, i.e. filters which reflect the light that is not transmitted rather than absorb it. The configuration shown here employs critical illumination: the light source is imaged in the object plane of the microscope objective, i.e. in the sample flow. Critical illumination provides somewhat higher intensity than is achieved with Kohler illumination, which is used in most microscopes. Kohler illumination has the advantage that the excitation intensity is less susceptible to lamp *flicker* - the sudden movements of the arc that is seen with some lamps.

used to focus the excitation light. Hence, excitation and fluorescence collection are automatically confocal with a common optical axis. This feature makes such instruments inherently much more rugged than laser-based ones. It also greatly simplifies the alignment, which is reduced to adjustment of the position of the lamp upon replacement and positioning of the flow chamber so that the sample flow runs through the focus. Alignment is facilitated by the fact that the sample flow can be seen through an ocular at high magnification, e.g. 500X.

The fluorescence can be split into various wavelength components by means of dichroic mirrors in the same way as in laser-based instruments. Light scattering is measured in a dark field configuration, as shown in Fig. 3. Low angle and large angle scattering is measured separately, but the large angle scattering is detected upward from 15° scattering angle rather than around 90° as it is in laser-based instruments.

Commercial arc lamp-based flow cytometers do not have provisions for excitation in two separate foci. One type (Pas II, Partec AG. Arlesheim, Switzerland) has facilities for sorting.

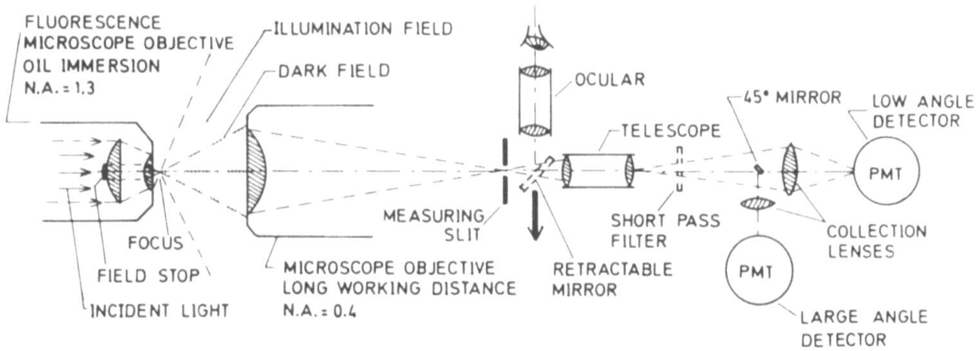

Figure 3. The optical configuration employed to measure light scattering in an arc lamp-based flow cytometer (Argus 100). A field stop in the primary microscope objective produces a conical shadow in the excitation light field. A secondary microscope objective situated with its aperture within this shadow, or *dark field*, thus *sees* only fluorescence and scattered light. The fluorescence, which is usually of insignificant intensity relative to the scattered light, can be eliminated by a suitable filter. The secondary microscope objective has a working distance of several millimeters to give room for the flow chamber. The scattered light is imaged on a *measuring slit* which covers only the image of the illuminated part of the sample flow and thereby eliminates the background of scattered light from other parts of the focus. A telescope situated behind this slit and a small 45° mirror separates the light scattered at low angles, i.e. upward from about 2°, from that scattered to larger angles, i.e. upward from about 15°, so that they are measured by two separate photomultiplier tubes (Steen, 1986).

***Light sources*:** The argon laser is the standard light source of laser-based flow cytometers. The large research instruments employ water-cooled lasers with a nominal output, that is, total power of all emission lines combined, of 2 to 5 W. These lasers can be tuned so as to emit at any of the emission wavelengths shown in Fig. 4A, although tuning to the lines in the UV, i.e. around 350 nm, may require some optional equipment. These lasers typically have a ripple below 1 % and good long term stability. However, they are expensive to buy and run. They consume 10 to 15 kW of electrical power and 2 to 4 gal/min of cooling water. With a plasma tube of about one meter in length they are susceptible to mechanical shock and vibration. Much smaller, air-cooled argon lasers are used in analytical instruments which are intended primarily for measurement of fluorescence conjugated antibodies on leucocytes and other mammalian cells (FACSscan, Becton Dickinson, San Jose, California, and Profile, Coulter Electronics, Inc., Hialeah, Florida). The lower emission power of these lasers, i.e. typically 10 to 50 mW, is partly compensated by the use of closed flow chambers and

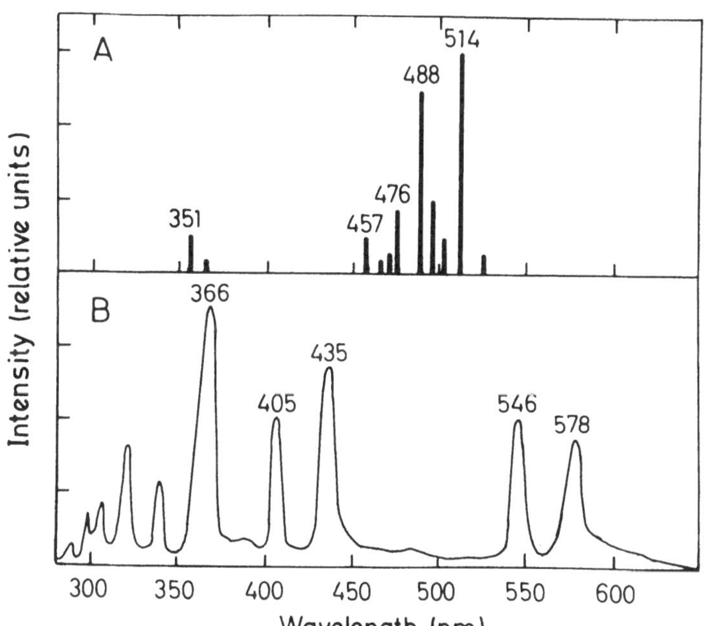

Figure 4. The emission lines of a tunable argon laser (A) and the emission spectrum of a high pressure mercury arc lamp (B). The scale of intensity is not the same in the two panels. The wavelengths of some of the most prominent lines are noted.

immersion optics. However, the output of these lasers have a ripple of a few percent which limits resolution, in terms of lowest cv. What may be more important in marine applications is that the lasers of such instruments are not tunable, but run at only one wavelength, namely 488 nm. This limits the range of applications. For example, chlorophyll is inefficiently excited, whereas dyes like Hoechst 33258, DAPI, and Chromomycin cannot be excited at all.

The alternative to lasers is the high pressure arc lamps containing either mercury, xenon or both. As shown in Fig. 4B, the Hg lamp, which is the most common, exhibits a number of strong emission lines superimposed on a continuum. Hence, it can be used for any dye which is excitable in the visible or near-UV. In particular, it has a strong line at 366 nm which is perfect for UV-excitable dyes like Hoechst 33258 and DAPI, and other strong lines at 436 and 546 nm, which coincides nicely with the excitation maxima of chlorophyll, phycocyanins and other pigments of marine organisms. The excitation intensity obtained with a 100 W Hg arc lamp in an instrument with epi-illumination and immersion optics is in the range 2 to 20 mW, depending on wavelength. The lowest power, about 2 mW, is in the region of excitation of FITC, i.e. around 490 nm. Nevertheless, the detection limit for FITC of such an instrument (Argus 100, Skatron AS, Tranby, Norway) is 2 000 - 3 000 FITC molecules, which is only twice the value achieved in the best laser-based instruments.

The 75 W Xe arc lamp, which produces a continuum covering all of the UV and visible, has a somewhat higher intensity around 490 nm than the Hg lamp, but has significantly less light at other wavelengths commonly used in the flow cytometry. It should be noted that no increase of excitation intensity is achieved by using lamps of higher power. Although larger lamps, say 200 or 500 W, emit more light, their larger arc exhibits a lower light density than that of the 100 W lamp. In addition, according to the laws of optics, a conventional light source cannot be focused to a higher light density than that of the source itself.

Provided a sufficiently stable power supply is employed, the short term stability of arc lamps is generally excellent. However, some lamps may exhibit *flicker*, that is, abrupt intensity changes of a few percent which is caused by arc movement and which occur with irregular intervals, typically of many minutes. On the long term, the intensity of Hg arc lamps decays at a rate of 1 - 2 % per day (8 hours). The effects of intensity fluctuations in arc lamp-based instruments can be eliminated by a monitoring system which measures the intensity and adjusts the signal amplifier gains so as to be inversely proportional to the intensity (Steen 1980). Under field conditions, arc lamps have an advantage in their low power consumption and resistance to mechanical shock and vibration.

Clearly, both lasers and arc lamps have their advantages and disadvantages. Lasers can produce higher intensities than arc lamps and have better long term stability. On the other hand, they have a limited number of emission wavelengths, which limits the selection of dyes that can be employed. The large lasers are not suited for field use, and they are expensive to buy as well as to operate. Arc lamps can be used for all dyes, they are rugged and inexpensive. Arc lamps have much shorter life times than lasers. (Typically the nominal life time is 200 hours for 100 W Hg lamp and 400 hours for a 75 W Xe lamp.) On the other hand, they are easy to replace and relatively inexpensive.

Can we expect that a light source which combines the best properties of lasers and arc lamps is going to be developed? The new laser diodes have a stable output with sufficient power, and they are rugged and inexpensive. However, so far the emission is only in the red and infrared, i.e. above 650 nm. Although there are now promising developments with regard to fluorescent dyes which can be excited by these wavelengths, such lasers cannot be used to excite many of the natural pigments of marine microorganisms. Just as today's laser diodes were developed primarily to satisfy the producers of mass produced products such as CD-players, a new generation of laser diodes, with emission in the blue and possibly UV, may result from efforts to develop more efficient optical memory devices for the computer industry. Such light sources are sure to have a major impact on flow cytometer design, performance and price.

Fluorescence detection optics: As evident from Eq. 2, optimal fluorescence detection requires that the numerical aperture of the light collection optics is as high as possible. In flow cytometers where the cells are measured in a jet in air, the numerical aperture of the detection optics is limited by the fact that the lens cannot approach the jet within a certain distance. Thus, the numerical aperture of the fluorescence collecting lens in such instruments is about NA = 0.6. The effective aperture is further limited by the *obscuration bar*, which is a strip of metal across the entrance pupil of the lens which prevents the intense reflection of laser light from the cylindrical surface of the jet from reaching the optics. Hence, only some 5 % of the fluorescence is actually reaching the detection optics in this type of instrument. This intensity is further reduced before it reaches the detector, primarily by optical filters used to isolate an appropriate portion of the fluorescence spectrum. Instruments with a closed flow chamber or a flow chamber of the JOOS type (see Figs. 8 and 9) may employ immersion optics for light collection. Oil immersion lenses can have NA = 1.3 -1.4, and since there is no need for an *obscuration bar*, collection efficiency may be as high as 25 %.

Efficient fluorescence measurement requires: a) that the lens which collects the fluorescence has the highest possible numerical aperture, and b) that the background light, caused by scattering by the flow chamber and fluorescence of optical components, is as low as possible. In order to reduce the level of background light it is important that the fluorescence and light scattering detectors *see* as little as possible outside the sample flow itself. This can be achieved by means of what is called spatial filtering, that is a measuring aperture or slit which is situated in the image plane of the detection lens (Figs. 2 and 3) and which covers only the image of the sample flow and thereby eliminates all light from other sources, i.e. from other parts of the excitation focus. Such a measuring aperture is sometimes termed the *pinhole*.

Such spatial filtering may also be used to reduce the volume of the excitation focus. Thus, an *excitation slit* is situated in the image plane of the excitation lens (Fig. 2). Ideally at least, the excitation and measuring apertures should be imaged in the excitation focus so as to overlap perfectly. Efficient spatial filtering depends on flow chambers that permit proper imaging of the cells and optics which produce images with high definition and contrast. In this regard, closed flow chambers having flat surfaces and the JOOS flow chamber are better than cylindrical designs or the jet in air, which act as a cylindrical lens and thereby distort the image. The detection optics should be a microscope objective or similar lens designed for imaging purposes. The aspherical lenses used as detection optics in some instruments are inferior in this regard and will provide less efficient spatial filtering and correspondingly higher levels of optical background, i.e. noise. In most flow cytometers the potential of spatial filtering is not fully utilized. Typically, the measuring aperture in front of the fluorescence detector is a circular pinhole which is much larger than the image of the exposed sample flow.

The reason may be that with a smaller pinhole alignment becomes more critical and difficult. Furthermore, if the optics do not exhibit sufficient contrast and resolution, a small measuring aperture will reduce not only the background, but the signal as well. Laser-based instruments generally do not employ spatial filtering of the low angle scattering. A significant increase of the signal to noise ratio can be achieved if this is introduced.

Light scattering measurement: The light scattering of the cells provides two types of information. The intensity of the light scattered at low angles relative to the excitation light, that is angles below about 5°, is primarily a function of cell size. As seen in Fig. 5, the intensity increases linearly with the square of the diameter, or the cross section, of particles in the size range upward from about 1 μm. For smaller particles, intensity falls off more rapidly with size: for particles less than 0.2 μm the intensity is proportional to the 6th power of the diameter (Van de Hulst 1957). However, direct calculation of cell size

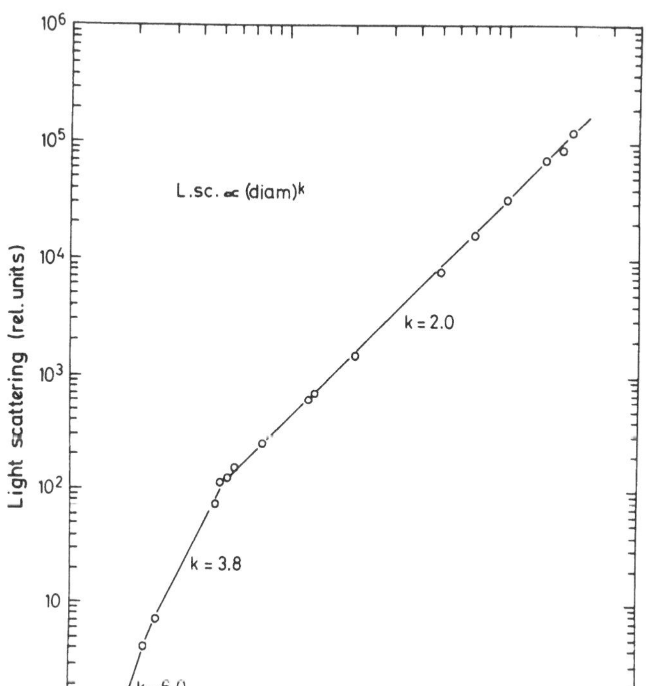

Figure 5. Low angle light scattering intensity as a function of particle size, as obtained with an Argus 100 arc lamp-based flow cytometer measuring monodisperse polystyrene particles of various diameters from 0,2 to 20 μm.

from light scattering is not possible in absolute terms. The light scattering intensity also increases strongly with the refractive index of the cell, which in turn is a function of its chemical composition. In practice, the size of cells of the same or similar types can be determined on a relative scale, which in turn has to be related to absolute values by an absolute sizing by means of electrical volume determination (Coulter volume) or microscopy. A further complication in size determination by light scattering measurement is that the scattering also depends on the shape of the cell, and the theory of light scattering is so complex that there is no practical way that the shape can be quantitatively accounted for.

The light scattering intensity, $N_l(\alpha)$, is also strongly dependent on the scattering angle, that is the angle, α, between the incident and the scattered light. As seen in Fig. 6, $N_l(\alpha)$ falls

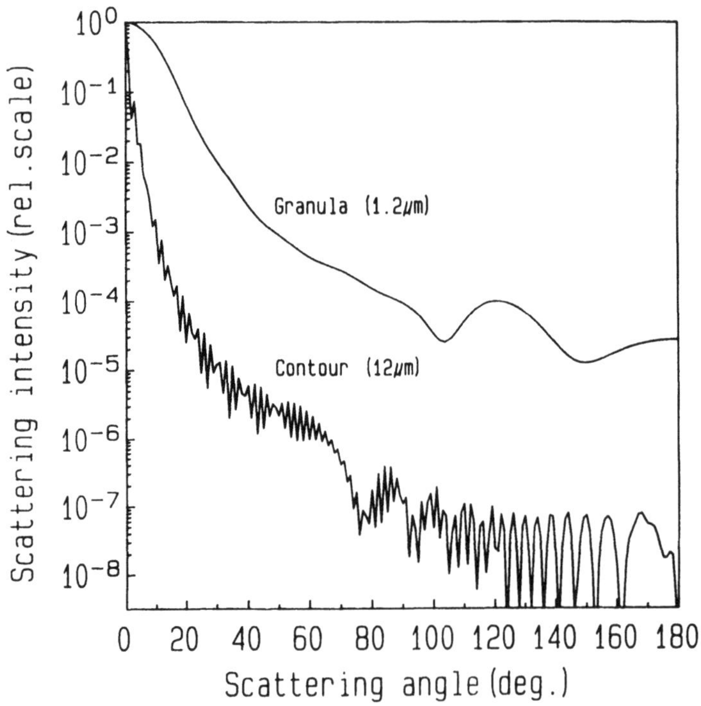

Figure 6. Light scattering intensity as a function of scattering angle as calculated from Mie theory for 12 μm spherical particles with a refractive index of 1,37 and for a gaussian size distribution of spherical particles with an average diameter of 1,2 μm and a refractive index of 1,39 (Steen, 1990). The curves are normalized to unity at zero angle. On an absolute scale the curve for the smaller particles would be shifted down by about 5,5 decades relative to the curve for the larger particles. The *fine structure* of the latter curve is the diffraction pattern.

off very rapidly with increasing scattering angle, and more so for larger particles. At small scattering angles, i.e. below 5°, $N_l(\alpha)$ depends primarily on cell size, whereas at higher angles, i.e. above 10°, $N_l(\alpha)$ is also significantly influenced by the shape and internal structure of the cell. Small structures, such as intracellular granules, will contribute more to $N_l(\alpha)$ at large scattering angles. Figure 7 shows the relative contribution of intracellular granules and the cell contour, i.e. size, as a function of scattering angle, as calculated for a spherical *cell* using Mie theory (Steen, 1990). It can be seen that this contribution is largely independent of α for α 15°. According to these theoretical data, large angle scattering, used as an indicator of intracellular structure, can be measured with approximately the same efficiency at any angle above 15°. This has been confirmed experimentally for various types of white blood cells (Steen and Lindmo 1985). On the other hand, $N_l(\alpha)$ falls off with increasing values of α by about three orders of magnitude between 15° and 90° (Fig. 6). Hence, in order to optimize

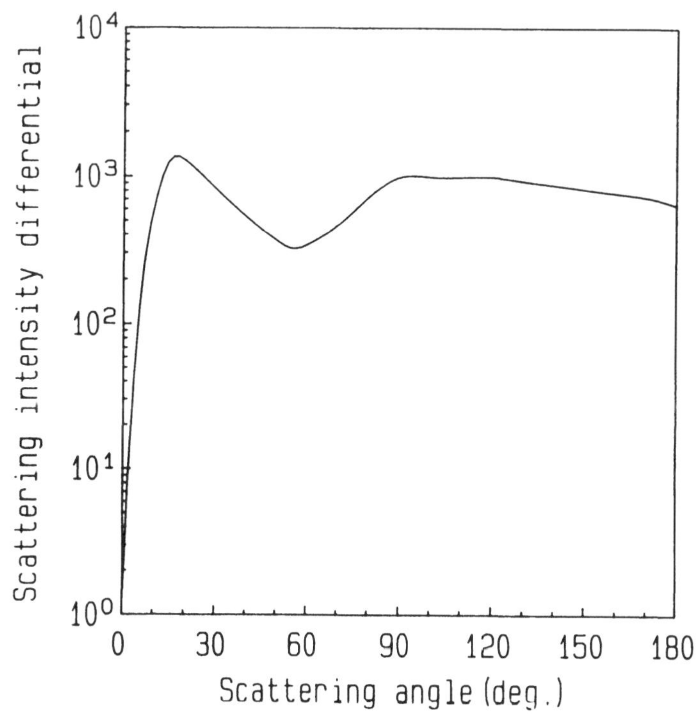

Figure 7. The ratio between the curves in Figure 6 after smoothing of the diffraction fine structure.

sensitivity large angle scattering should be measured at relatively low scattering angles, say around 20°, rather than around 90° which is standard in current laser-based instruments. This is one reason why an arc lamp-based instrument, which detects large angle scattering upward from about 15° (Argus 100) exhibits a higher light scattering sensitivity than any of the laser-based instruments that are commercially available.

***Flow chambers*:** The flow chamber is a critical part of the flow cytometer. It is an important factor in the overall performance of these instruments, and probably the most common source of trouble in daily routine. There are several types of flow chambers (see Figs. 8 and 9). Common to all is that they have what is called hydrodynamic focusing, which means that the sample is introduced into a much larger flow of water, the so called "sheath flow", at a point where the cross section of the flow is much larger than what it is when it passes through the excitation focus. Since the flow is laminar and there is essentially no mixing between the sample and the sheath fluid, the cross section of the sample flow relative to that of the total flow is maintained as the flow converges toward the excitation focus. Hence, the sample is confined to a narrow core of the flow. This is done to allow a smaller focus of excitation light and a correspondingly high excitation intensity. As evident from Eq. 9, there is a proportionality between the diameter of the sample core, d, and the diameter of the flow, D. Since the rate of sample flow, w, is always much smaller than that of the total flow, W, typically by a factor of 100 or more, W is approximately equal to the sheath flow rate

$$d = (w/W)^{1/2} \cdot D \qquad (9)$$

In order to achieve a sufficiently high excitation intensity to measure very small and/or weakly fluorescent cells with adequate sensitivity, the width of the excitation focus should be as small as possible. On the other hand, since the intensity profile of such a focus is approximately gaussian, all cells must pass through the central part of the focus in order to be exposed to the same intensity and therefore to be measured reproducibly. For example, for all cells to be exposed to the same excitation intensity to within 2%, they must pass through the center of the focus to within 10 % of its width. (Focal width is defined as the distance between points having $1/e^2$ of the peak intensity). If one allows a 10 % variation in excitation intensity, the diameter of the sample flow, d, can be increased to 23 % of the focal width.

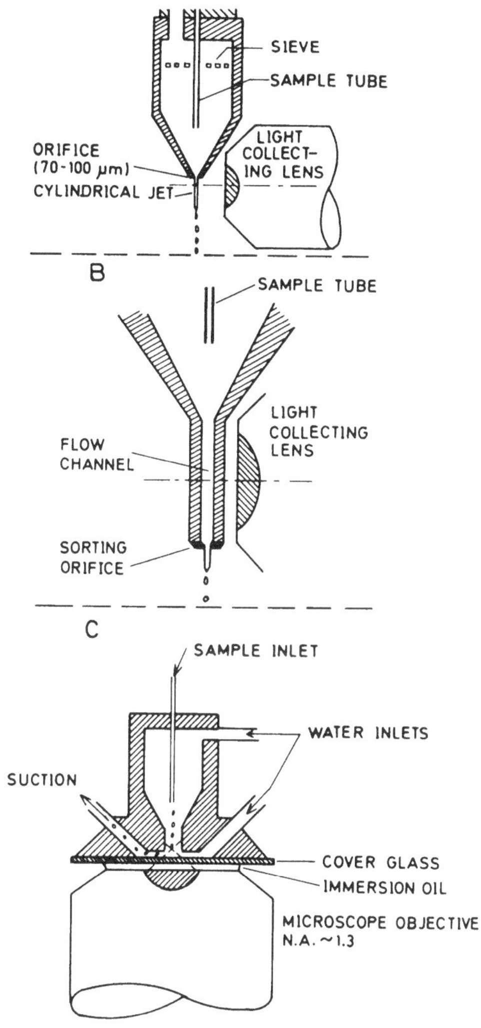

Figure 8. Schematic drawings of different types of flow chambers: A) *jet in air*, B) closed flow chamber used in laser-based instruments, C) closed flow chamber used in an arc lamp-based instrument. Note that all types have hydrodynamic focusing, which is to say that the sample is introduced into a much larger flow of water which converges laminarly into a narrow stream that is intersected by the excitation light.

According to Eq. 9, d can be reduced by reducing the sample flow, w, and/or by increasing the sheath flow, W, in order to allow a smaller focus and a correspondingly higher excitation intensity. However, there are practical limits to this. A smaller focus calls for higher precision in the alignment of the excitation light relative to the flow and makes the instrument more susceptible to mechanical shock and vibration. Reduction of the sample flow means that for a given concentration of cells, fewer cells can be measured per unit time. Increasing the sheath flow means that the flow velocity increases, which in turn means reduced sensitivity (Eq. 1) and increased consumption of water. In any case the width of the focus in the direction perpendicular to the flow must be considerably larger than the largest cell diameter. Obviously, one must find a practical compromise between the need for high sensitivity, i.e. high excitation intensity, on the one hand, and practical values for sample-and sheath flow on the other.

For instruments designed primarily to measure mammalian cells, which are usually less than 20 μm in diameter, the typical compromise is a focal width of about 100 μm. A jet in air (Fig. 8a) usually has about the same diameter and a velocity around 10 m/s. That means a flow rate of approximately 5 ml/min. For precision measurements, d should not exceed 10 μm, which, according to Eq.9, implies that the sample flow, w, must be within 50 μl/min. For a sample with 1000 cells/ml this means a measuring rate of 20 cells/min.

Closed flow chambers typically have a cross section of 200 by 300 μm. With a sheath flow of 10 ml/min and d = 10 μm this means that the sample flow must be within 16 μ l/min; that is, significantly less than with a jet in air system. So why not reduce the cross section of the closed flow chamber? The answer: increased susceptibility to clogging, and larger background due to scattering from the flow chamber itself.

In order to scatter as little light as possible, the flow chamber should have a minimum number of surfaces, since scattering arises mainly from impurities and imperfections on surfaces. Ideally all exposed surfaces should be flat and perpendicular to the direction of the excitation light. Furthermore, the exposed volume of glass and other materials should be as small as possible. A jet in air has only two exposed surfaces (the front and the back of the jet), which are both perfectly clean. However, with its cylindrical shape, the jet in air reflects an enormous intensity of excitation light over the entire plane perpendicular to the jet, that is, the plane of the detection optics in laser-based instruments. Hence, the need for *obscuration bar* to protect the detection optics. Another factor which may reduce the sensitivity of a jet in air flow cytometer is that in order to maintain a laminar jet, the flow velocity cannot be reduced below a certain limit, i.e. a few m/s.

A closed flow chamber (Fig. 8b) with a rectangular cross section is much better in this regard although it exhibits four reflecting surfaces, i.e. the air/glass and the glass/water interface on each side of the flow. Such flow chambers are used with excitation and fluorescence detection at perpendicular directions. Although sorting can be performed in instruments with closed flow chambers, it needs more fine tuning and is less stable than with jet in air instruments. Furthermore, closed flow chambers are cumbersome and time consuming to clean.

The JOOS flow chamber (Fig. 9), which is used with epi-illumination optics, has only two flat surfaces oriented perpendicular to the direction of the excitations, i.e. the glass/water and the water/air interface. Furthermore, it is easy to clean because of its open configuration. Altogether, it is probably the flow chamber with the lowest light scattering background. The thickness of the flow (on the cover glass) of the JOOS flow chamber is only 10 to 20 μm, which means a smaller measuring volume than that of closed flow chambers. The JOOS flow chamber does not allow sorting.

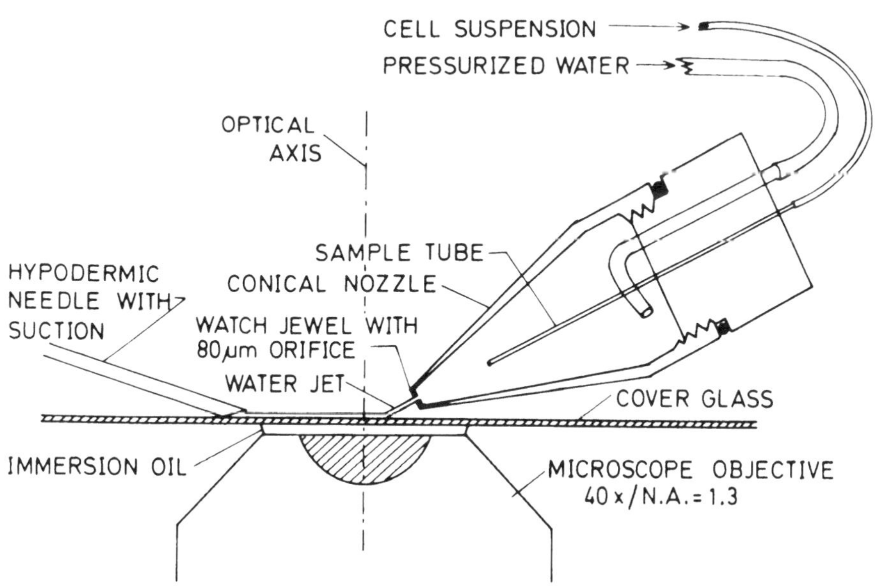

Figure 9. *Jet on open surface* (*JOOS*) flow chamber used with an arc lamp-based instrument.

***Sample injection systems*:** The sample injection of most flow cytometers is of the differential pressure type (Fig. 10 a). That means that the sample is put under a pressure which is slightly higher than that of the sheath fluid. The pressure differential thereby drives the sample from the sample tube into the flow chamber or nozzle. This is a simple method which has the advantage that it allows rapid and efficient flushing of the sample tubing simply by removing the sample tube. For many applications this advantage is offset by the fact that with this method it is not possible to set the sample flow to a constant calibrated value. This means that the number of cells per unit sample volume cannot be determined directly, but must be estimated by counting some reference particle which is added to the sample in a known concentration (and which must not adhere to or be phagocytized by the cells). Furthermore, this method does not facilitate very low sample flow rates, which may be required for samples with high cell densities, because the pressure differential becomes so small.

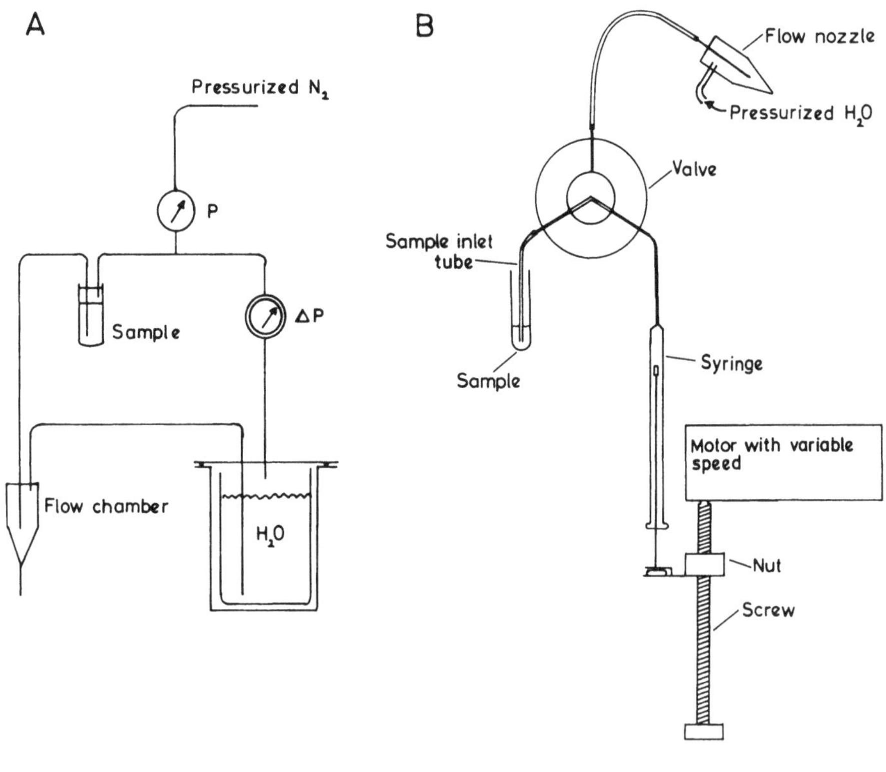

Figure 10. Systems for sample injection in flow cytometers. A) Pressure differential system. Sheath pressure is set by means of the pressure regulator P, whereas the sample flow is determined by the pressure difference between the sample container and the nozzle, which is set by the pressure regulator P. B) Volumetric sample injection. The sample flow is determined by the velocity of the syringe piston.

A better alternative is volumetric sample injection by which the sample is injected into the flow chamber or nozzle by means of a syringe driven by a motor running at an adjustable, calibrated speed. An example of such a system is shown in Fig. 10b. Using this method, the sample flow rate may readily be reduced to 0.1 μl/min, and the cell density of the sample can be calculated directly from the number of cells measured during a given period of time and the rate of sample injection as given by the motor speed. The disadvantage of volumetric sample injection is that flushing of the system may be less efficient, although this depends on the design of the system.

***Data systems*:** The flow cytometer generates a considerable amount of data which the data system has to handle. For example, five thousand cells/s measured in four parameters, each of which turns out a 8 bit number, means 20 kbytes/s. Until a few years ago this task required dedicated computer systems. With the recent advances in computer technology it can now performed by a PC computer.

The data system of the flow cytometer has four major tasks: 1) To control and record the various measurement parameters, such as sheath and sample flow rates, electronic amplifier gains, etc. In the previous generation of instruments, these parameters were set by turning knobs and recorded by pencil in the lab journal, to the extent that the operator remembered to jot them down. In most current instruments the measurement parameters are set from the keyboard of the computer and automatically recorded in its memory. With future instruments the operator may be able to leave some of the control to the computer, which will supervise the instrument and automatically optimize the parameters for a given task and level of performance.

2) A large part of the computer capacity is used to accumulate and store the measuring data. The data may be stored either as histograms, or in *list mode*. In histogram mode, the data are used to generate histograms, either single parameter histograms or two or more parameters are combined into multiparameter histograms, whereas the data themselves are not stored. In list mode, all the measuring data are stored consecutively in the computer memory so that histograms may be generated by running the data through the histogram generating program of the system. List mode data acquisition occupies much more computer memory. For

example, 100 000 cells measured in two parameters with 8 bit resolution consumes 200 kbytes, whereas a typical 64 by 64 channels dual parameter histogram takes up 8 kbytes. However, list mode acquisition has the big advantage that the data can be rerun through the computer in order to generate histograms representing any combination of parameters and with any selection of acquisition gating. The list mode memory also contains the complete time sequence of the data, which may be utilized in kinetic studies or to disclose measuring artifacts such as sedimentation of cells. With the inexpensive optical memories that are now coming on the market, memory capacity will be no concern and list mode acquisition will be the standard mode of data acquisition, especially in multiparameter studies.

3) The data generated by the flow cytometer in many cases are so complex that there is no way the information can be presented as a limited set of simple numbers. For such tasks the human eye and brain is still indispensable. To exploit this *utility*, it must be possible to visualize the data in all kinds of combinations and perspectives. For example, isometric representations of dual parameter histograms, which can be rotated freely, may be quite useful. Multiparameter histograms can be visualized by the use of colors. The new graphical software which is now coming to the PC market is certainly going to extend our facilities in this regard.

4) Nevertheless, eventually the data have to be boiled down to simple numbers in order to be communicated and utilized. Quantitative data analysis is an essential function of the flow cytometer data system. Except for DNA histogram curve fitting for cell cycle analysis, this a field which is not well developed at the present time. Hopefully, the near future will bring software to do cluster analysis and other types of data reduction which will help us to utilize our flow data more fully.

***Some aspects of a flow cytometer for measurement of marine microorganisms*:** Marine microbiology presents a challenge to the designers of flow cytometers. What the marine biologist would like is an instrument which is small enough to be carried into the field, and sufficiently rugged to withstand the vibration of a relatively small vessel in rough seas. It should still have all the measuring capabilities of a large laboratory instrument, such as five measuring parameters and sensitivity to measure submicron bacteria.

One big problem is the large variation in the size of marine microorganisms. The very large cells in particular call for a flow chamber and tubing of larger cross section than is currently used in commercial instruments. To avoid turbulence and clogging, the smallest cross section should be at least twice that of the largest organisms passing through the system. Some phytoplankton may reach 500 μm (Drebes, 1974), which means that the smallest cross section of the fluidics system, including the flow chamber, should be no less than 1 mm. That in turn means that the water consumption will be roughly one order of magnitude higher than in current instruments. More importantly, it means that the width of the excitation focus must be several millimeters in order to achieve homogeneous excitation intensity across the cells. This is hardly compatible with the high sensitivity required to measure the organisms at the lower end of the size range, which calls for a narrow focus to achieve a sufficiently high excitation intensity, that is, a focus width of about 100 μm. One could of course compensate for a large focus width by means of a very powerful laser. However, that means a large water cooled device which is so susceptible to mechanical shock and vibration that it is not a true field instrument. An instrument taking such considerations into account has recently been developed (Peeters *et al.*, 1989; Dubelaar *et al.*, 1989). In conclusion, it appears that based on the present technology, an instrument which is sufficiently compact and rugged to be used in the field will not be able to measure simultaneously organisms covering the entire size range from sub-micron to several hundred microns. However, it is possible to design an instrument which can be switched between two different focal widths, for example 100 μm and 2 mm. Thus, to cover the entire size range, one would have to perform a separate acquisition cycle with each of the focal widths.

Another problem which is particular to marine organisms is that most of these organisms are non-spherical and highly asymmetrical, and therefore will produce fluorescence and light scattering signals which depend on the orientation of the cell in the flow. As noted above, this type of measuring artifact is most pronounced in laser-based instruments with a near-parallel beam of excitation light. With such an instrument, the fluorescence signal of flat, disc-formed cells has been found to vary by more than a factor of two, depending on the orientation of the cells in the flow (Pinkel *et al.*, 1982). It has been demonstrated that a flow exhibiting asymmetrical convergence can orient flat cells (Kachel *et al.*, 1977; Kay and Wheeless, 1977), for example, when a cylindrical flow runs into a flow having a rectangular or flattened cross section. This can be done by using a wedge-tipped sample inlet tube in the flow chamber (Fulwyler, 1977) or by means of a nozzle which has a cylindrical cross section and a rectangular orifice (Pinkel, 1982). As discussed above, closed flow chambers having a flow channel with rectangular cross section may be preferable to nozzles producing a jet in air. It may be that by choosing the right combination of cross section profiles for the flow channel and the convergent part of the flow chamber leading into that channel, one may achieve sufficient orientation of the cells. Further elimination of the orientation artifact will be obtained by the use of optics which produce a highly convergent beam of excitation light, the type of optics used in arc lamp-based instruments. A combination of these two approaches, for example, an arc lamp-based instrument having a flow chamber which orient non-spherical cell in the flow, may solve this problem.

Organisms larger than 10 - 20 μm may sediment so rapidly that it affects the cell density of the sample and produce some degree of selection of small cells versus large ones. The *swimming* of some cells may conceivably cause similar artifacts, for example with phototropic species. Hence, it may be necessary to stir the sample continuously while it is being injected into the flow chamber. This can be done by means of a glass coated iron rod in the sample tube or sample syringe, which is moved by an exterior magnet.

Large cells in particular are exposed to considerable mechanical stress as they pass through the fluidic system. Narrow tubing with diameters approaching that of the cell produce shear forces which may tear the cell apart. This effect may be alleviated by reducing the length of the sample tube and by increasing the bore of this tubing as much as possible without creating an unacceptably large dead volume. The tubing should preferably be made from a hydrophobic material, such as Teflon, or its inside should be coated with a hydrophobic material.

Cell disruption may also be caused by accelerations produced in the flow chamber or nozzle. Thus, the velocity gradients which result when the flow converges into the channel of the flow chamber or the orifice of the nozzle may constitute a considerable tear on the cells. The flow chamber or nozzle should be designed to minimize this effect by having a cross section which decreases gradually over a relatively long distance in such a manner that the velocity gradient is nearly constant (Dubelaar *et al.*, 1989).

Finally, with regard to compact size and shock resistance required for a true field instrument, the arc lamp-based instruments with epi-illumination optics are probably superior to laser-based systems, and in particular the systems which comprise large water-cooled lasers. Still, we will probably have to wait a few years before we can carry our flow cytometer on board the research vessel in a rucksack.

REFERENCES

Arndt-Jovin DJ, Ostertag W, Eisen H, Klimek F, Jovin TM (1976) Studies of cellular differentiation by automated cell separation.Two model systems:Friend virus transformed cells and *Hydra Attenuata*. J Histochem Cytochem 24:332-340

Drebes G (1974) Marine phytoplankton. Georg Thieme Verlag, Stuttgart

Dubelaar GBJ, Groenewegen AC, Stokdijk W, van den Engh GJ, Visser JWM (1989) Optical plankton analyzer: A flow cytometer for plankton analysis II: Specifications. Cytometry 10: 529-539

Fulwyler MJ (1977) Hydrodynamic orientation of cells. J Histochem Cytochem 25: 881-883

Kachel V, Kordwig E, Glossner E (1977) Uniform lateral orientation, caused by flow forces on flat particles in flow-through systems. J Histochem Cytochem 25: 774-780

Kay D B, Wheeless LL (1977) Experimental findings on gynecologic cell orientation and dynamics for three nozzle geometries. J Histochem Cytochem 25: 870-874

Lindmo T, Steen HB (1977) Flow cytometric measurement of the polarization of fluorescence from intracellular fluorescein in mammalian cells. Biophys J 18:173-187

Peeters JCH, Dubelaar GBJ, Ringelberg J, Visser JWM, Optical plankton analyzer: A flow cytometer for plankton analysis I: Design considerations. Cytometry 10: 522-528

Pinkel D, Stovel R (1985) Flow chambers and sample handling. In Van Dilla MA, Dean PN, Lærum OD, Melamed MR (eds) Flow cytometry:Instrumentation and data handling. Academic Press, London

Pinkel D, Lake S, Gledhill BL, Van Dilla MA, Stephenson D, Watchmaker G (1982) High resolution DNA content measurement of mammalian sperm. Cytometry 3: 1-9

Salzman GC, Singham SB, Johnston RG, Bohren CF (1990) Light scattering and cytometry. In: Melamed MR, Lindmo T, Mendelsohn ML (eds) Flow cytometry and sorting. Wiley-Liss, New York, p 81

Steen HB (1980) Further developments of a microscope-based flow cytometer: light scatter detection and excitation intensity compensation. Cytometry 1: 26-31

Steen HB, Lindmo T (1985) Differential light scattering detection in an arc-lamp-based epi-illumination flow cytometer. Cytometry 6: 281-285

Steen HB (1986) Simultaneous separate detection of low angle and large angle light scattering in an arc lamp-based flow cytometer. Cytometry 7: 445-449

Steen HB (1990) Light scattering measurement in an arc lamp-based flow cytometer. Cytometry 11: 223-230

Van de Hulst HC (1957) Light scattering by small particles. Dover Publications, Inc, New York

Wheeless LL (1990) Slit-scanning. In Melamed MR, Lindmo T, Mendelsohn ML (eds) Flow cytometry and sorting. Wiley-Liss, Inc New York pp 109-125

DIGITAL MICROSCOPY AND IMAGE ANALYSIS: WHERE DOES IT IMPACT ON OCEANOGRAPHY?

D. J. Arndt-Jovin

Department of Molecular Biology
Max Planck Institute for Biophysical Chemistry
Postfach 2841
D-W-3400 Goettingen
Fed. Rep. of Germany

ABSTRACT

This paper presents the *coming of age* of digital microscopy as it relates to the study of marine organisms in and with their environment. In particular, recent advances in microscopic components as well as image analysis techniques are reviewed.

Confocal and non-confocal systems are compared for their unique capabilities as well as the choice of camera systems. Commercial confocal microscope systems have been optimized for their optical performance and are undergoing a second wave of development in their image acquisition systems which should make them much more useful in the future for biologists in all areas of research, including oceanography. The high resolution, linearity and extreme sensitivity of scientific 2-dimensional array charge-coupled devices (CCDs) make them the best low-light level detectors for all but extremely fast acquisition times.

The range of marine biological problems which may be addressed by digital microscopy is very broad. These can be divided into the following areas: a) the identification of objects on the basis of size and structure; b) the use of endogenous fluorescence molecules (e.g. chlorophyll) to both identify species and to assess viability; c) the characterization of bioluminescence organisms; d) the measurement of physical and physiological properties of marine organisms which provide information on water quality.

For many of the above problems a variety of software systems are already available which may be directly applied to marine questions. In other cases, the combination of biological information with image processing algorithms from a variety of disciplines will be necessary to implement the desired analysis. It follows from the high resolution of digital microscopy and image analysis that only small sample volumes can be processed at any one time, excluding high statistical precision in measurements of particle numbers but allowing precise discrimination between biological particles and debris.

INTRODUCTION

Although diatoms have long been beloved by microscopists for demonstrating the resolution of their optical systems, digital imaging microscopy (DIM) is only beginning to make a significant contribution to the study of marine organisms in and with their environment. Heterogeneity in size, abundance and species of the plant and invertebrate animal populations found in the oceans presents an enormous challenge to the description of the interdependent communities and to the assessment of their metabolic activity which finally determines the

NATO ASI Series, Vol. G 27
Particle Analysis in Oceanography
Edited by S. Demers

health of the ecosystem. The recent advances in DIM hardware and software suggest that oceanography will experience a *bloom* in this discipline in the imminent future. The first eight references in this paper are books or reviews on DIM which cover in detail both instrumentation and specific applications of the technique in cell biology.

HARDWARE

Microscope systems

Light microscopes can be divided into two types, confocal and non-confocal. The former require that only light emitted or reflected from the focal point of the objective lens be collected by the detector thus increasing the resolution of the microscope to the diffraction limit (super-resolution instrumentation will not be discussed here). Resolution of two point objects in the optical microscope is limited, as shown by Rayleigh, by a distance at which the peak intensity of the Airy disc of one object overlaps the first minimum of the Airy disc of the second, a value of about 0.2 μm for visible light (Taylor and Wang, 1989; Jovin and Arndt-Jovin, 1989; Pawley, 1990). The z-axis resolution can be equivalent for phase-contrast, interference contrast and polarized light microscopy (see Inoué in 6) but is not better than 0.7 μm for fluorescence (Taylor and Wang, 1989; Jovin and Arndt-Jovin, 1989). Much smaller objects can be detected if these objects are separated from one another by the requisite distance indicated above. For example, 5-10 nm gold particles are detectable in the reflection mode or 50 nm microtubules in phase contrast or immunofluorescence modes (see discussion of Inoué in 3). Thus, resolution of marine microorganisms from the smallest phytoplankton to the largest algae is realizable with optical microscopes. The range of confocal laser scanning microscopes (CLSM) can be extended to small macroscopic and multicellular species as well. The optical superiority of the CLSM is, at present, offset somewhat by their image acquisition systems which are slow and limited to successive single-parameter measurements. The lessons from flow cytometry of simultaneous, multi-wavelength excitation and emission will be applied increasingly to CLSM in the future to provide multiparameter measurements in real time. Such capabilities with the enhanced spatial resolution of these systems will make them highly competitive to present 2-dimensional array microscopes. The analysis of multi-cellular or complex unicellular species requires analysis by 3-D image cytometry, a

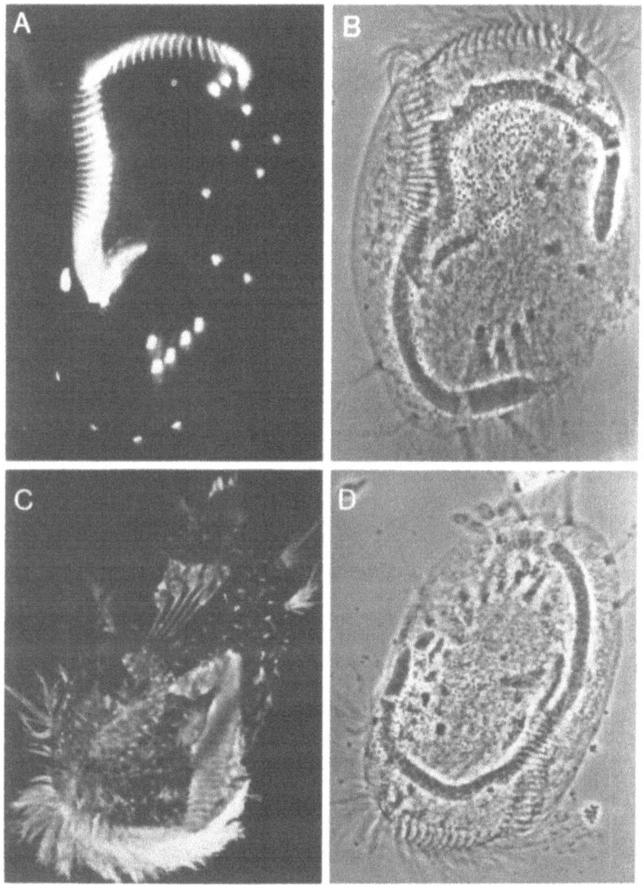

Figure 1. Confocal immunofluorescence optical sections of a fresh-water protozoa, *Euplotes eurystomus*. The cells were permeabilized and probed with monoclonal antibodies specific for post-translational modified carboxy terminal tryosine-tubulin (A) or unmodified carboxy terminal glutamic-tubulin (C). A and C are CLSM immunofluorescence images; B and C, phase contrast images of the respective animals. The anti-tyr tubulin is found at the base of the cirri and membranelles whereas the anti-glu tubulin epitope is found throughout the cirri and membranelles and surrounding the macro- and micro nuclei.

rapidly growing field. In Figure 1 are shown confocal fluorescence images and phase contrast laser scan images of a fresh-water unicellular *Euplotes eurystomus* stained with monoclonal antibodies to alpha-tubulin isotypes (Olins *et al.*, 1989). This example demonstrates the necessity of spatial as well as quantitative fluorescence information for structural and physiological analysis of such organisms.

Detectors

Three types of detectors are generally employed in image microscopy; (a) photomultipliers for point sources such as CLSM, or 2-dimensional cameras such as (b) charge-coupled array devices (CCD) and charge induction devices (CID), and (c) video cameras (Wang and Taylor, 1989; Jovin and Arndt-Jovin, 1989; Inoué, 1986). Photomultipliers of high sensitivity and good linearity exist and the optical properties of the point source microscopes ensure a high spatial resolution. However, the scanning mechanisms of all of the available commercial microscope systems are very slow compared to 2-dimensional detection systems. Assuming that many measurements are low-light level applications (fluorescence) a variety of 2-dimensional cameras have been assessed under these circumstances. Their resolution, linearity, dynamic range, sensitivity and speed have been compared by Tsay *et al.* and Bookman (in ref 8). A summary of this and other data available in the literature shows that slow scan CCD cameras are superior in resolution, linearity and sensitivity but cannot be used for rapid kinetic studies. The best choice for the latter is the multichannel plate intensified CCD although some fast kinetic studies can be performed on subarrays of the slow-scan CCD cameras as well (Jovin *et al.* in 8). The newest slow-scan CCDs contain 7 micron square pixels in arrays of 10^6 or more elements providing spatial resolution in the X-Y plane equivalent to the optical components of visible microscopes. New CID cameras are now available which will probably overcome the disadvantages of the slow read-out CCD arrays since they provide nondestructive and direct subarray read-out capability (Sims and Denton, 1987; Kaplan, 1990) and may take over the detector field in the near future.

BIOLOGICAL PROBLEMS

The identification of objects in oceanography encompasses a broader spectrum of size and diversity than any other biological discipline. One need only to reflect on the interdependence of the world's largest creatures, the cetaceans, and their food source, plankton, to grasp the problem of identification and classification. It has been recently estimated (Giovannoni *et al.*, 1990) that $<1\%$ of all microbial species in the world have been characterized so far. Thus, identification alone, not to speak of description of whole ecosystems, requires complex analysis not afforded by simple DNA or macromolecular dyes, or even by enzymatic tests and a few monospecific antibodies. The loss of spatial discrimination in flow cytometry compounds the limitations for analysis and argues strongly for increased use of DIM. One approach currently being developed employs specific labelling techniques and oligonucleotide probes for *in situ* characterization of microbial ecosystems (Giovannoni *et al.*, 1990; DeLong *et al.*, 1989) . We have tested a method similar to that of DeLong and colleagues (16) using direct coupling of fluorophores to synthetic oligonucleotide rDNA probes to identify various microbiol populations and analyzing the samples with a DIM-CCD camera system in our laboratory (Suematsu *et al.*, 1987 and N.B. Ramsing, unpublished data). An example of these data is shown in Figure 2. A mixed culture of *Saccharomyces cerevisiae* and *E. coli* was probed with a eukaryotic oligonucleotide sequence, labeled with rhodamine, and a prokaryotic sequence, labeled with fluorescein. Both populations are simultaneously visible by DNA staining or phase contrast but completely distinguished by their respective fluorescent emissions.

Clearly, the inherent multiparameter nature of digital microscopy affords an integrated approach to solving the problems of identification and characterization of microscopic and small macroscopic marine populations. One can summarize the obvious advantages of DIM over bulk measurements and flow cytometry as follows: (a) precise discrimination between biological particles and debris, (b) direct assessment of both size and structure, (c) the mechanism for metabolic determinations, and (d) established procedures for reproducible *in situ* hybridization techniques. However, DIM excludes high statistical precision in the measurement of particle numbers due to the low sample volume that can be measured.

Digital microscopy should be much better able to exploit endogenous fluorescence molecules than flow cytometry for assessment of the physiological properties of marine organisms since (a) the range of excitation wavelengths is unlimited, (b) spatial parameters are essential to correct identification, and (c) very low-light levels can be measured by signal integration over time without loss of resolution. The sensitivity of digital microscopy has reached the range where individual bioluminescence organisms can be detected (Suematsu *et al.*, 1987; Seliger, 1979). The combination of present spectroscopic techniques with microscopy will certainly provide new understanding of the physiology of these organisms in the very near future.

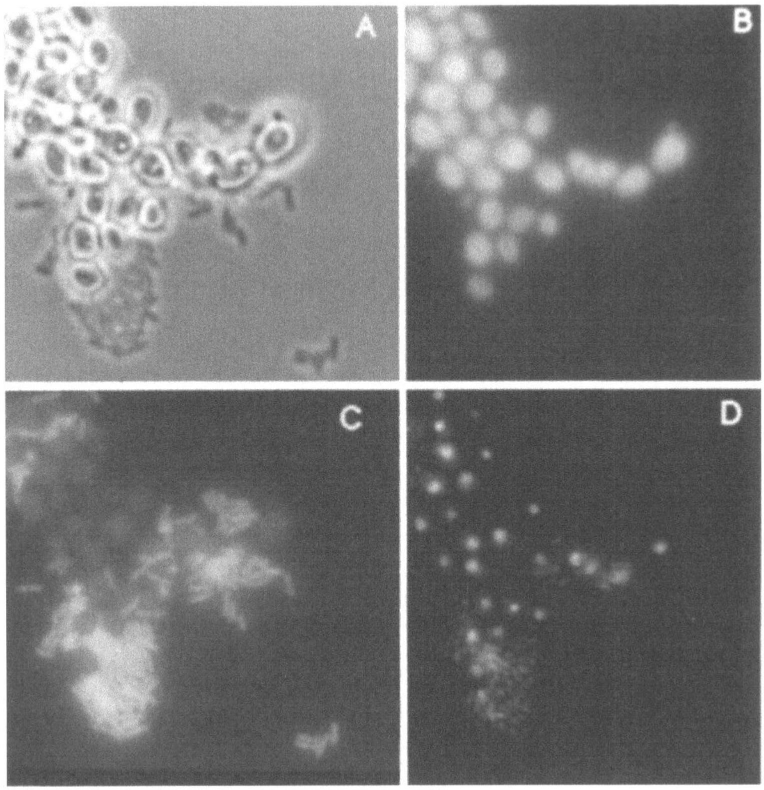

Figure 2. Phylogenetic identification of single microbial cells using *in situ* hybridization with fluorescently labeled oligonucleotide probes. A mixed culture of *Saccharomyces* and *E. coli* hybridized with a rhodamine-labeled eukaryotic probe and a fluorescein-labeled prokaryotic probe and counterstained with (Hoechst) BBI-33342 DNA dye at 1 μM. Images acquired with a Photometrics 2100 cooled CCD (Kodak 1035 x 1280 pixels) camera interfaced to a MacIntosh Cfx computer, and processed by TCL-image and NIH-image software. **A.** phase contrast; **B** rhodamine channel: excitation 546 nm, emission >590 nm; **C.** fluorescein channel: excitation 440-480 nm, emission 510-550 nm; **D.**BBI channel: excitation 365 nm, emission > 450 nm.

An overriding concern to the future of life on our planet is water quality. The composition as well as the physiological state of marine populations will certainly be some of the criteria used to measure water quality. Here quantitative cytometric measurements are the key to these questions. Again, I expect a multi-parameter approach to be the most successful, combining microsensors for elemental and pH analysis and DIM for physiological measurements of the organisms.

Software: the analysis problem

The texts (Wang and Taylor, 1989; Taylor and Wang, 1989; Jovin and Arndt-Jovin, 1989; Inoué, 1986; Castelman, 1979; Gonzalez and Wintz, 1987) given in the reference list are only a small number of those available on image processing related to cytometry. Many commercial microscopes and 2-D cameras can be purchased with accompanying analysis as well as acquisition software. The utility of any one of these systems for oceanographic problems will relate directly to the goal of those studies. Some software packages are more adaptable than others to user-specific modification or addition of specialized programs. One such program which is widely disseminated in the image cytometry field is TCL-Image (Multihouse, Amsterdam). An important consideration involving both hardware and software for anyone entering the field of image cytometry is an adequate capability to deal with the large data base which is generated by each experiment. Mass storage devices and networking between acquisition and analysis systems are a must.

Another important consideration in the application of DIM to oceanography is the proper statistical analysis of the generated data. Here oceanographers already have wide-ranging experience and can be expected to contribute direction to the field.

In conclusion, I believe the application of DIM to oceanographic research is at the beginning of an exponential growth phase. As an outsider to the discipline, I have tried to indicate what I think are the obvious advantages of the technique for solving problems in a number of areas but have clearly neglected a plethora of equally probable applications. I hope that this article stimulates a number of readers in oceanography to apply DIM to their problems and look forward to future conferences where I can profit from their successes.

ACKNOWLEDGEMENTS

I thank my colleagues Thomas Jovin, Niels Ramsing and Michel Robert-Nicoud for stimulating discussions and permission to use data collected by them in this laboratory.

REFERENCES

Amann RI, Krumholz L, Stahl DA (1978) Fluorescent-oligonucleotide probing of whole cells for determinative, phylogenetic and environmental studies in microbiology. J Bacteriol 172:762-770

Castelman KR (1979) Digital Image Processing Prentiss Hall: Englewood Cliffs NJ pp 429

DeLong EF, Wickham GS, Pace NR (1989) Phylogenetic stains: ribosomal RNA-based probes for the identification of single cells. Science 243:1360-63

Giovannoni SJ, Britschgi TB, Moyer GL, Field KG (1990) Genetic diversity in Sargasso Sea bacterioplankton. Nature 345:60-63

Gonzalez RC, Wintz P (1987) Digital Image Processing, Second Edition. Addison-Wesley: Reading MA pp 503

Herman B, Jacobsen, K (1990) ed. Optical Microscopy for Biology Wiley-Liss: New York

Inoué, S (1986) Video Microscopy. Plenum: New York

Jovin TM, Arndt-Jovin DJ (1989) Digital imaging microscopy. Ann Rev Biophys Chem 18:271-308

Jovin TM, Rippe K, Ramsing NB, Klement R, Elhorst W, Vojtiskova M (1990) Parallel stranded DNA, structure and methods. in Vol 3: RNA and DNA, Eds. R.H. Sarma and M.H. Sarma, Adenine Press: Albany NY 155-74

Kaplan H New jobs for charge-transfer devices. Photonics Spectra Nov:86-87 1990

Kohen E, Hirschberg JG (1989) eds. Cell structure and function by microspectrofluorometry. Academic Press: San Diego, pp 465

Olins DE, Olins AL, Robert-Nicoud M, Jovin TM, Wehland J, Weber K (1989) Differential distribution of alpha-tubulin isotypes in *Euplotes eurystomus* determined by confocal immunofluorescence microscopy. Biol. Cell 66:235-46

Pawley J (1990) ed. The handbook of biological confocal microscopy. Plenum Press: New York.

Seliger HH The measurement of light emitted by living cells in ref 1 p 417

Shotton DM (1989) Confocal scanning optical microscopy and its applications for biological specimens. J Cell Sci 94:175-206

Sims GR, Denton MB (1987) Characterization of a charge-injection-device camera system as a multichannel spectroscopic detector. Opt Eng 26:1008-1019

Suematsu M, Oshio C, Miura S, Tsuchiya M (1987) Real-time visualization of oxyradical burst from single neutrophil by using ultrasensitive video intensifier microscopy. Biochem Biophys, Res Commun 149:1106-1110

Taylor DL, Wang, Y-L (1989) eds. Methods in Cell Biology, Vol. 30. Fluorescence microscopy of living cells in culture. Part B. Quantitative Fluorescence microscopy-imaging and spectroscopy. Academic Press: San Diego pp 503

Wang, Y-L, Taylor DL (1989) eds. Methods in Cell Biology, Vol. 29. Fluorescence microscopy of living cells in culture. Part A. Fluorescent analogs, labelling cells, and basic microscopy. Academic Press: San Diego pp 333

Ward DM, Weller R, Bateson MM (1990) 16S rRNA sequences reveal numerous uncultured microorganisms in a natural community. Nature 345:63-65

Williams B, Carta D (1989) CID cameras: more than an alternative to CCDs. Adv Imag Jan:48-50

APPLICATIONS OF MICROPHOTOMETRY TO OPTICAL OCEANOGRAPHY

R. Iturriaga

Department of Biological Sciences
University of Southern California
Los Angeles, California - U.S.A.

ABSTRACT

The study of individual particulates through microphotometry provides an accurate measurement of the specific contribution of phytoplankton and detrital particles to the spectral absorption of submarine light. Thus, a link between individual particles and bulk particulate can be analyzed and integrated with physical and biological data. Such information provides the basis for developing appropriate and linked models of variability from individual particles to mesoscale particulate suspended matter in the ocean.

Results obtained using the present methodology are quantifiable and are consistent with standard macrophotometric methods. Microphotometry has been successfully applied to natural samples of marine particulates, enabling taxonomic, geometrical and optical data to be collected simultaneously. In addition, the determination of the absorption and fluorescence of individual phytoplankton cells has been valuable in characterizing individual phytoplankton cells by their excitation spectra, in determining their photoadaptive levels, and in assessing their intrasample variations.

Other applications of microphotometric analysis to marine optics and cell biology of phytoplankton cells are being investigated with a highly sensitive video image system interfaced to the photomicroscope system. We hope image processing will facilitate the quantification of morphological and optical properties of the particulate suspended matter. This will enable us to assess the contributions that different particle types and sizes make to particulate absorption of submarine light in aquatic environments.

INTRODUCTION

A thorough characterization of the optical properties of the particulate suspended matter is essential to understanding the spectral absorption of submarine light and its sources of variability. A common approach used to determine the spectral absorption of marine particulates has been the concentration of samples on filters. However, these determinations only provide the absorption of the bulk particulates indiscriminately, making it difficult to interpret such measurements in terms of taxonomy, physiological adaptation of species, or relative contribution to light attenuation by specific categories.

A microphotometric procedure to directly measure the spectral absorption and fluorescence properties of phytoplankton cells and detrital matter has proved valuable for determining

NATO ASI Series, Vol. G 27
Particle Analysis in Oceanography
Edited by S. Demers

optical properties at the level of individual cells and detrital particles in aquatic environments (Iturriaga *et al.*, 1988; Iturriaga and Siegel, 1989). This technique has allowed the identification of the components that are major contributors to attenuation within the particulate assemblage. In addition, because of its capability to perform simultaneous particle identification and optical characterization, microphotometric analysis has proven valuable for understanding optical properties and variability at the level of individual cells and detrital particulates in aquatic environments.

METHODOLOGY

Microphotometric system

The system is comprised of a universal microscope (Carl Zeiss) equipped with a type 03 photometer (Zeiss) interfaced with a tungsten-halogen light source (150W, Xenophot) and a scanning monochromator (Type H-10, Instruments SA, N.J.). The photomicroscope records continuous transmission or fluorescence spectra from 400 to 750nm (Fig. 1).

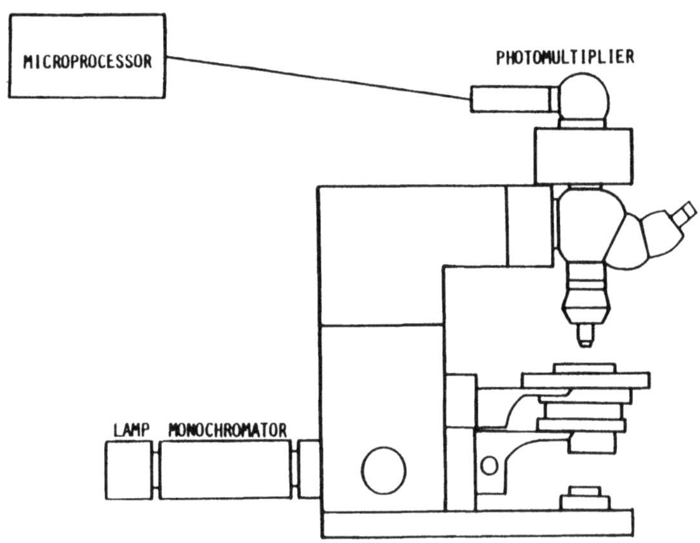

Figure 1. The photomicroscope system records continuous transmission or fluorescence spectra from 370 to 750 nm (100 discrete measurements per nm). Fluorescence spectra require a dark field condenser and objectives equipped with an iris. A second monochromator is placed before the photomultiplier to run emission spectra at a fixed wavelength, or to set the emission for excitation spectra.

Sample preparation

Water samples (100ml approx) are concentrated onto 25 mm diameter Nuclepore filters (0.4 μm) by filtration at low vacuum pressure. Particles collected on the filter are transferred onto a gelatin-coated microscope slide. The filter is removed and the transferred sample is covered with few drops of a 30% glycerol aqueous solution and a cover slip. Samples are kept frozen at -20°C until analysis. The absorption and fluorescence properties of cells remain unaltered when samples are stored frozen at -20°C indefinitely and do not change their properties when brought back to room temperature for analysis.

ABSORPTION

The technique has a spectral resolution comparable to traditional spectrophotometric measurements allowing the simultaneous determination of particle type, cross-sectional area (G), and the absorption efficiency factor ($Q_a(\lambda)$) for a single particle. The application of microphotometry to field samples enables variations of absorption properties within the particulate assemblage to be directly assessed, allowing discrimination of major absorption bands corresponding to the algal pigments: chlorophyll *a* at 435 and 675 nm, chlorophyll *b* at 470 and 650 nm, chlorophyll *c* at 465 and 630 nm, and fucoxanthin and peridinin from 470 nm to 550 nm. The spectra may be used to fingerprint taxonomic differences between algal cells or characterize detrital particulates (Fig. 2).

Microphotometric determinations

Measurements of the spectral transmittance of individual particles are determined by focusing on the targeted particle and selecting an appropriate optical slit so that a representative cross-sectional area of the particle is sampled. The minimum practical target area that can be sampled by the present system is in the range of 3 μm diameter.

The spectral transmittance of individual particles is determined by focusing on the targeted particle. An adjacent particle-free area is used as the blank transmittance. Measurements of

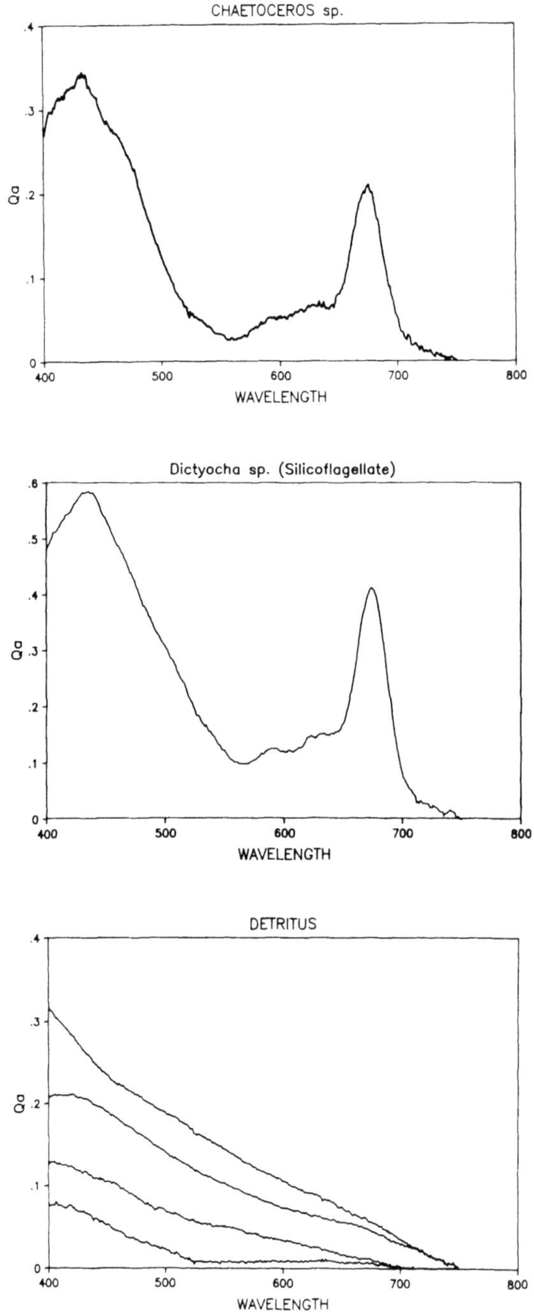

Figure 2. Absorption efficiency factor (Qa(λ)) for field samples of individual phytoplankton cells and detritus.

the sample transmission spectra, I_s, and the blank spectra, I_b, enables the determination of the absorption efficiency factor for the particle:

$$Q_a(\lambda) = \frac{I_b(\lambda) - I_s(\lambda)}{I_b(\lambda)}$$

Partitioning of the particulate absorption spectra

In order to link individual optical properties to bulk optical properties,a procedure was developed (Iturriaga and Siegel, 1989). The particulate absorption spectrum (*ap* (λ)) can be calculated from the optical and geometric properties of the particulate assemblage with:

$$ap(\lambda) = \sum_{i=\lambda}^{N_t} N_i \cdot G_i \cdot Q_{ai}(\lambda)$$

where N_t is the number of different particle types, N_i is the number of a specific type, G_i is the cross-sectional area for the ith type, and $Q_{ai}(\lambda)$ is its absorption efficiency factor.

A simpler way to partition the particulate absorption spectra is proposed. We assume that there are only two types of particles in the ocean: *phytoplankton cells and detrital matter.* Thus, the particulate absorption can be partitioned into phytoplankton and detrital components.

$$ap(\lambda) = aph(\lambda) + adet(\lambda)$$

or

$$ap(\lambda) = N_{ph} \cdot G_{ph} \cdot Q_{aph}(\lambda) + N_{\det} \cdot G_{\det} \cdot Q_{adet}(\lambda)$$

We use linear regression to decompose the total particulate absorption spectra *ap* (λ), spectra using the mean $Q_{a,ph}$ (λ) and $Q_{a,det}$ (λ) spectra for each depth.

$$ap(\lambda) = AQ_{aph}(\lambda) + BQ_{adet}(\lambda)$$

$$= aph(\lambda) + adet(\lambda)$$

The quality of fit is evaluated by comparing reconstructed *ap* (λ)to the measured *ap* (λ) (Fig. 3).

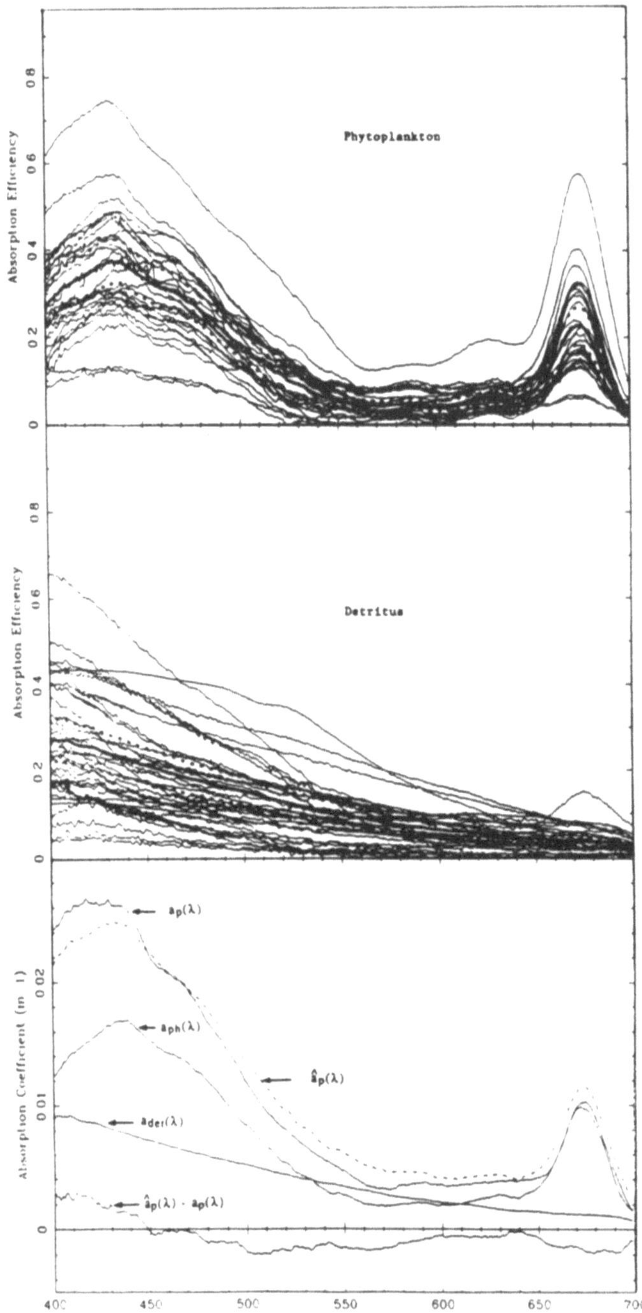

Figure 3. Absorption efficiency factors for phytoplankton cells and detrital materials sampled at a depth of 20 m in the Northern Sargasso Sea, and results of the particulate absorption coefficient partition. The dashed line corresponds to the reconstructed particulate absorption spectrum ap(λ), and the line about the zero axis is the deviation between the sampled ap(λ) and the reconstructed spectra, Δ ap(λ).

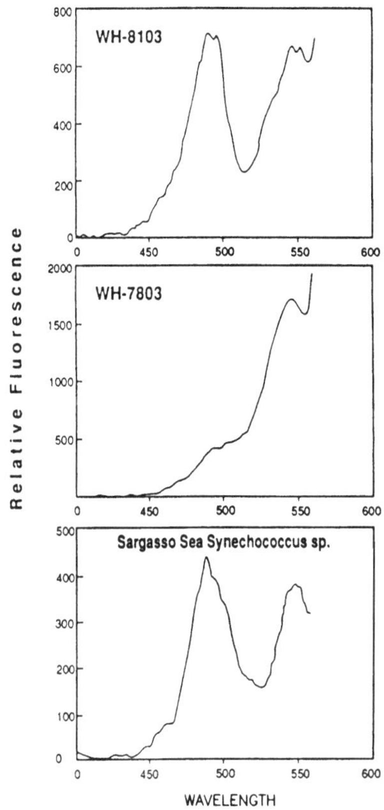

Figure 4. Fluorescence excitation spectra (uncorrected) of individual Synechococcus from cultured cells and natural cell types. Emission measured at 575 nm (uncorrected spectra) indicates natural strains in the Sargasso Sea contain high ratios of phycourobilin to phycoerythrobilin.

In addition, the abundance of phytoplankton and detrital particles (N_{ph} and N_{det}) may be estimated using regression coefficients (A and B) and the relationship between Q_a and ap.

$$ap(\lambda) = N_{ph} \cdot G_{ph} \cdot Q_{aph}(\lambda) + N_{det} \cdot G_{det} \cdot Q_{adet}(\lambda)$$

Thus,

$$N_{ph} = A/G_{ph}$$

and

$$N_{det} = B/G_{det}$$

FLUORESCENCE

Individual cell fluorescence spectra

Determination of the fluorescence properties of phytoplankton cells(excitation and emission spectra) is possible using monochromatic dark field illumination (Campbell and Iturriaga, 1988). A special dark field condenser (e.g. Ultracondenser (Zeiss)) and objectives with iris diaphragms are required. A second monochromator placed before the photomultiplier is used to scan or set the emission wavelength. Samples are prepared as described for absorption measurements.

Fluorescence excitation spectra has been useful in discriminating between clones types of picoplanktonic cells such as coccoid cyanobacteria (*Synechococcus* spp.) which can be differentiated by their phycourobilin to phycoerythrobilin ratio (PUB:PEB). It has been found that in the oligotrophic ocean and in most oceanic regions, the PUB-rich type predominate (Fig. 4). In addition, this type of analysis has been useful for determining photoadaptive states of individual planktonic microalgae.

ACKNOWLEDGEMENTS

The present research is supported by the Ocean Optics Program of the Office of Naval Research (contract N00014-89-1047). The author thanks the NATO Advanced Study Institute on Individual Cell and Particle Analysis in Oceanography Organizing Committee for their support.

REFERENCES

Iturriaga R, Mitchell BG, Kiefer DA (1988) Microphotometric analysis of individual particle absorption spectra. Limnol Oceanogr 33:128-135

Campbell L, Iturriaga R (1988) Identification of *Synechococcus* spp. in the Sargasso Sea by immunofluorescence and fluorescence excitation spectroscopy performed on individual cells. Limnol Oceanogr 35:1196-1201

Iturriaga R, Siegel D (1989) Microphotometric characterization of phytoplankton and detrital absorption properties in the Sargasso Sea. Limnol Oceanogr 34:1706-1726

CALIBRATION AND QUANTITATIVE ANALYSIS

K. A. Ault M.D.

Maine Cytometry Research Institute
190 Park Avenue
Portland ME 04102

INTRODUCTION

Flow cytometers have traditionally been used primarily for semi-quantitative measurements. The earlier flow cytometers were mainly research instruments and were so completely adjustable that standardization and calibration outside of the boundaries of a single experiment were both difficult and not really needed. Only recently in the medical applications field has there been a necessity for quantitation of flow cytometric results, and for standards that can be applied in many laboratories. The history of attempts to provide standards for quality control and calibration of flow cytometers in the medical field has been long and difficult. Clinicians and researchers have long recognized the clear need for standardization but only in the past few years have there emerged, largely by default rather by careful planning, some fairly widely recognized standards.

Much of the problem in the field has resulted from confusion about terminology and fuzzy thinking about what types of controls are needed and what is being controlled for in any given case. In this overview I will attempt to provide an outline by which it may be possible to discuss issues of standardization and calibration. I will use some examples drawn from the medical field in the hope that they may serve as models for applications in marine biology and oceanography. Unfortunately I must acknowledge that all of the information I will offer here deals with the use of exogenously applied fluorochromes. To my knowledge the problem of standardization and calibration of intrinsic fluorescence ("autofluorescence") which is a major issue in marine biology has not been seriously addressed in the flow cytometric field and this, I believe, will be a major challenge for the future.

There are four types of quantitative data that can be obtained from a flow cytometer. First is the determination of proportional populations, i.e. "percent positive". This requires a minimum of calibration because the measurement is entirely relative and is only meaningful within the context of a single sample. The only requirement that must be met is that the population of interest must be resolved from the other cells. If they are not, then the

NATO ASI Series, Vol. G 27
Particle Analysis in Oceanography
Edited by S. Demers

determination of their proportion must be based upon some type of model or on assumptions about the reason for the lack of resolution. This type of determination is by far the most common one in medical flow cytometry today and is the basis for such tests as "T4/T8 ratios", phenotyping of leukemias, etc.

Second is the analysis of fluorescence intensity distributions. In this case some, information, usually the mean or modal fluorescence value for a subpopulation, must be obtained from a fluorescence distribution. An example is the analysis of a DNA distribution in which the various populations at different stages of the cell cycle are identified within a continuous distribution. This type of analysis requires that the response characteristics of the flow cytometer be known - usually the requirement is that the instrument be linear. It usually involves measurement of fluorescence in "channels" in arbitrary units and not in any absolute terms.

Third is the measurement of absolute fluorescence levels. This requires at a minimum that the flow cytometer be calibrated and that it be stable from one assay to another. Examples include the measurement of auto-antibodies and cross-matching for organ transplantation. The results are usually expressed as a channel shift relative to a control or as mean fluorescence, again in arbitrary units. In those cases where it is desirable to express the measurement in terms of non-flow cytometric units, such as molecules of fluorochrome, moles of antigen, etc., special problems are encountered that we will discuss below.

Fourth is the measurement of absolute counts of cells, i.e. cells per unit volume. For reasons that are somewhat obscure, none of the commercially available flow cytometers offer the easy determination of absolute cell counts. This has become very important medically in the determination of the absolute numbers of CD4 lymphocytes in AIDS. The solutions usually involve obtaining a proportional count in the flow cytometer and then determining the absolute count with a different instrument. Other solutions include using calibration particles in the sample at a known concentration, injecting a known volume of sample or injecting the sample at a known volume rate. Each of these solutions involves special problems that we will discuss.

The general topic of calibration of a flow cytometer can be broken down into several separate considerations.

ALIGNMENT

First is that of alignment of the flow cytometer. This usually requires very stable particles that may have properties very different from the biological particles which are ultimately to be analyzed. Alignment is usually done using plastic beads of uniform size and very high levels of fluorescence over a wide spectral range. The necessity for a brightly fluorescent particle arises because, if the flow cytometer is seriously out of alignment, it is necessary to see the weak signal from the bright particle in order to begin the process of alignment. As flow cytometers have improved, this step of alignment has become less important. The stability of the particles is important in order to detect changes from day to day that may reveal gross instrument malfunction. However, since the particles may not resemble biological particles and are usually very bright, they may not reveal subtle malfunctions. For example, if their emission spectrum is not the same as that of the biological particles they may not reveal defects in optical filters.

SENSITIVITY AND LINEARITY

Second is the determination of the sensitivity and linearity of the now optimally aligned flow cytometer. This is best done with particles that fluoresce with a known intensity. Usually multiple particles of differing intensities are used. They must be stable, but again need not bear any relation in terms of fluorochrome or emission spectrum with the biological samples to be analyzed. Perhaps the greatest difficulty with the determination of sensitivity is the need to determine the "noise" level which ultimately limits sensitivity. This is usually done with a "non-fluorescent" particle which, when run through the instrument gives a signal corresponding to "noise". The determination that a particle is non-fluorescent is not easy. In addition, in the absence of fluorescence, there may nevertheless be a signal due to light scatter from the particle which penetrates the barrier filter(s). Thus the determination of instrument noise is complex and includes both true noise, i.e. noise in amplifiers and photomultiplier tubes, as well as noise due to photon statistics, and aberrant or irrelevant signals that are operationally included with the noise.

One widely used set of particles for determination of sensitivity and linearity are offered commercially by the Flow Cytometry Standards Corporation in Research Triangle Park, North Carolina U.S.A. They offer a set of five beads, including "non-fluorescent" beads and four others of known fluorescence intensity using fluorescein as the fluorochrome. Beads containing other fluorochromes may also be available from this source. These beads are run on the flow cytometer and a regression of the known mean fluorescence for each bead versus the corresponding mean channel number obtained on the flow cytometer is plotted (Fig. 1). The intersection of this regression with the mean channel of the "non-fluorescent" beads is

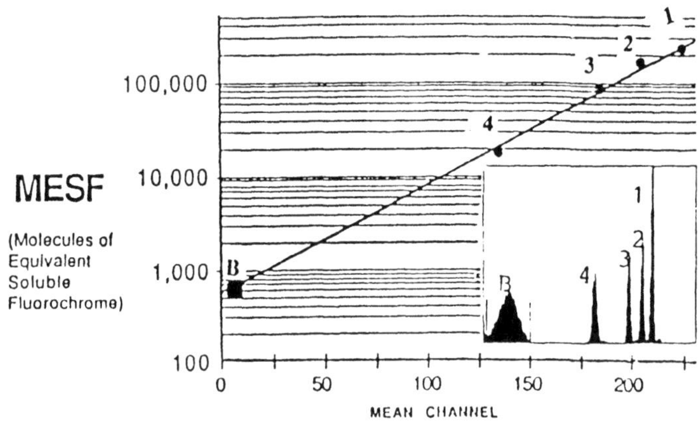

Figure 1. Example of a calibration curve using beads of known fluorescence intensity (Flow Cytometry Standards Corporation). A regression line is calculated based on the mean fluorescence of beads 1 through 4 and this line is then extrapolated to the mean of non-fluorescent beads (B). This intercept is reported as the instrument sensitivity. Linearity is calculated from the goodness of fit of the regression.

used as a measure of sensitivity. The goodness of fit of the linear regression (r^2) is used as a measure of linearity.

This approach has proven to be feasible and has been applied in at least two informal surveys of large numbers of laboratories using different instruments. The results can best be summarized by saying that there are remarkably wide variations in sensitivity, even within the same model of flow cytometer. For example, in one published survey (1) the variation in sensitivity in all 110 instruments surveyed was from about 200 molecules of fluorescein to more than 16,000 molecules. (The units of fluorescence used by the company are "molecules of equivalent soluble fluorochrome" which we will discuss later.) In this survey the most frequently tested instrument was represented 39 times and the sensitivity measurements for that instrument varied from 200 to about 1200. Whether these variations are due to poor maintenance in the laboratory, poor alignment by the users, or to artifacts in the way "sensitivity" is being determined is not clear. I suspect that all three explanations are true. One major problem with this method is the relatively long extrapolation needed to determine sensitivity. The dimmest bead has a fluorescence value of about 18,000 units and the extrapolation down to the level of fluorescence of the non-fluorescent beads spans more that one decade of fluorescence. It is unlikely that most log amplifiers are "linear" in the lowest decade. Thus very small changes in the position or resolution of the non-fluorescent beads can lead to large differences in the estimation of sensitivity.

The results are much more encouraging for determination of linearity, which many would argue is really more important. In the published survey, all the tested instruments proved to have r^2 values of greater than 0.99 and there was very little variation between instruments or within instrument models. Thus the basic tenant of flow cytometry, that the amplitude of the fluorescent signal is proportional to the amount of the fluorochrome, seems to be a good one in practice. Again this is limited by the caveat that the instruments are being tested primarily in the range of greater than 10,000 fluorochrome equivalents and linearity almost certainly does not extend down to the limits of fluorescence detection, at least using log amplifiers.

Another, more economical approach to the question of linearity can be made by using particles, usually beads, that are partially aggregated. When this is done, one obtains a series of peaks which should be spaced uniformly corresponding to singlets, doublets, triplets, etc. If the coefficient of variation of the beads is sufficiently small, these peaks can be resolved cleanly and can be used to determine linearity over a limited range.

For the purposes of determining both sensitivity and linearity, it is not necessary for the beads to be calibrated in terms of any absolute fluorescence standard. Thus the fact that the commercial beads provide a result in terms of "molecules of equivalent soluble fluorochrome" is not really relevant to our current discussion. The special problems introduced by this type of calibration will be discussed below.

COMPENSATION

Another instrument variable that must be checked in the process of obtaining quantitative data from a flow cytometer is that of compensation. By this is meant the correction of the measured fluorescence signals for signals spilling from another fluorochrome. This problem arises because we generally identify each fluorochrome by measuring its emission through a band pass filter. Some fluorochromes have wide emission spectra and may "spill" their emission into the band pass of a filter designed to measure another fluorochrome. The correction for this is usually done in hardware by subtracting a fraction of the measured signal in one channel from the measured signal in another. The fraction subtracted is set empirically by the operator so that each fluorochrome, when analyzed by itself, gives a signal only in its own channel. The setting of these fractions is often critically dependant upon the gain setting for the various fluorescent channels. Thus small setting in photomultiplier high voltage or amplifier gains can result in large changes in the necessary compensation.

The proper sequence for setting compensation is first to run the "negative" sample, placing it in the desired position by adjusting gain in all fluorescence channels. Next the brightest possible sample of each fluorochrome is run, and gains are further adjusted so that these are on scale. Having thus established the required gain settings to visualize both the dimmest and the brightest samples, these setting are fixed and must not be changed during compensation. Then, bright examples of each fluorochrome alone are run sequentially and adjusted so that the mean fluorescence signal obtained in the non-relevant fluorescence channels is the same as the mean for the negative control in that channel. The result is that each fluorochrome gives a signal only in its appropriate channel.

One major drawback to this procedure is that although one is subtracting two signals, their errors (or the uncorrelated noise contained in each) are being added. Thus as one adds increasing amounts of compensation the coefficient of variation of the cluster of particles increases. This is sometimes a dramatic effect which significantly interferes with resolution of closely spaced populations (Fig. 2). For this reason and others, it has been argued that it

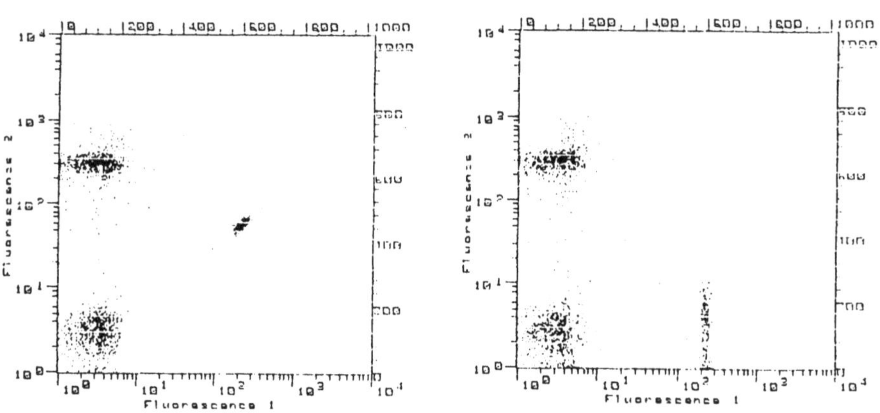

Figure 2. The panels show dot plots of FL1 (green fluorescence) versus FL2 (red fluorescence) for a mixture of three beads. One bead is unlabelled, one has only fluorescein, and the other only phycoerythrin. In the left panel the data has no electronic compensation and there is significant FL2 signal from the fluorescein beads. In the right panel compensation has been added so that the FL2 mean of the fluorescein beads is identical to the FL2 mean of the unlabelled beads. Note that the cluster of fluorescein beads is considerably expanded along the FL2 axis by the compensation procedure.

would be preferable to work with uncompensated data. If compensation is required later for data presentation, it could in theory be done with software. Compensation is not really necessary except to permit correlation of each fluorescence channel on the flow cytometer with a single fluorochrome. No information is added (in fact some is lost) by compensation.

When systematically applied, this procedure is not difficult. However, failure to adhere to the rules about not changing gain settings during the procedure, or failure to run each fluorochrome individually, can result in great confusion. When setting compensation for more than two fluorochromes, the conflicting requirements and the possibility of multiple "spillovers" can be very confusing. Usually compensation is set by running appropriately stained biological samples, but bead kits for this procedure are commercially available. It must be stressed that in order to use beads it is absolutely necessary that they have exactly the same emission spectrum as the corresponding fluorochrome in the biological sample.

CONTROLS

Achieving stable beads with correct emission spectra is technically difficult. In order to insure stability it is preferable to place the fluorochrome within the bead to protect it from environmental degradation. However, the emission spectrum is very likely to be different inside the plastic of the bead that it is on the biological sample. Conversely, placing fluorochromes on the outside of beads makes them subject to instability. Most available particles with correct emission spectra are not as stable as one would like.

It might be argued that beads need not have the same emission spectrum as the biological particle if the relationship between the two spectra is known and is constant. This may be true, but it assumes that filters in the instrument(s) have not changed. The possibility of high intensity laser light damaging delicate interference filters in a flow cytometer is not remote, and thus most laboratories have decided that they need to have particles with emission spectra similar to the particles of interest.

The next topic that must be addressed is that of controls. Controls are designed to allow one to detect errors within data or an experiment. Controls for flow cytometry are not different than controls for any scientific endeavour and thus should not require discussion here. However, there is frequently considerable confusion about controls in flow cytometery. This usually stems either from a misconception about the need for controls or from fuzzy thinking about what is being controlled for. There is obviously no magic in a flow cytometer that relieves one of the necessity of running controls, but there are a number of levels at which controls can be instituted and these must be clearly stated so that one can be certain that the controls are appropriate. Examples are: controls for the instrument, i.e. is the flow cytometer functioning correctly?; controls for the procedure, i.e. were the correct stains used?; controls for the sample, i.e. was the sample obtained correctly?; controls for the reagents, i.e. have they gone bad since last used?, etc.

There are three types of controls that are relatively unique to flow cytometry that we might discuss. First is that of an internal standard which may be added to each sample and which, if everything is correct, has known characteristics. This is a very good method of controlling for a number of variables. Such internal standards must have properties that are similar to, but clearly distinct from, the biological particles of interest. An example of this will be given below.

Another type of control which has not been widely used in flow cytometry but which is ideally suited to flow cytometry data is that of testing the shape of the distribution of some parameter, either over a population of particles or over a population of samples. The distribution shape is often very sensitive to errors in the instrument, the procedure, or the sample. Relatively little software is available for this type of testing but it will likely be exploited to a greater extent in the future.

Additionally, it is frequently possible, due to the multi-parameter nature of flow cytometry data, to find ways of testing for internal consistency in a data set. If the same particles are counted in different ways or in different samples they should give the same results. Tests for this type of internal consistency are now being applied to great advantage in medical flow cytometry.

ABSOLUTE CALIBRATION

Finally we come to the most difficult aspect of quantitating flow cytometry data: expressing the results in terms of quantities that are not intrinsically flow cytometric, i.e. in absolute values. This is can be done in terms of quantity, e.g. the conversion of arbitrary fluorescence to molecules of fluorochrome or moles of antigen, or the conversion of light scatter to size. Alternatively, conversion to absolute concentration (particles per unit volume) may be required. Both of these types of conversions are difficult with current flow cytometers.

ABSOLUTE COUNTS

Determining absolute concentration is difficult because no commercial flow cytometers presently provide this measurement. The most direct way of adding it would be to build the flow cytometer with an injector mechanism that either injects a fixed volume or injects the sample at a known volume flow rate. Although some instruments have such injectors, one must be careful about how accurate they are. It is a common problem that not all of the injected sample passes through the measurement point. It is usually necessary to chase the sample in with an air bubble or a blank sample to be certain that all of the injected sample was analyzed. In order to use injection rate as the basis for calculating concentration, it is necessary to insure that the sample is uniformly mixed, or that the average particle flow rate is used in the calculation. For various technical (and perhaps economic) reasons these approaches have not yet been implemented.

A method that does not require instrument modification is that of mixing distinctive particles into the sample at a known concentration. The relative numbers of these particles and the biological particles of interest can then be used to calculate concentration. This approach has been used successfully in a number of laboratories. One must to be sure that the added particles do not aggregate with themselves or with other particles in the sample, that they are added accurately, that the sample is well mixed, and that the concentration of added particles is comparable to the concentration of particles of interest. Although this sounds difficult, it is not hard in practice to find beads that, if added immediately prior to running the sample, meet these criteria. If the beads are allowed to remain in the sample for a longer period, they are very likely to interact with the sample in some way that makes the procedure invalid.

ABSOLUTE FLUORESCENCE

A similar approach may be taken in order to obtain calibration of fluorescence in absolute terms by adding beads to the sample which bear a known amount of antigen on their surface. The beads are labeled at the same time as the cells during the staining procedure. During subsequent analysis the level of fluorescence on the beads can be used to calibrate the fluorescence scale. This procedure also requires beads that are similar to, but distinguishable from, the biological particles. An example is shown in Figure 3. In this example, beads were treated with various amounts of antigen (human IgG) which had been trace radiolabeled. After the IgG had bound to the beads, the mean amount of IgG per bead was calculated using the radioactive tracer. The beads were then added to preparations of human platelets and the mixture was labeled with fluorescent antibody to human IgG. During analysis, separate gates were placed on the platelets and the beads, and fluorescence histograms generated. In addition, unconjugated beads of the same type were also added. The difference in fluorescence

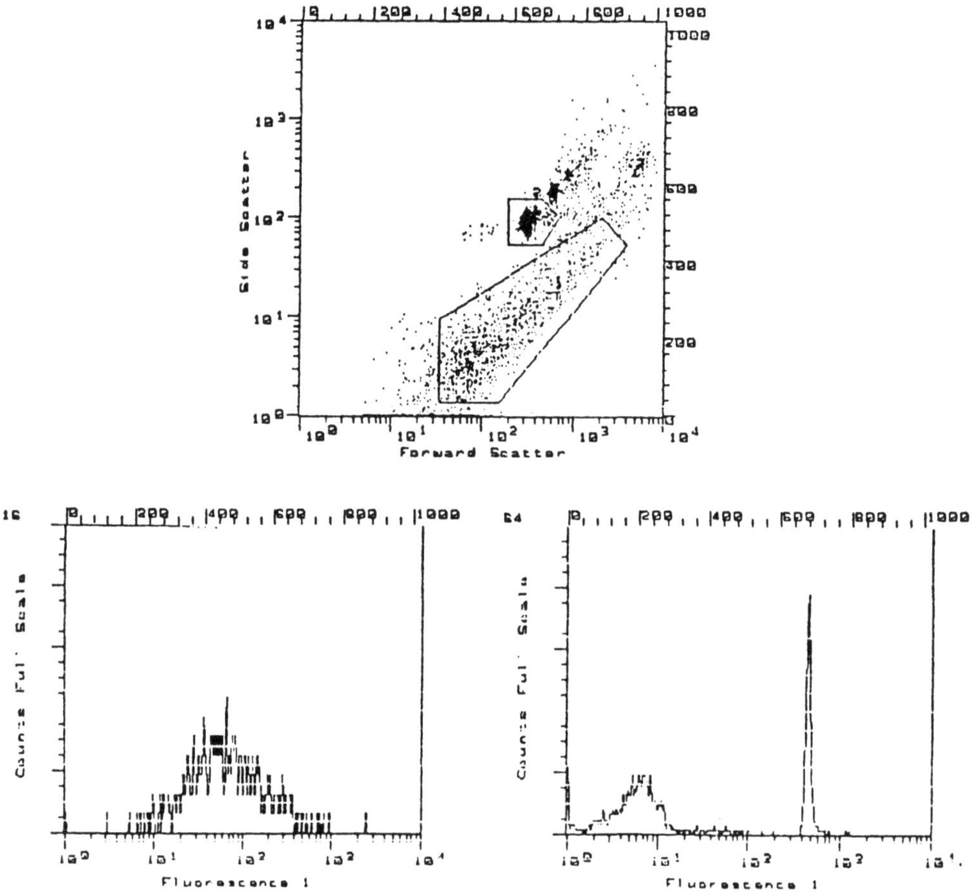

Figure 3. A dot plot showing forward scatter versus side scatter illustrates a mixture of platelets and beads. Gating on the platelets yields a FL1 histogram with a broad distribution. Gating on the ead singlets yields an FL1 histogram with two peaks corresponding to unlabelled beads and beads containing a known amount of antigen (see text).

levels between the unconjugated beads and the labeled beads was used to calibrate the fluorescence scale; the amount of IgG per platelet could be calculated from the mean fluorescence of the platelets.

This procedure involves a number of assumptions. First, that the beads and platelets are labeled in the same way in the mixture as they would be if labeled separately. This is easily tested, and proved to be the case in this example (Figure 4).

Second is the assumption that the antigen on the bead surface is labeled in the same way by the antibody as is the antigen on the platelets. This is a weak assumption and may well be the most difficult question in any similar quantitation attempt. The antigen on the bead is likely

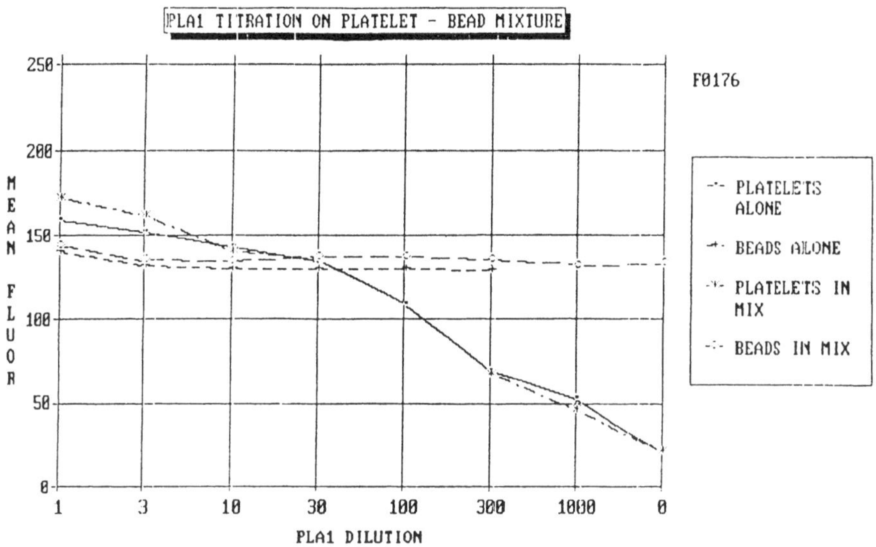

Figure 4. An experiment using the system described in Figure 3 in which the labelling of beads alone, platelets alone, and beads and platelets in the mixture is compared over a titration of anti-platelet antibody. The data shows that the beads and platelets in the mixture are labelled as if they were alone.

to be oriented differently and subject to much different steric conditions than the antigen on the cells. Thus the ratio of antibody to antigen may be very different and is difficult to determine. In addition, beads with organic molecules exposed on their surface are very likely to be unstable and subject to degradation over time. Therefore they frequently do not make very good standards. The same may be said for other similar standards, such as freeze dried or otherwise stabilized cells.

Our experience, and that of others who have attempted quantitation of antigen binding in flow cytometers, has shown that it is possible with considerable effort, but that the degree of uncertainty in the result seldom justifies the effort.

ENDOGENOUS FLUORESCENCE

All of the preceding discussion applies to exogenous fluorochromes. The problem of quantitation of endogenous fluorescence has not been seriously addressed. There are, however, a few possible approaches which should be mentioned here. First is the use of fluorescence parameters to obtain information about the physical state, local environment, or binding of ligands to the fluorochrome.

Many fluorochromes change their properties with changes in their environment. This can be exploited in a quantitative sense if it is possible to calibrate the change. Changes in fluorescence intensity are most difficult to quantitate because the intensity is obviously a function of both the amount of fluorochrome and its physical state. In the case of endogenous fluorochromes, the amount is not easily controlled; in the case of exogenous fluorochromes, the ability of the dye to penetrate the cell or be metabolized may dramatically alter its amount in each cell.

More widely used is the measurement of fluorescence intensity ratios at two different wavelengths, either of excitation or emission. These ratio measurements are independent of the amount of fluorochrome and have been well exploited for measurements of calcium concentration, ph, etc. in cells. The ratio can be calibrated either with free fluorochrome *in vitro* or in the cells using various drugs to obtain equilibration between the cellular compartments and the outside medium.

Another property of fluorochromes that changes with their environment is the fluorescence lifetime. Direct measurement of this is difficult and cannot be done in flow cytometers. However, it may be possible to indirectly measure the same phenomenon by using the phase shift of emitted light when the fluorochrome is excited by a modulated excitation source. This phase fluorescence is being explored in non-flow cytometric applications and there are a few laboratories working on measuring phase shift in a flow cytometer. It is too early to tell if this approach is technically feasible, but if it proves to be, it may be a very good method of quantitating changes in endogenous fluorochromes.

REFERENCES

Flow Cytometry Standardization Forum Vol. 2, No. 1. Published by Flow Cytometry Standards Corporation P.O. Box 12621, Research Triangle Park NC 27709. A. Shwartz Ed.

Pigments and fluorescence

ALGAL PIGMENT FINGERPRINTS: CLUE TO TAXON-SPECIFIC ABUNDANCE, PRODUCTIVITY AND DEGRADATION OF PHYTOPLANKTON IN SEAS AND OCEANS

W.W.C. Gieskes

University of Groningen
Department of Marine Biology
PO Box 14
9750 AA Haren (Gn),
The Netherlands

INTRODUCTION

A principal assignment of the Pelagic Ecosystem Research Group established at the Netherlands Institute for Sea Research in the early seventies was to contribute to the existing body of knowledge of the primary production of seas and oceans. We soon realised that incubations of natural phytoplankton to assess ^{14}C incorporation are time-consuming and can only be performed at a few stations, which should be as representative as possible for the survey region. The phytoplankton in the North Sea coastal areas that we were to survey appeared to be distributed very irregularly both in space and in time. Also in the open ocean patchiness was reported to be the rule rather than the exception (Lorenzen, 1967a). The only way to arrive at reasonable estimates of primary production in the marine environment is to establish the relation between photosynthesis and a few parameters determining the growth potential of algal cells that can be measured quickly and therefore frequently. Chlorophyll *a*, the pigment that is central in light capture for photosynthesis, was selected by us as one of these because reliable methods for its measurement have been presented by Holm-Hansen *et al.* (1965) and Lorenzen (1967b) just a few years before we started our work in the Dutch part of the North Sea. The great seasonal and regional variability in phytoplankton biomass in these waters is expressed in the chlorophyll distribution (Fig. 1), assuming that chlorophyll is an indicator of biomass. The other major parameter that is important for aquatic primary production, underwater light availability, was simply determined from surface irradiance and Secchi disc observations, the attenuation coefficient k being calculated from $k = f/S$, with f ranging from 1.4 in turbid coastal waters, 1.7 in offshore neritic waters, and 2.2 in oligotrophic ocean waters - a variability due to reasons unknown to us at that time.

NATO ASI Series, Vol. G 27
Particle Analysis in Oceanography
Edited by S. Demers

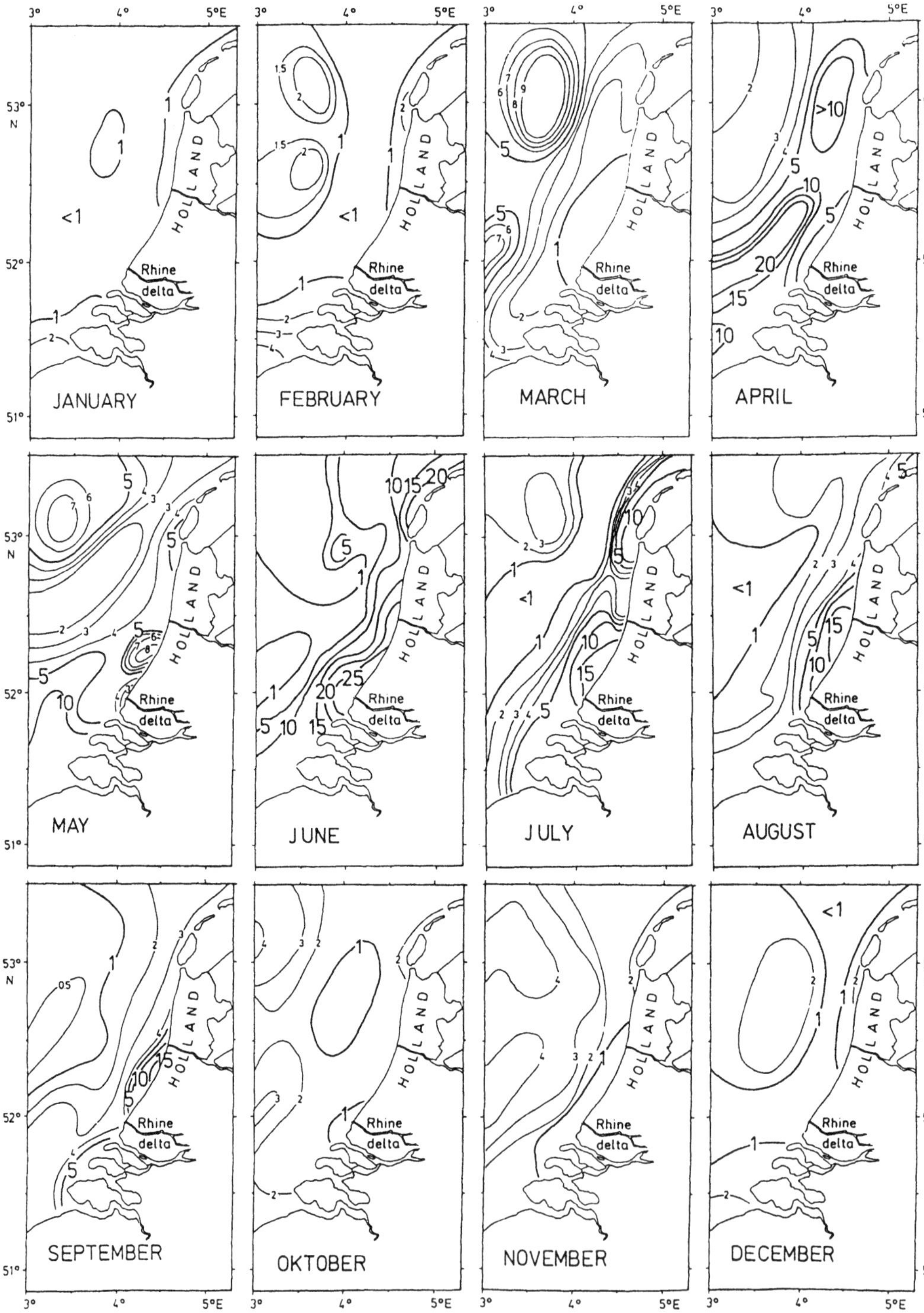

Figure 1. Monthly chlorophyll distribution in the Southern North Sea, showing the rapid temporal changes and patchiness. Concentrations are mean values (μg liter^{-1}) for the usually well-mixed water column. Based on surveys in the late seventies.

Correlations between our chlorophyll, irradiance, and primary production estimates were less than satisfactory. It occurred to us that in order to accurately model production rates one should take into account not only the intensity but also the spectral composition and angular distribution of underwater light. In addition, not only the light absorbed by chlorophyll *a* should be considered, but also absorption by the other pigments that are directly active in photosynthesis through efficient light energy transfer to chlorophyll *a* (Chl *b*, the fucoxanthins, the Chl *c*'s, (Jeffrey, 1989); violaxanthin, peridinin (Owens *et al.*, 1987)), or indirectly by having a protective function through interaction in photochemically induced oxidation reactions at high light intensities (lutein, zeaxanthin (Bidigare *et al.*, 1989)). The remarkable abundance of the photosynthetically inefficient zeaxanthin recorded by us in the open tropical and subtropical ocean (Gieskes and Kraay, 1983; Gieskes *et al.*, 1988) obviously called for a new approach in the thinking of absorption of light by pigments. Indeed, over much of the world's oceans there appeared to be as much zeaxanthin as chlorophyll!

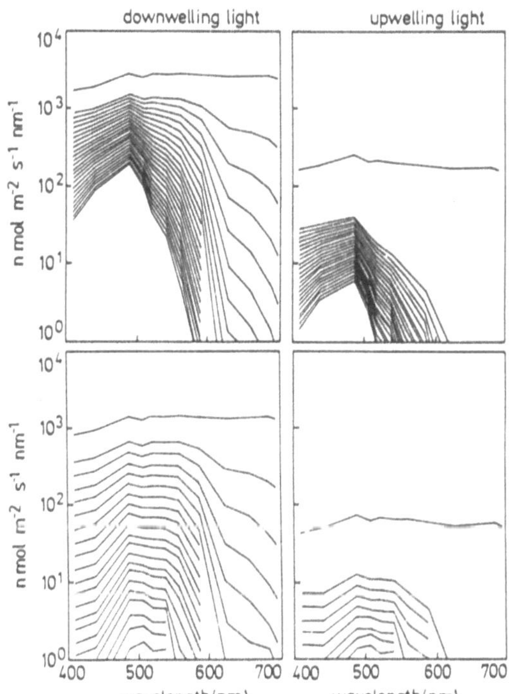

Figure 2. Spectral distribution of light in the euphotic zone of the Weddell Sea. Courtesy M.M. Tilzer (Konstanz).

The design and subsequent use of an underwater light recording instrument by my colleagues Spitzer and Wernand (1979) extended our simple Secchi disc observations to measurements of scalar irradiance and its spectral distribution in the euphotic zone. Today, the best available and highly user-friendly equipment for this purpose is available from Rocky Booth's

Biospherical Instruments in San Diego, California (Fig. 2; courtesy M.M. Tilzer). Our effort to expand our sets of Chlorophyll *a* measurements to the detection of other chlorophylls and carotenoids was stimulated by G.W. Hallegraeff, then at Amsterdam University, who routinely used thin-layer chromatography, albeit not to eventually relate pigment

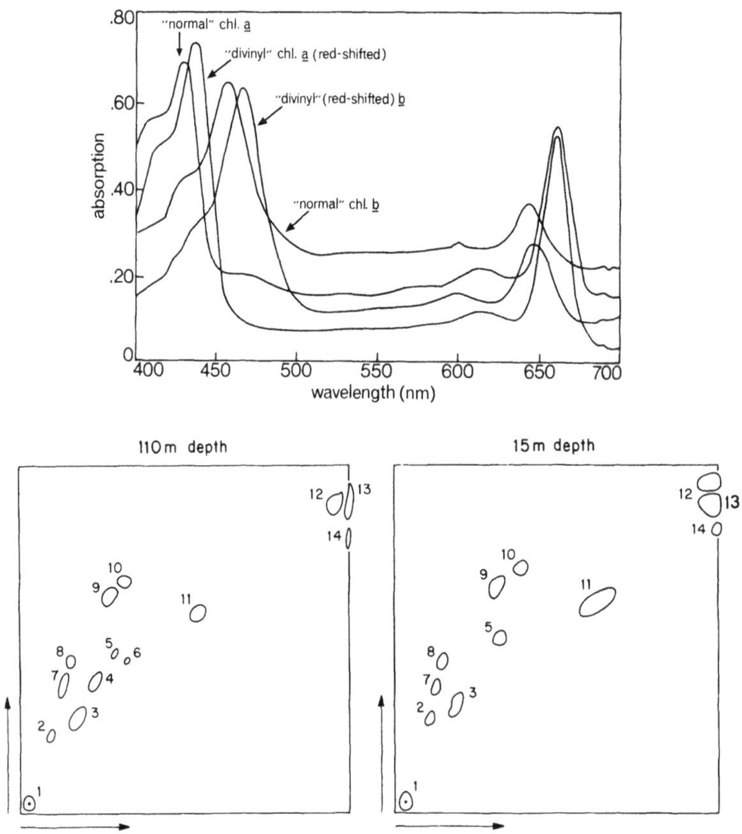

Figure 3. Absorption spectra (in acetone) of Chl *a*, Chl *b*, and the picoplanktonic Prochlorophyte's "divinyl-Chl *a*" and "divinyl- Chl *b*", both red-shifted. 2-d thin-layer chromatograms are also shown (samples taken near surface and at 110 m, at the depth of the deep chlorophyll maximum in the central tropical Atlantic, November 1977). 1. Origin (olive-brown) 2. Butanoyloxyfucox (yellow). 3. Fucoxanthin (yellow). 4. Hexanoyloxy-fucoxanthin (yellow). 5. Unknown. 6. Unknown (yellow-orange). 7. Divinyl-Chl *b* (brown-green). 8. Chl *b* (olive-green). 9. Divinyl-Chl *a* (green). 10. Chl *a* (blue-green). 11. Zeaxanthin. 12. Carotene (orange). 13. Secondary carotenoid (bright yellow). 14. Secondary carotenoid (pink).

concentrations to wavelength-dependent absorption and photophysiological properties of phyto-plankton, which was our objective, but merely to study succession in the taxonomic composition of natural freshwater lake populations (Hallegraeff, 1977).

In the sixties, the pioneer of pigment analysis for chemotaxonomic purposes in marine phytoplankton studies had been S.W. Jeffrey (1961, 1968). She convinced most biological oceanographers of the promises of this approach by publishing her finding of high chlorophyll *b* concentrations in the open ocean, which, as she argued (Jeffrey, 1976), revealed the abundance of a group of phytoplanktonic algae that was until that time not known to occur there at any numbers of importance: green algae. This was the only taxonomic entity of microalgae characteristically containing Chl *b* according to the literature of the time (Weber and Wettern, 1980; Van den Hoek, 1978). Jeffrey's report drew much attention, but it can now be regarded as an example of the danger of neglecting, more conventional methods for detection of natural phytoplankton components, notably microscopy. Green algae may be present in the open ocean (Takahashi and Hori, 1984), but we now know that except in estuarine regions (Lubian and Establier, 1980) or coastal upwellings (Gieskes and Kraay, 1986b; Gieskes *et al.*, 1988), where green algae can be prominent, most chlorophyll b in the ocean is associated with picoplanktonic Prochlorophytes, not green algae (Chisholm *et al.*, 1988; Gieskes *et al.*, 1988; Olson *et al.*, 1990; Veldhuis and Kraay, 1990) - *a fortiori*, most so-called Chl *b* is in fact not even normal Chl *b* but a derivative, possibly containing a divinyl bond (Goericke, pers. comm.), that gives it a red-shifted absorption spectrum (Gieskes *et al.*, 1979, their Fig. 2; see also Neveux *et al.*, 1989; and see Fig. 3). Prochlorophytes do not contain violaxanthin (Burger-Wiersma *et al.*, 1986) while green algae do (Van den Hoek, 1978), so Chl *b* presence without a trace of violaxanthin indicates that Prochlorophytes contribute to a phytoplankton population - a straightforward way to detect and quantify this group's abundance (Gieskes *et al.*, 1988). Such subtle pigment analysis is best performed after separation by high-performance liquid or thin-layer chromatography (HPLC, HPTLC). Nevertheless, Jeffrey's early TLC work has given the interest in pigment analysis its momentum, and it continues to do so even now that modern HPLC has replaced paper chromatography (Jeffrey, 1961; Jensen and Sakshaug, 1973), thin-layer chromatography (Jeffrey, 1968, 1981), HP-thin-layer chromatography (Wright and Jeffrey, 1987), and the conventional ways of HPLC (Abaychi and Riley, 1979).

As I have argued, pigment analysis should not be applied indiscriminately in studies of taxonomy and succession of natural phytoplankton populations, but it is still valuable as a tool for population structure analysis because manipulation, fixation and preservation of samples needed for alternative methods (ranging from conventional microscopy and Confocal Laser Scanning microscopy to flow-cytometry) tend to destroy an unknown proportion of the cells. Also, the presence of cells associated with aggregates of particulate matter, with marine snow, or with other organisms in symbioses (e.g. Cox *et al.*, 1985; Knight and Mantoura, 1985), including alga-alga symbioses (e.g. in dinoflagellates: Schnepf and Elbrächter, 1988; cf.

Jeffrey, 1975) can only be readily revealed by the detection of taxon-specific pigmentation (e.g. Gieskes and Kraay, 1983a). In the following I will present a short review of the methodology used for the measurement of chlorophylls, carotenoids, and their breakdown products, describe how knowledge of their concentrations can be applied to unravel the taxonomic composition of natural microalgal populations and the ecophysiological and photophysiological properties of the various taxonomic groups, and outline how insight into the distribution of pigments can in the future be used as an aid in the interpretation of results obtained with the newest and most promising instruments available for the investigation of plankton distribution and ecology: remote sensing and flow-cytometry. The more conventional methods of pigment analysis will also be discussed, as they remain valuable in phytoplankton biomass and productivity research, can be put to unexpected use, and are more accessible to field workers or to those who have no means to afford the excessive technical and financial resources needed for the sophisticated approaches just mentioned. I will start with a treatise on the single most important pigment, chlorophyll *a*, and gradually expand the ideas developed here to a discussion of other pigments: the other chlorophylls, the carotenoids, the mycosporine-like pigments absorbing ultraviolet light, and light absorption by detrital material.

DISTRIBUTION OF CHLOROPHYLL IN THE SEA: PROS AND CONS OF DETECTION METHODS

Chlorophyll concentrations are usually measured because the pigment is considered to be indicative of algal biomass. However, it is not the most appropriate measurement available, as the amount per cell and the phytoplankton-to-carbon ratio is highly variable (Eppley *et al.*, 1977; see also Table I), depending on the physiological condition and degree of light adaptation (Olson *et al.*, 1990a; Olson *et al.*, 1990b), and the state of degradation and lysis of a population (Daley, 1973). Nevertheless, it is often a better biomass index than conversions from cell counts based on microscopic observations (Strathmann, 1967), especially when, as is normally the case, nano- and picoplankton (0.5 - 5.0 μm) are abundant. These cells are easily overlooked, and their structure is often altered beyond recognition due to sample treatment. This is probably the reason why in many reports the carbon:chlorophyll ratio is unrealistically low, often less than 10. Counting particles with electronic particle counters (e.g., Coulter Counters) is not practical; in most size classes, algal cells are usually only a fraction of the total particle load. A solution to this problem is to count the various components of natural phytoplankton by flow cytometry. However, only small volumes of water can be processed. Concentration of cells before counting causes considerable

Table I. Carbon-to-Chlorophyll *a* ratios in batch cultures of a few algal species. Highest values are recorded at the post-stationary growth phase.

Species	C:Chl *a*.
Biddulphia regia	8 - 100
Biddulphia sinensis	13 - 100
Thalassiosira sp.	7 - 120
Phaeocystis colonies	70 - 162
Thalassiosira rotula	20 - 150
Thalassiosira excentrica	20 - 150
Isochrysis sp.	40 - 80

cell loss with practically all methods because of adherence to filters and cell destruction due to manipulation. Large cells may even lose their chloroplasts and may then be recorded by a flow cytometer as eukaryotic picoplankton, or coccoid prokaryotes when chlorophyll fluorescence has faded. There can be no doubt that bulk chlorophyll measurements, either *in situ* or in extracts of suspended matter collected on fine glass fiber filters (Lorenzen and Jeffrey, 1980), remain a necessity for the realistic reconstruction of phytoplankton biomass distribution in natural waters.

Recording Chlorophyll in situ

During the last 20 years equipment has become available for *in situ* fluorometric measurement of chlorophyll (Fig. 4). The instruments that have been used with the most success are VarioSens (Herman and Denman, 1977), AquaTracka (Fasham *et al.*, 1983), and Almondbury (first used by Jim Aitken of the Continuous Plankton Recorder Team of Plymouth Marine Laboratory). The instruments all measure red light emitted from chloropigment molecules excited with blue light. In natural waters, the chloropigment pool may consist of chlorophylls *a*, *b* and *c*, their derivatives, and phaeopigments. Not all these pigments have the same fluorescence excitation and emission maximum. The *in situ* recording instruments have been equipped with blue and red filters that match with the excitation and emission spectra of individual pigments (Lewitus and Broenkow, 1985), such as chl *a*, the pigment that is of greatest interest as a phytoplankton biomass indicator. However, the variable contribution of chl *a* to the total amount of phorbins in natural waters remains a problem. Moreover, fluorescence per unit of chl *a* depends on the physiological condition of cells and on chloroplast structure, and is therefore not constant in living cells (Kiefer, 1973; Loftus and Seliger, 1975). The decrease of fluorescence near the surface, often observed in vertical profiles made during the daylight hours (Fig. 4), can be caused by overexposure of cells to surface light (Papageorgiou and Govindjee, 1968) due to a low rate of vertical mixing. The relation between cellular photoadaptation and vertical mixing has been explored by Lewis *et al.* (1984) in order to calculate mixing rates. In general, *in situ* measurements of chlorophyll fluorescence should only be used as a method to obtain a quick look of the distribution of phytoplankton and to get an impression of patchiness on a smaller scale than sampling usually permits. Profiles should be calibrated frequently with measurements of chlorophyll made in a different way, preferably by methanol or acetone extracts of suspended matter samples at discrete depths.

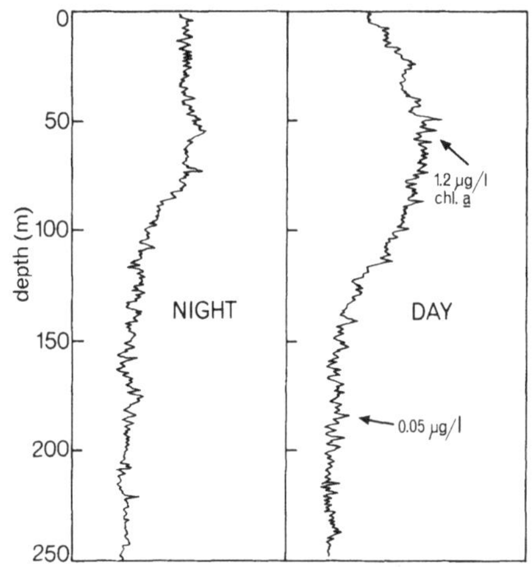

Figure 4. *In situ* records of vertical chl *a* distribution (AquaTracka), Weddell Sea, November 1986. a: Mid-day cast. b: Night cast.

Chlorophyll detection from space

The Coastal Zone Color Scanner (CZCS) on the Nimbus-7 satellite has provided comprehensive views of pigment distribution near the surface of large ocean areas surveyed between 1978 and 1986. Airplane remote sensing can also provide rapid scans of pigment distribution over large areas. Again, these observations are limited to the sea's surface - even when obtained by LIDAR (as practiced, for example, at Oldenburg University). NASA's Earth Science and Application Division recently claimed that the technique of ocean color imaging based on CZCS records has allowed oceanographers to see the global distribution of photosynthetic organisms and their seasonal variation in abundance for the first time, but this claim is somewhat exaggerated: in the early seventies Russian scientists published global maps of primary production (Koblentz-Mishke *et al.*, 1970) that reveal a pattern of distribution remarkably similar to the (admittedly much more colourful, glossy and spectacular) composite images of chlorophyll concentration produced by technicians of the Goddard Space Flight Center and the University of Miami (USA). In these early reports, geographic differences were ascribed to variability in solar radiation, upwelling conditions along continents, and monsoonal wind direction, and along fronts. Also, features such as the development of the spring bloom in the North Atlantic had been identified in the sixties by the CPR survey (Glover, 1967).

Other conventional techniques, based on compilations of nutrient depletion, light measurements, and primary production estimated with Steemann-Nielsen's ^{14}C method in seas and oceans all over the world, have been used by Berger *et al.* (1987) to reveal globe-wide patterns of phytoplankton distribution and productivity. However, CZCS pictures are extremely detailed geographically, and, even more importantly, near-synoptic over enormous areas. Moreover, the frequency of observations of large regions inaccessible by other means (especially ships) is unique. Time series of CZCS images have revealed that the dynamics of the pelagic ecosystem are much greater than expected, and governed to a large extent by physical forcing. Conversely, physical oceanographers can use the complicated patterns of pigment distribution as markers of small-scale and large-scale water movements (Holligan, this volume). A major disadvantage of the CZCS was that the signal due to the presence of phytoplankton cannot be distinguished from the signal due to the abiotic light absorption component. In neritic regions, much light is absorbed by Gilvin (*Gelbstoff* or *yellow substance*) and particulate detritus, and absorption by phytoplankton pigments in such areas is relatively insignificant at most wavelength ranges in the visible part of the spectrum at the 4 spectral bands (443, 520, 550 and 670 mm) that are used for the CZCS images. The persistent chlorophyll abundance in the shallow North Sea throughout the year on CZCS series (Fig. 5) is not in agreement with Seatruth observations (Fig. 1), and merely reflects non phytoplanktonic detrital suspended matter that may become particularly dense after winter storms. A successor to CZCS, SeaWiFS, which uses twice the number of spectral bands and much higher radiometric sensitivity in order to allow discrimination between phytoplankton and detritus, possibly even between different algal groups absorbing light at different bands/wavelengths, will be launched in 1992. Another ocean color sensor will be on Japan's space agency NASDA's ADEOS satellite starting in 1996. Images derived from NOAA-n satellites and AVHRR will continue to provide distribution maps of coccolithophorids, a group of algae that is of major importance in the global CO_2 cycle, which scatter light intensely when present in blooms (*milky waters*: Ackleson *et al.*, 1988).

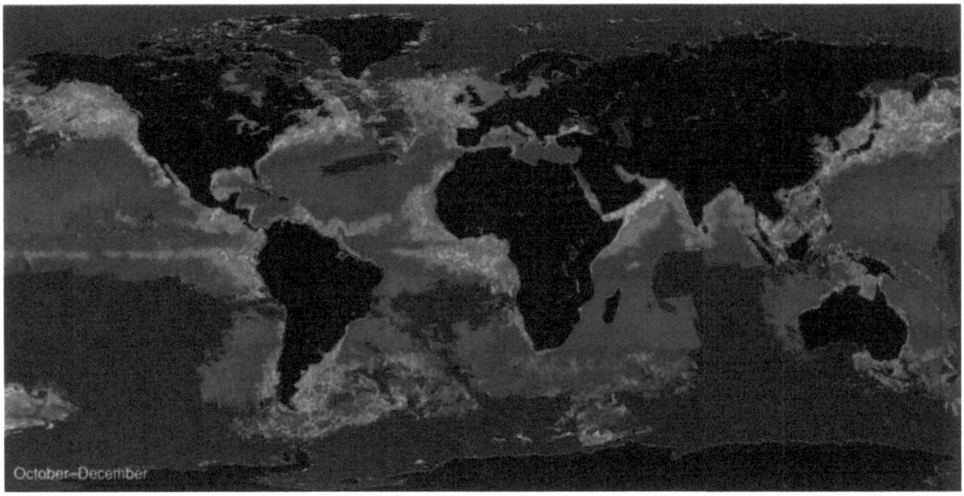

Figure 5. CZCS image of "chlorophyll" distribution during winter in the North Sea; in reality, mostly detrital pigment (compare with Fig. 1).

Seatruth observations of light absorption at all wavelengths to be covered by the new satellite generation remain a necessity, not only because phytoplankton pigmentation (both chlorophylls and carotenoids) and detrital abundance vary enormously in space and time, but also because space imagery only yields estimates of plankton presence in the upper ocean layer, thereby omitting the occurrence of a significant proportion of the global biomass that is concentrated in the deep chlorophyll maximum characteristic of all oceans between 40°N and 40°S.

Fluorometric measurements in extracts

The method used most often in marine research has been described by Holm-Hansen *et al.* (1965). The measurements are usually done with a Turner 111 fluorometer before and after acidification of an acetone extract, the latter preferably obtained by cell homogenising with glass beads and CO_2 cooling. We have also used a Perkin Elmer Hitachi fluorescence spectrophotometer, choosing a very high acid factor to reduce as much as possible the interference of phaeopigments when measuring chlorophyll *a*. Phaeopigment concentrations measured with Holm-Hansen *et al.*'s method are not reliable because not all phaeopigments (of which there are many) have the same fluorescence characteristics. Moreover, their concentration can be grossly overestimated in natural waters where chlorophyll *b* is abundant (Loftus and Carpenter, 1971), which occurs frequently (Jeffrey, 1976; Gieskes and Kraay, 1986a,b; Chisholm *et al.*, 1988; Olson *et al.*, 1990). In Fig. 6, we show the extent of this overestimation. In fact, the abundance of phaeopigment measured with the conventional fluorometric method near the bottom of the ocean's euphotic zone cannot simply be used as an indicator of the presence of detritus or of grazing. It will often merely reflect the abundance of Chl *b*-containing phytoplankton. Using HPLC, we have never detected more than traces of phaeophorbides and phaeophytins at this depth.

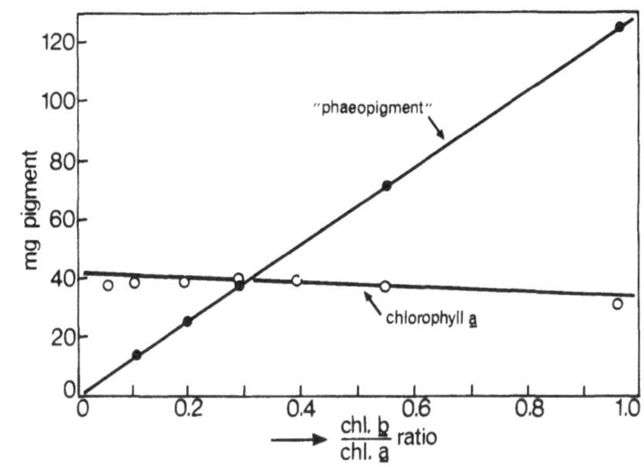

Figure 6. Overestimation of phaeopigments with the fluorometric method in the presence of Chl *b*.

A further problem concerns the equations for calculating chl *a* and phaeopigment: phaeopigment concentrations are expressed in chl *a* equivalent weight, i.e. on a molar basis. Phaeopigments in natural waters, however, may be several phaeophorbides (Gieskes and Kraay, 1986a) and several phaeophytins, possibly all of different molecular weights, so errors in the concentration of phaeopigments may be appreciable when using the Strickland and Parsons equations (Engelkes, 1985; Roy, 1986).

Absorption measurements in extracts

Other common methods for the measurement of chl *a* and bulk phaeopigment are based on absorption of light. The absorption of light by algal pigments depends on the arrangement of pigments in the chloroplasts and on their location in the algal cells. The cell's organisation (chains, cluster-like colonies, or single cells) also influences light absorption. Light absorption further depends on characteristics of the algal cell wall, especially structure and composition (e.g., calcium-carbonate containing coccolithophorids are strong light scatterers, though mainly in the blue range of the spectrum). All these problems are avoided when chloropigment light reabsorption is measured in a methanol or acetone extract of suspended matter. Chlorophyll *a* and its phaeopigments dominate light absorption in the red part of the spectrum;

measurements in acetone extracts are usually made at 665 nm before and after acidification (Lorenzen, 1967). Chlorophyll *b* and *c* have absorption spectra that overlap with the spectrum of chl *a*. Trichromatic equations, such as those designed by Richard and Thompson (1952) for calculation of the concentration of Chlorophyll *a*, *b* and *c* separately after absorption readings at 665, 645 and 630 nm, can hardly be reconciled with Beer's law and have been criticized heavily by Rott (1980). Improved trichromatic equations have been given by Jeffrey and Humphrey (1975), but even these cannot always be used without caution (Rott, 1980; Lorenzen and Jeffrey, 1980).

The specific absorption coefficients of the major phaeopigments (phaeophorbide *a* and phaeophytin *a*) are known, and so are their absorption spectra. However, unknown phaeopigments may also be present (Gieskes and Kraay, 1986a, b), perhaps with different spectral characteristics, and a choice of a fixed acid factor (1.7: Lorenzen, 1967b) for all natural phytoplankton populations may therefore occasionally give questionable results. The acidification step in the method described by Lorenzen is quite critical (Moed and Hallegraeff, 1978): too much acid causes shifts in the absorption maximum to wavelengths below 660 nm and increases the 750 nm blank reading, e.g. when fucoxanthin breaks down to products absorbing in that spectral region (Riemann, 1978).

ABSORPTION SPECTRA OF FRESHLY COLLECTED SUSPENDED MATTER

In recent years, interest has been renewed in an old, very simple method for the measurement of phytoplankton pigments, and this time not only chl *a* (foregoing sections) but also other pigments. The procedure typically involves filtration of suspended matter onto a glassfiber filter (Whatman GF/C or GF/F) which is then placed, while still wet, in the light path of a spectrophotometer to obtain an absorption spectrum of any wavelength range between 220 and 720 nm (Shibata *et al.*, 1954; Trüper and Yentsch, 1967; Shibata, 1969; Gieskes and Kraay, 1990). When the spectrum is recorded quickly one doesn't even need a glass support for the filter. Second and higher derivative analysis of such spectra (Fig. 7) provides much information on the chlorophylls and phaeopigments in the sample, and also of different carotenoids (Faust and Norris, 1985; Stolte, 1988; Bidigare *et al.*, 1989). An advantage of absorption spectra made of material that is not extracted in acetone is also that the absorption by water-soluble phycobilin pigments is visible (Steenbergen and Korthals, 1982; Zevenboom, 1986). These pigments are present in Cryptophyceae and Cyanobacteria, algal classes that may dominate natural phytoplankton in tropical seas (Waterbury *et al.*, 1979) but are also present in temperate waters (Gieskes and Kraay, 1983a).

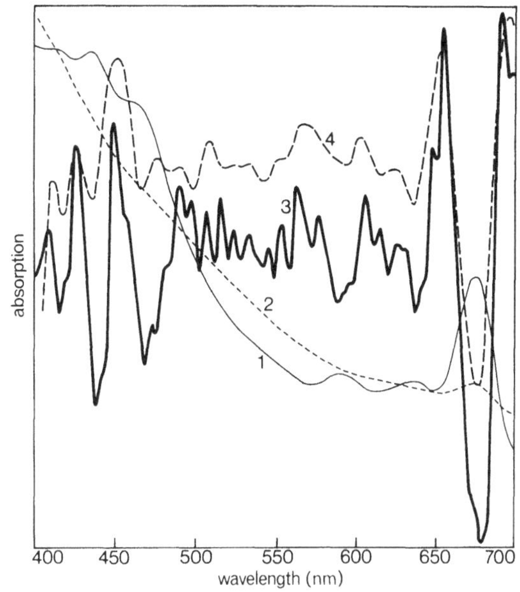

Figure 7. Derivative analysis of *in situ* absorption spectra for recognition of chlorophylls and carotenoids. The chl *a* peak at 673 nm is correlated with the chl *a* concentration measured in an acetone extract (r = 0.8). 1 and 2: Absorption of suspended matter in the North Sea at two stations; sample 2 consists mostly of detritus. 3 and 4: second derivative spectra of 1 and 2.

The major difference between absorption spectra of pigments in acetone, ether, methanol or ethanol extracts and *in vivo* spectra is that the latter are flattened (Duysens, 1956), and that absorption peaks are shifted to longer wavelengths, especially in the blue part of the spectrum, where carotenoids of the fucoxanthin family cause peak shifts of up to 40 nm. This does not diminish the advantages of the method: scans of fresh material are made within minutes, and on-board derivative analysis of the absorption spectra for detection of taxon-specific carotenoids and chlorophylls may give a quick impression of taxonomic composition of phytoplankton without the need of concentrating samples, microscopy, or other time-consuming alternatives for population structure analysis. If, for example, we examine Fig. 7, the height of the absorption maximum that can be identified without derivative analysis, the 673 nm *in vivo* maximum due to chl *a*, is directly proportional to the chl *a* concentration in acetone extracts, and can thus be used as a rough index of algal biomass.

Finally, little-studied pigments absorbing ultraviolet light can be identified and quantified readily with the *in vivo* method: the mycosporine-like compounds that were found in coral reef organisms including algae by the inventor of the method, Shibata (1969), and were recently found by us in ice algae and in phytoplankton of the Weddell Sea. In a recent article (Gieskes and Kraay, 1990), we have suggested that the relatively high absorption peaks at 270 and

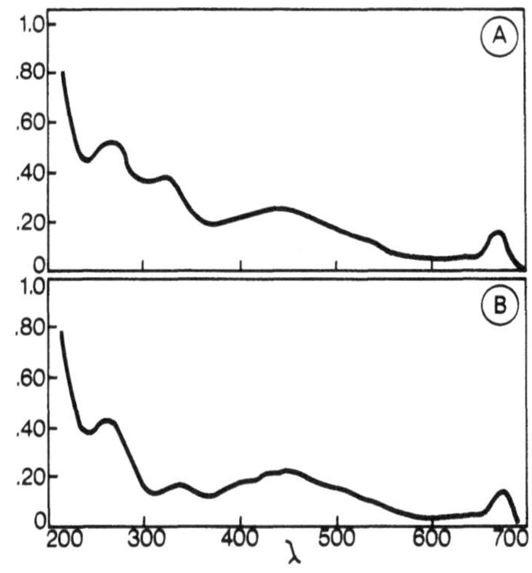

Figure 8. Absorption spectra of ice algae (a) and of phytoplankton (b) in the Weddell Sea, Nov. 1988. Notice pigments absorbing light in the U.V. part of the spectrum (peaks at 270 and 330 nm).

330 nm (and thus probably the concentration of these compounds) recorded by us in November 1988 in Wedell Sea ice algae (typically exposed to high UV-B and UV-A irradiance in the snow-ice intertistium or under thin sea-ice) may be indicative of adaptation to the increased UV irradiance underneath the ozone hole (Fig. 8).

CHROMATOGRAPHY: COMPARISONS WITH OTHER METHODS

In the previous section, I have described the simple but fairly effective methods that allow rough quantitative analysis of individual pigments (e.g. chl *a*, mycosporine-like amino- acids) and qualitative analysis of pigment groups (e.g. the fucoxanthins, other carotenoids). However, the best way to measure pigments in mixtures is to separate the individual components, identify each one, and measure their concentrations. The latter is possible because the specific absorption coefficient and other spectral characteristics of the most important phytoplankton pigments are known (Foppen, 1971; Davies, 1976; Abaychi and Riley, 1979; Mantoura and Llewellyn, 1983). Very good separation can be achieved by paper (Jensen and Sakshaug, 1973) and thin layer chromatography (Jeffrey, 1981). Even better is high-performance liquid chromatography is even more accurate because concentrations can be quantified with great precision. HPLC (Robinson, 1979) can easily be performed at sea, and a complete pigment fingerprint can be obtained in less than 10 minutes, although we usually limit the retention time of the slowest pigments (carotene α and ß, phaeophytin *a*) to 35 minutes.

A great variety of HPLC equipment is now available, and the number of methods described in the literature has been increasing steadily (Roy, 1987). With our own methods, which resemble those of Abaychi and Riley (1979) and Mantoura and Llewellyn (1983), we obtain chromatograms of very high quality (Fig. 9). Recent versions of our methods (one isocratic, the other reversed-phase) have been described in Gieskes and Kraay (1983b), Gieskes and Kraay (1986a) and Gieskes *et al.* (1988). A major difference with the method of Mantoura and Llewellyn (1983) is that we did not use an ion-pairing reagent: buffering of solvents with a phosphate buffer (ion strength 0.05, pH 7) and extraction of the sample in 2% ammonium acetate in methanol was found to be sufficient to achieve good resolution of all chloropigments, including Chl *c*1+2, Chl *c*3, chlorophyllide a, etc. (see Fig. 9). Zeaxanthin and lutein separation is notoriously difficult (Hager and Meyer-Bertenrath, 1967), but co-occurrence in natural phytoplankton is in our experience exceptional. Ben-Amotz *et al.* described a method for separation in 1982, in J. Phycol. 18:529-537. The two pigments have a different absorption spectrum in all organic solvents (we prepare lutein standard from ferns, zeaxanthin from maize), and this can be the test for which is present in a sample.

Our measurements of chl *a* after separation by HPLC are normally well-correlated with chl *a* concentrations measured with the conventional fluorometric and absorption acidification methods (see previous sections), both in cultures and in natural populations, in a wide variety of water types (Fig. 10a). The regression equation is never far from Y = 0.67 X, so the conventional methods give a mean overestimate of 30%. This can normally be ascribed to the presence of chl *a* allomers and isomers, chlorophyllide *a*, and unknown substances detected as tiny but sometimes numerous peaks in the chromatograms. The large discrepancy between HPLC and the conventional methods reported by Jacobsen (1978), Sartory (1985) and Mantoura and Llewellyn (1983) has never been found by us, not even in the turbid Wadden Sea, the polluted Rhine river, or Northwest European coastal waters (see Fig. 10a). On the other hand, Trees *et al.* (1985) stated that HPLC measurements overestimate chl *a* concentrations! Our own comparison between methods of TLC (Gieskes and Kraay, 1978), HPLC (Gieskes and Kraay, 1983b, 1984, 1986, 1988) and conventional methods was always quite satisfactory.

Nevertheless, it is clear that intercalibration remains a necessity. An improve intercalibration exercise is presently being coordinated by R.R. Bidigare, now at Hawaii University, Honolulu. The JGOFS program (Joint Global Ocean Flux Studies) for which this intercalibration was designed, tries to define and estimate of the extent of oceanic carbon transformations. Phytoplankton plays a vital role in the global carbon budget; the synoptic mapping of these organisms by satellite which is considered essential for this program,

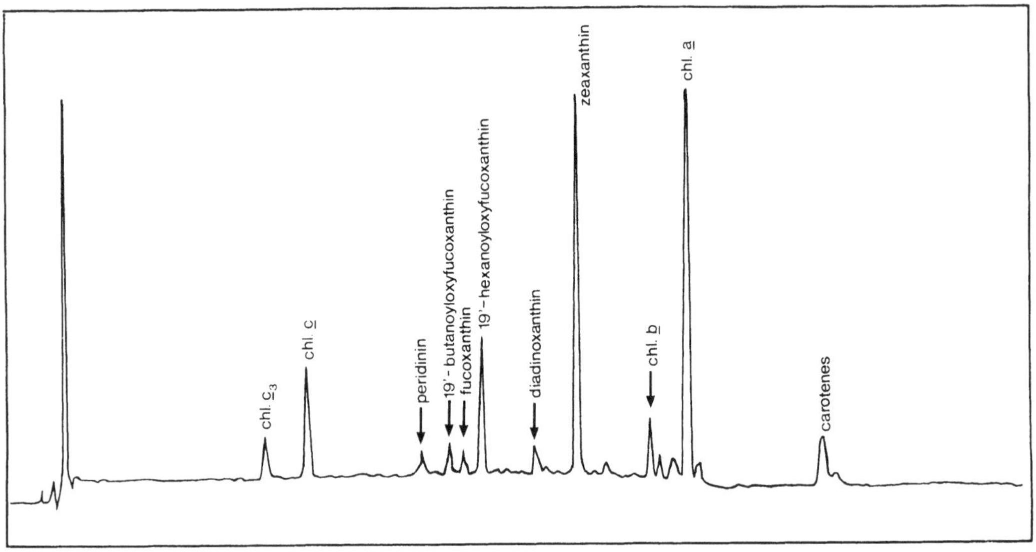

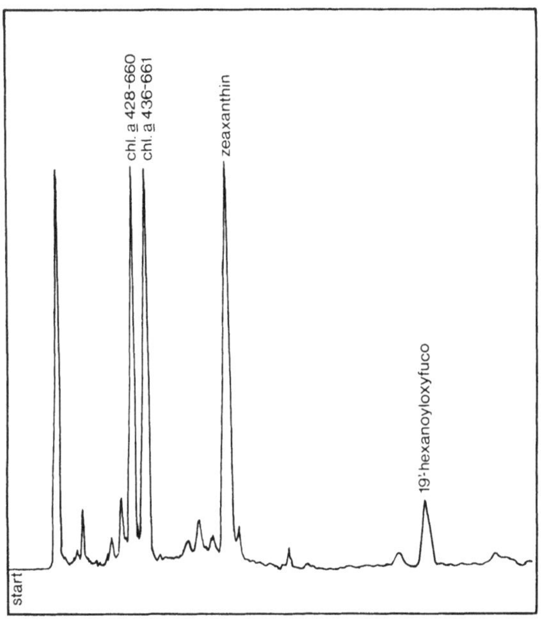

Figure 9. High-performance liquid chromatograms: isokratic (lower) and reversed-phase (upper diagram).

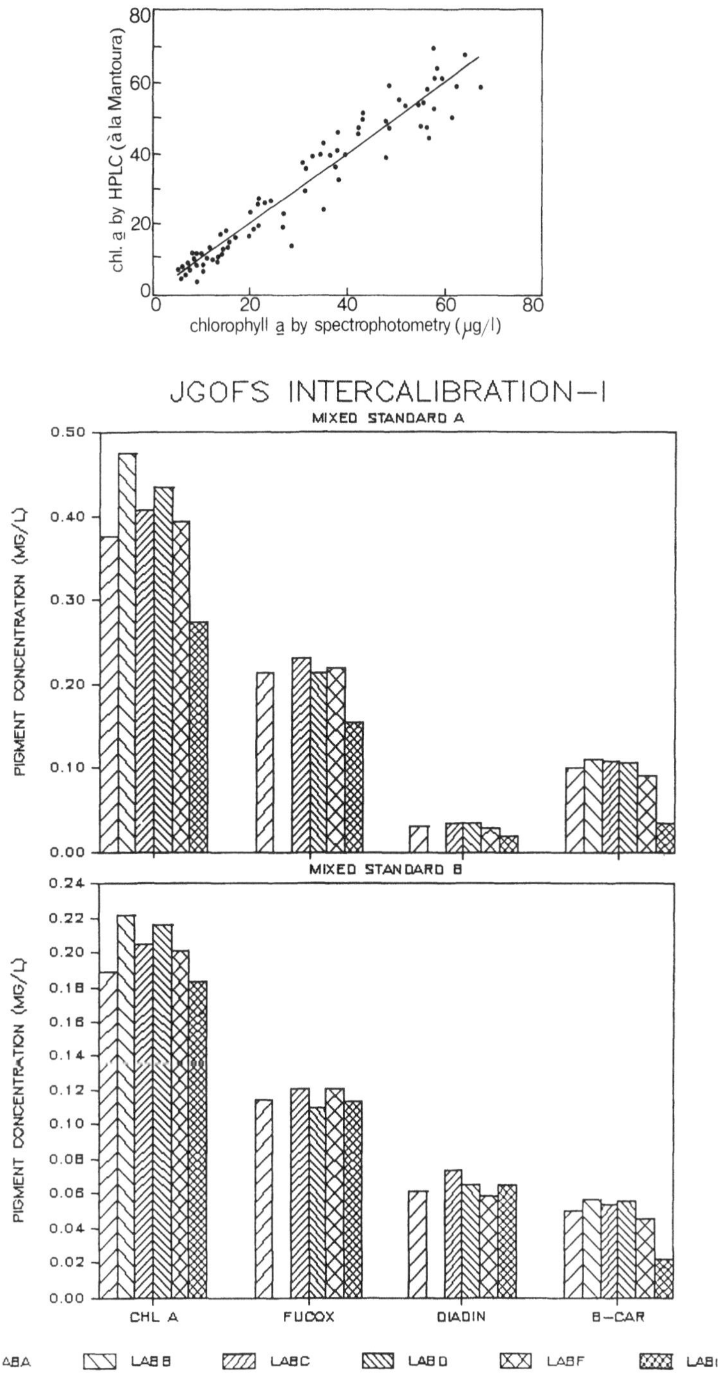

Figure 10. a) Correlation of chl *a* measured after HPLC separation (method of Mantoura and Llewellyn) and concentrations measured with conventional spectrophotometric methods. b) Example of Dr. R. Bidigare's intercalibration exercise, 1989 - 1990 (JGOFS)

requires a complete set of seatruth observations for the interpretation of the satellite signal received on earth. Early intercalibrations for chl *a* alone often showed enormous discrepancies (Schlittenhardt, 1980; Larsson *et al.*, 1978; Nusch, 1984), but progress has apparently been made and the results of the participants in Bidigare's intercalibration exercise are surprisingly close to each other (Fig. 10b)). One of the tests conducted by Bidigare reveals that it is possible to keep the concentration of pigments unchanged for many months after sealing and refrigeration at liquid nitrogen temperatures. We have the experience that pigment loss is kept to a minimum (at most 20% after 1 year) under normal deep-freeze conditions of -30°C. Freeze-drying is not allowed (Lenz and Fritsche, 1980).

APPLICATION OF PIGMENT ANALYSIS IN PLANKTON ECOLOGY

Adaptation of phytoplankton populations to environmental conditions

The adaptation of phytoplankton to environmental conditions is often expressed in the pigment composition. For example, zeaxanthin is produced in green algae that are nitrogen-limited (Czygan, 1968; Weber and Wettern, 1980) and in *Synchococcus* (cyanobacteria),

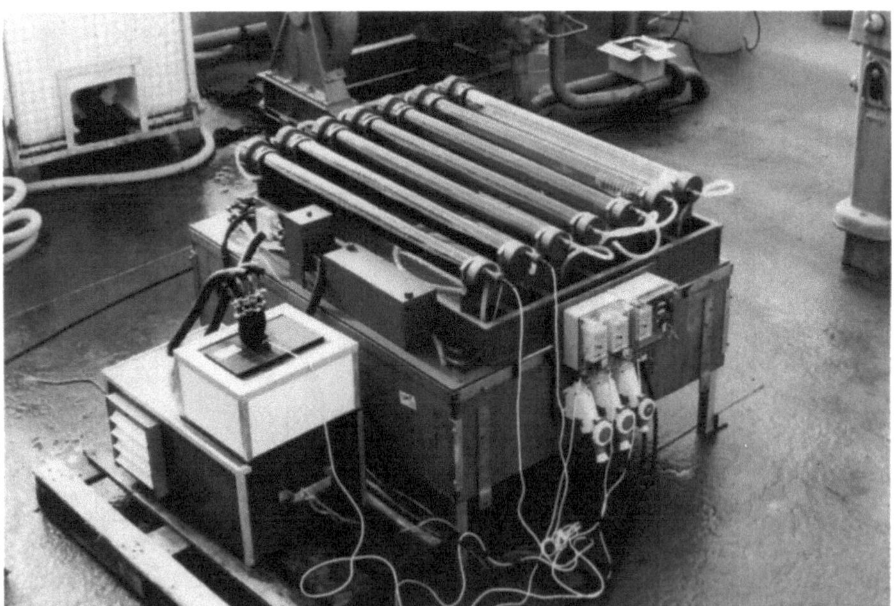

Figure 11. Blue-light incubator for simulation of underwater light conditions in the ocean.

phycoerythrin is used as a nitrogen store in oligotrophic waters (Wyman *et al.*, 1985). However, light is the most obvious condition that influences the pigment contents of cells. The chl *a*, Chl *b*, Chl *c*, fucoxanthin, violaxanthin and peridinin concentration can be changed to enhance light absorption for photosynthesis in the aquatic environment (see Introduction), where light is not only variable in quantity but also in quality (Fig. 2); zeaxanthin and lutein, on the other hand, are photosynthetically incompetent (Bigidare *et al.*, 1989) but do provide protection against light-induced, photosensitized oxidation. Indeed, the high zeaxanthin concentration near the tropical ocean's surface (as much zeaxanthin is found here as chl *a*) suggests adaptation of the omnipresent coccoid Cyanobacteria (Waterbury *et al.*, 1979) and other bluegreen algae to bright light (Pearl *et al.*, 1983; Gieskes and Kraay, 1986).

New picoplankton, new ideas on light adaptation

The scarcity in the ocean's deep chlorophyll maximum layer of violaxanthin (a shade-adaptation pigment of green algae and abundant in Halosphaera, a representative of the shade flora, Sournia, 1982) suggests that green algae are less prominent there than previously thought (Jeffrey, 1976). Interestingly, the picoplanktonic Prochlorophytes commonly occurring near the bottom of the euphotic zone of oceans (Gieskes *et al.*, 1988; Chisholm *et al.*, 1988; Olson *et al.*, 1990; Veldhuis and Kraay, 1990) seem to be better adapted to shady conditions than green algae, which have been reported to contain relatively high amounts Chl *b* in dim light (French *et al.*, 1959; Brown and Richardson, 1968; Ramus *et al.*, 1976; Titlyanova *et al.*, 1977). The Prochlorophytes do not contain normal chl *a* and *b*, but red-shifted versions (Fig. 3), with an *a : b* ratio of 1 : 1 (Olson *et al.*, 1990; Chisholm *et al.*, 1988). This enables increased absorption of blue light, the only spectral region of sunlight irradiance available at this depth in the ocean (Wood, 1985; Glover *et al.*, 1986, and see Fig. 2 and 11), and also in the wavelength range emitted by luminescent bacteria associated with the decaying organic debris that tends to accumulate near the bottom of the euphotic zone of the stratified open ocean (Sieburth, 1988).

The red-shifted chl *a* and *b* of the deep-chlorophyll-maximum's Prochlorophytes, which we detected in Atlantic Ocean samples by TLC in 1977 (Gieskes *et al.*, 1979; see their Fig. 2), disappears rapidly upon transfer to shallower depths where more surface light penetrates (Fig. 12b). This also suggests that this picoplankton group is adapted to life at depth. The red shift of about 5 nm in the absorption spectrum in acetone and ether (Fig. 3) is probably due to a divinyl bond in the chlorophyll molecule (Goericke, pers. comm.). Bazzaz (1981) and Bazzaz and Brereton (1982) found a divinyl chlorophyll in maize mutants with absorption and fluorescence characteristics that resembled those of our picoplanktonic chl *a*. It was the only chl *a* that the Maize plants had - exactly as is the case with oceanic Prochlorophytes, which do not contain a trace of normal chl *a* (Chisholm *et al.*, 1988; surprisingly enough, freshwater Prochlorophytes and Prochloron sp. do not contain divinyl-Chl *a*: Burger-Wiersma *et al.*, 1986; Foss *et al.*, 1987). There are other reports of algae photosynthesizing in spite of a lack of normal chl *a* (Kowalewska *et al.*, 1987; Kowalewska, 1988).

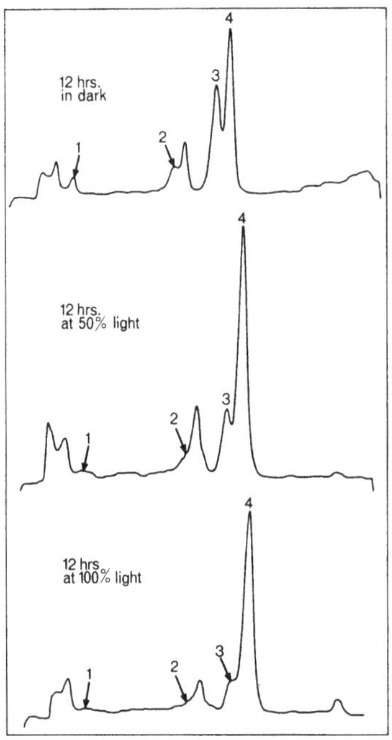

Figure 12. Results of transfer experiments with Sargasso Sea samples containing the red-shifted "divinyl-Chl *a*": photobleaching. Fluorescence scans of thin-layer chromatograms developed in one direction. Peak 2: Divinyl-Chl *b*. Peak 3: Divinyl-Chl *a*.

Our own attempts to characterize the marine Prochlorophyte's chl *a* and *b* by NMR, including COSY, have not been successful due to difficulties of sampling enough material (Van Dijken, 1990). We intend to use Fast Atom Bombardment mass spectroscopy and Time-of-Flight mass spectroscopy (as developed by J. Grotemeyer, Germany) for further characterization of the red-shifted chlorophylls *a* and *b* of oceanic picoplanktonic Prochlorophytes, in cooperation with Dr. W. Wierenga, Dept. of Chemistry, Groningen University, and Prof. J.J. Boon, F.O.M. Institute, Amsterdam.

Fingerprinting of taxon-specific pigments and recognition of water masses

Several pigments or pigment combinations are, as we have seen, only found in certain algal groups or classes. Van den Hoek (1978) gave a list of phytoplankton classes and their

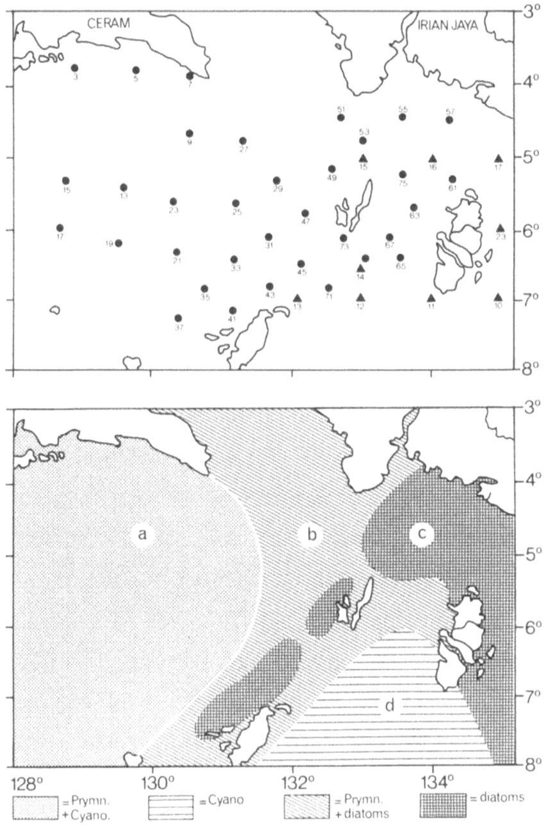

Figure 13. Distinction of water masses, southeast monsoon (upwelling season), in the Banda Sea (Indonesia) by their fingerprints of algal pigments. a) Mainly Prymnesiophyceae and Cyanobacteria. b) Mainly Prymnesiophyceae and Diatoms. c) Mainly diatoms. d) Mainly Cyanobacteria.

pigments known at that time. Since then, Hooks *et al.* (1988), Wright and Jeffrey (1987), Jeffrey (1989), Olson *et al.* (1990) and Gieskes and Kraay (1983, 1986, 1988) have reported on other taxon- specific pigments: red-shifted chl *a*, red-shifted Chl *b*, Chl *c*3, Prasinoxanthin (Foss *et al.*, 1986), 19'-hexanoyloxyfucoxanthin, 19'-butanoyloxyfucoxanthin, etc.

In different water masses, the population composition of phytoplankton is usually different, and the pigment fingerprint of the suspended matter in a sample can therefore be characteristic for the water mass from which it was taken. Spectral fingerprinting of phytoplankton populations by two- dimensional fluorescence and Fourier-transform-based pattern recognition was attempted by Oldham *et al.* in 1985. A much simpler method, cluster analysis of HPLC-derived pigment fingerprints measured in large numbers of samples taken during surveys in the North Sea and in the Indonesian Banda Sea, revealed that the border front between water masses can often be better described by differences in phytoplankton species composition than by hydrographic parameters (salinity, temperature) (see Fig. 13). Of course, detailed microscopic analysis of populations would have made us conclude the same, but the disadvantages of microscopy are numerous (see Introduction). Moreover, microscopy is time-consuming, while a series of chromatograms can be developed rapidly, even on board, for cluster analysis.

^{14}C Labelling of pigments: assessment of phytoplankton-specific biomass and taxon-specific productivity

In order to estimate the contribution of different algal groups to total phytoplankton production, we have measured how much ^{14}C was incorporated into different taxon-specific pigments during incubation of samples of Banda Sea water (eastern Indonesia). Redalje and Laws (1981) described a method to estimate algal biomass and algal-specific growth rates by incubating phytoplankton and measuring the amount of ^{14}C that they found in particulate matter and in chl *a*, which they separated by TLC. We have expanded this method to measuring ^{14}C in all major chlorophylls and carotenoids separated by HPLC. By using HPLC, it is easier to avoid coelution of pigments with other labeled cell compounds than with TLC. We were able to show (Gieskes and Kraay, 1989) that the specific growth rate of zeaxanthin-containing groups (*in casu* Cyanobacteria) was usually highest, except in some areas where the specific growth rate calculated on the basis of ^{14}C labelling of fucoxanthin was as high; diatoms were apparently the main producers here. In other areas, the specific growth rate based on ^{14}C labelling of hexanoyloxyfucoxanthin was highest, suggesting that here the contribution of Prymnesiophyceae (mainly Coccolithophorids) was highest. An alternative method for estimating growth rates of individual algal groups in natural populations using flow cytometry has recently been used by Veldhuis. His (net) growth rates indicate that the high (gross) rates registered by us are balanced by high rates of consumption.

R. Goericke (1990: Ph.D. thesis; a number of papers will be published in the course of 1991) has examined the method of ^{14}C- labelling the various pigments of algal families in great detail, and he concludes that it should be used with caution. The high, differential turnover rate of some pigments may be a significant source of error, leading to false estimates of primary production rates of some classes. High rates of pigment production can, for example, be due to rapid interconversions related to photo-adaptation of algal cells, e.g. those in the diadinoxanthin / diatoxanthin system. Measurement of such rates may be useful as an indicator of the rate of vertical water movements through a light gradient (Welschmeyer and Hoepffner, 1986).

Prof. E. Laws (University of Hawaii at Manoa, U.S.A.), one of the originators of the *Redalje-Laws Method*, also recently pointed out some (minor) flaws in the calculations of Gieskes and Kraay (1989). He objected to their statement that Since chl *a* is the pigment common to all three classes, its specific activity should be ... equal to the mean of the specific activities of the class-specific carotenoids.; G. and K. then went ahead and averaged the growth rates calculated from the carotenoids and compared that average with the chlorophyll *a* - specific growth rate (Gieskes and Kraay, 1989). Laws argues as follows: first, the chl *a* specific activity is a weighted average of the specific activity of the chl *a* in the components of the phytoplankton community. The weighting factor is the fraction of chl *a* contributed by each component. For example, in a two- component system

$$\text{total chl } a = (\text{chl } a)_A + (\text{chl } a)_B$$
$$\text{activity chl } a = (\text{activity chl } a)_A + (\text{activity chl } a)_B$$

where A and B are the two components. The specific activity of the chl *a* is then (activity chl *a*)/(total chl *a*). It is easy to show that the specific activity of the chl *a* is therefore equal to f_A (specific activity chl $a)_A$ + f_B (specific activity chl $a)_B$ where $f_A = (\text{chl } a)_A/\text{total chl } a$ and $f_B = (\text{chl } a)_B/\text{total chl } a$. This conclusion is easily generalized to show that

$$\text{specific activity chl } a = \Sigma f_K \, (\text{specific activity chl } a)_K$$

where the summation is performed over all components in the population. Given information on the carotenoids in the population, one would need to know the ratio of carotenoid to chl *a* for each component in order to calculate the weighted average of the component specific activities. What one can rigorously say is that the specific activity of the total chl *a* will lie somewhere between the extreme specific activities of the components. The total chl *a* specific activity will lie close to the minimum component specific activity if most of the total chl *a* is associated with the slowest-growing component and close to the maximum component specific activity if most of the total chl *a* is associated with the fastest-growing component.

Even if one knew the appropriate weighting factors, the comparison of growth rates could not be made by averaging the component growth rates. *The reason is that growth rate and specific activity are related in a logarithmic manner*. The average value of a logarithm is not equal to the logarithm of the average value of the argument of the logarithm. One can, however, show that the carbon-specific growth rate of the entire population is equal to a weighted average of the carbon-specific growth rates of the population components. The reasoning is similar to that used in the foregoing example of chl *a* specific activities. Hence one can say that the carbon-specific growth rate of the entire population should lie somewhere between the extreme carbon-specific growth rates of the population components, being close to the minimum if phytoplankton carbon is dominated by the slowest-growing component and close to the maximum if phytoplankton carbon is dominated by the fastest- growing component.

Calculating the abundance of different algal taxa using pigment concentrations

Multiple regression analysis

Recently, Gieskes et al (1988) have used a mathematical analysis of algal pigment fingerprint series described earlier (Gieskes and Kraay, 1983b) by which it is possible to estimate the contribution of the major taxonomic groups to the total chl *a* contents of natural phytoplankton populations. This method may suffer from the instability of the carotenoid-to-Chl *a* ratio, which is in several classes determined by the physiological condition of cells (e.g., Klein, 1988) and the level of light adaptation (see previous section). Bidigare *et al.* (1989a) summarized existing literature on light adaptation by stating that in the ocean's phytoplankton the ratio of non- photosynthetically active pigments-to-Chl *a* decreases with depth while the ratio of photosynthetically-active pigments-to-Chl *a* increases with depth. Gieskes *et al.* (1988) avoided the ratio problem by first arranging similar populations by cluster analysis, then doing multiple regression analysis on each group separately.

Another approach, and application to picoplankton populations from oligotrophic oceans

Another method developed with G. Kraay and M. Veldhuis, which calculates biomass of individual taxonomic groups from pigment data is more straightforward. Errors made with this approach are minimal when cell size accounts for variations in pigment concentration per unit biomass. A few examples will be presented to illustrate how effective this method is. I use the chlorophyll and zeaxanthin concentrations per cell of picoplankton Prochlorophytes given in Chisholm *et al.* (1988). Concentrations recently published by Veldhuis and Kraay (1990) are somewhat different: 0.91 fg divinyl chl *a* per cell Prochlorophyte near the surface, 5.4 near the bottom of the euphotic zone.

At Station 53, (Banda Sea, Indonesia, March 1985, upper 25 m) (Fig. 13), the zeaxanthin concentration was 0.0487 μg/l and the chl *a* concentration 0.132 μg/l. Exactly half of this chl *a* (0.065 μg/l) was *divinyl-Chl a* - the red-shifted chl *a* (Gieskes and Kraay, 1983b) of picoplanktonic Prochlorophytes (Chisholm *et al.*, 1988). According to Chisholm *et al.* (1988), such Prochlorophytes contain 2.15 fg chl *a* per cell, so at this station there must have been 30,000 Prochlorophyte cells per ml. These cells contained 0.0207 μg zeaxanthin per liter (0.69 fg/cell; Veldhuis and Kraay reported 1.97 fg per cell), so the rest of the zeaxanthin, 0.028 μg/l, must have been associated with another zeaxanthin-containing group - coccoid Cyanobacteria (Guillard *et al.*, 1985; Kana *et al.*, 1988), numbering 15,500 cells per ml. In *Synchococcus,* the zeaxanthin:Chl *a* ratio is 1.2 (Fig. 1D in Kana *et al.*, 1988), so 0.065 - 0.023 = 0.042 μg chl *a*/l at this station was associated with other algae. The chromatogram at this station shows clear peaks of fucoxanthin, hexanoyloxyfucoxanthin and butanoyloxyfucoxanthin, indicative of the presence of these other algae, in casu diatoms, Prymnesiophyceae and Chrysophyceae.

The same reasoning can be applied for Station 7 (Fig. 13). In surface samples taken here in March 1985, 69.3% of the chl *a* (0.069 μg/l) was red-shifted *divinyl-Chl a* (= 0.048 μg/l). For Prochlorophytes, 0.048 (red-shifted) chl *a* corresponds with 0.016 μg zeaxanthin/l. The rest of the zeaxanthin, 0.0378 μg/l, must have been associated with *Synchococcus* cells, numbering (0.0378 x 10^9): 1.8 = 2.1 x 10^4 cells per ml (each containing, according to Kana *et al.*, 1988, 1.8 fg zeaxanthin). The number of Prochlorophytes at Station 7 (surface) was (0.016 x 10^9) : 0.69 = 2.3 x 10^4 cells per ml (each containing 0.69 fg zeaxanthin). According to Kana *et al.* (1988), each *Synchococcus* cell may have contained 1.5 fg chl *a*, so 0.0315 μg chl *a* was associated with this group. Apparently, all normal Chlorophyll a present in the sample was contained in *Synchococcus* cells; the contribution of other taxonomic groups was clearly nil - a conclusion supported by the chromatogram, that did not show any carotenoid peaks except zeaxanthin.

At Station 25 (Fig. 13) samples taken at the deep chlorophyll maximum (60-80 m) contained 0.263 μg chl *a*, 0.118 μg Chl *b*, and 0.034 μg zeaxanthin per liter. The percentage of *divinyl-Chl a* was 48.6%, i.e. 0.128 μg/l - nearly as much as Chl *b*. Chisholm *et al.* (1988) also reported a 1:1 ratio in Prochlorophytes. Apparently, all Chl *b* at Station 25 (60-80 m) was associated with Prochlorophytes. 0.128 μg chl *a* corresponds to 0.043 μg zeaxanthin, so all zeaxanthin present in the sample was Prochlorophyte zeaxanthin (59,500 cells/ml), leaving no room for *Synchococcus* here. On the other hand, surface samples at Station 25 (0, 20 and 30 m) contained mostly *Synchococcus*: the chl *a* concentration was 0.0688 μg/l, *divinyl-a* was 65.3% or 0.0449 μg/l, Chl *b* was 0.0069 μg/l, and zeaxanthin 0.061 μg/l. The divinyl-*a*:Chl *b* ratio here was 6.5 - far higher than the 1:1 ratio at depth, where the Prochlorophytes were clearly shade-adapted. 0.049 μg *divinyl-Chl a* corresponds with 20,883 cells/ml Prochlorophytes, which contributed 0.0135 μg zeaxanthin (Chisholm *et al.*, 1988). Therefore, zeaxanthin from the *Synchococcus* cells contributed 0.061 - 0.0135 = 0.048 μg/l, i.e. (0.048 x 10^9) : 1.8 = 26,666 cells/ml.

On the basis of the simple exercise presented above, I conclude that at Banda Sea Station 25 (chosen because it is a representative station for the oligotrophic ocean: Gieskes *et al.*, 1988) *Synchococcus cells were most abundant near the surface, Prochlorophytes near the bottom of the euphotic zone*. This is also the conclusion of workers who surveyed oligotrophic tropical and subtropical regions elsewhere (Olson *et al.*, 1990a,b; Veldhuis and Kraay, 1990). It is remarkable that the cell numbers recorded by these colleagues with flow cytometry are not dramatically different from the results derived by simply using taxon-specific pigment concentrations. Apparently, the differences in the pigment content per cell, which is most variable in a vertical direction (Bidigare *et al.*, 1989), are not always so large in natural populations that calculations such as those presented above are not feasible. The constancy in picoplankton cell numbers in the euphotic zone in all the globe's oligotrophic ocean regions (Atlantic and Pacific) implies that conditions determining the apparently delicate balance between production and consumption are the same everywhere in the ocean's upper 100-150 m.

In Table II, I summarize the results of the exercise presented above. A similar treatment of sample series in the Gulf of Guinea (West of Africa) and in the tropical Atlantic along 20°N yielded approximately the same cell numbers.

Grazing, degradation

Grazers such as copepods do not feed indiscriminately on algal cells of all types (Paffenhöfer and Knowles, 1978). Species selectivity can readily be estimated by HPLC analysis of class-specific pigments in waste products of grazers (Kleppel and Pieper, 1984; Burkill *et al.*, 1987), and carotenoids have often been used in food chain studies (e.g., Partali *et al.*, 1987). However, a quantitative assessment of grazing on phytoplankton is still a problem because of loss of pigments, i.e. incomplete recovery, between ingestion, digestion and egestion (Lopez *et al.*, 1988; Gieskes et al, 1991; Fig. 14). Even the concentration of phaeopigments (of which there are many, e.g., Gieskes and Kraay, 1986a) cannot always be used as a measure

Table II. Cell numbers of *Synchococcus*, Prochlorophytes, and eukaryotes in samples taken in February - March 1985 (Banda Sea, Indonesia). DCM = Deep Chlorophyll Maximum layer (60 - 80 m).

Station	Total Chl *a*	% Divinyl Chl *a*	Total zeaxanthin	Number Prochlorophyte cells per ml	Number Synechococcus cells per ml	%Contribution to total Chl *a* by Eukaryotes
SURFACE						
53	0.130	50.0	0.049	30.200	15.466	32.15
7	0.069	69.0	0.054	22.100	21.400	0
5	0.070	63.9	0.055	20.800	22.580	0
7	0.070	69.3	0.063	22.500	26.350	0
7	0.069	69.3	0.054	22.200	21.474	0
9	0.057	65.5	0.059	17.300	26.121	0
13	0.054	60.1	0.044	12.500	19.622	0
15	0.056	63.2	0.055	16.400	24.245	0
17	0.086	66.3	0.047	26.500	15.945	5.89
19	0.100	60.9	0.051	28.300	17.475	12.89
23	0.065	57.2	0.032	17.200	11.148	17.07
29	0.126	60.4	0.046	35.400	11.968	25.33
41	0.077	63.5	0.053	22.700	20.726	0
65	0.043	57.1	0.027	11.400	10.622	5.85
23	0.208	25.2	0.025	24.300	4.543	71.52
47	0.215	45.0	0.027	45.000	0	56.57
61	0.494	30.1	0.092	69.100	24.599	62.43
65	0.225	43.1	0.036	45.100	2.709	55.09
DCM						
5	0.256	54.9	0.033	65.300	0	49.04
7	0.342	45.1	0.046	71.700	0	55.83
9	0.243	37.2	0.022	42.000	0	65.20
13	0.104	47.3	0.016	22.800	118	52.53
15	0.207	47.5	0.034	45.700	1358	51.52
19	0.246	37.5	0.029	42.900	0	62.71
25	0.264	48.6	0.034	59.400	0	3.62
31	0.272	42.3	0.032	53.500	0	59.21
35	0.141	58.3	0.029	38.200	1454	40.51
41	0.163	64.5	0.063	48.900	16	20.54

of grazing, certainly not when protozoa are the most important grazers: when this is the case very little chl *a* is degraded to phaeophytin and none to phaeophorbide (Klein *et al.*, 1986); carotenoids are digested to colorless residues.

Engelkes (1985) and Gieskes *et al.* (1991) recorded the appearance of a fucoxanthin derivative when copepods grazed diatoms (Fig. 15), and peridinin and other carotenoid derivatives were reported by Gieskes and Kraay (1984) and Nelson (1989). Superfluous copepod feeding can

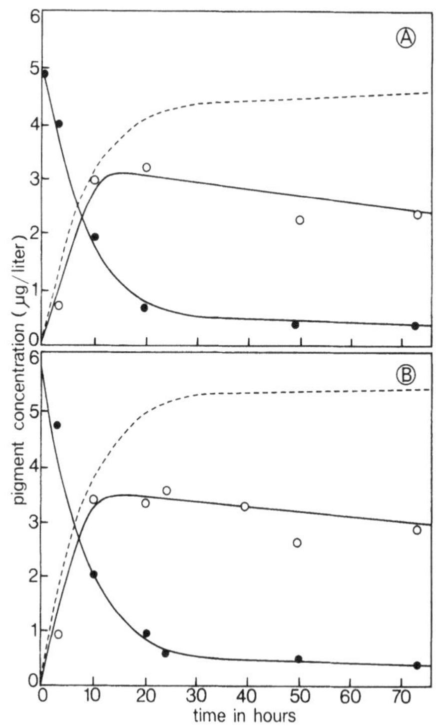

Figure 14. Recovery of chlorophyll after grazing of algal cells by a copepod: measurements by HPLC and fluorometer.

be inferred from the appearance of chlorophyllide a (Gieskes and Kraay, 1980), formed when chloroplasts liberated from cracked cells (Gieskes and Elbrächter, 1986) come in contact with seawater (Hallegraeff and Jeffrey, 1984).

Interpretation of pigment patterns in sediments is more difficult because they tend to be very complicated and confounded with, a.o., bacteriopigments (Züllig, 1985). Knowledge of the preservation and the conditions of diagenesis of the various pigments (Daley, 1973) can increase the value of the sedimentary pigment record as a tool in paleoceanography (Leavitt and Brown, 1988; Züllig, 1989). Organic geochemical, as well as taxonomic and food web

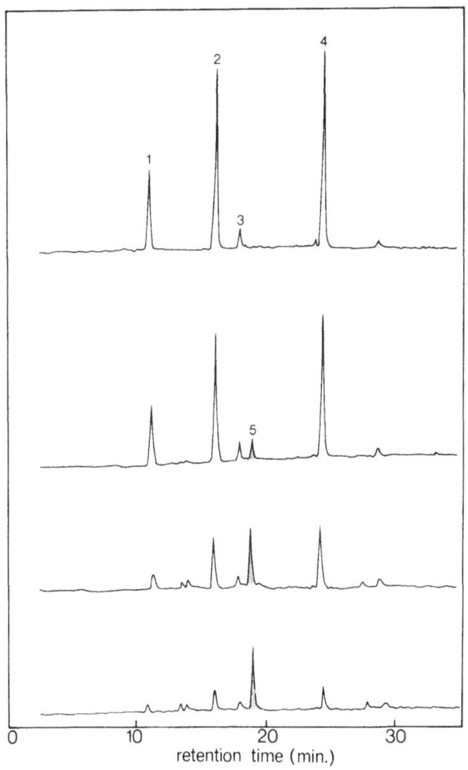

Figure 15. Appearance of a fucoxanthin derivative during copepod grazing.

studies, may further benefit from the analysis of the composition of other compounds, such as sterols and fatty acids (Nichols *et al.*, 1987), which may be more taxon- specific than pigments.

Absorption of light by detrital pigments

Models rooted in physiology to obtain vertical profiles of primary production normally make use of empirical relationships between photosynthesis, quantum yield, and photosynthetically absorbed radiation. The photosynthesis-irradiance parameters should preferably be wavelength-dependent. Therefore, not only chlorophyll *a* light absorption is of importance, but also that of all other pigments absorbing in the wavelength range of photosynthetic action (400-700 nm). An important additional source of variability in underwater light absorption is

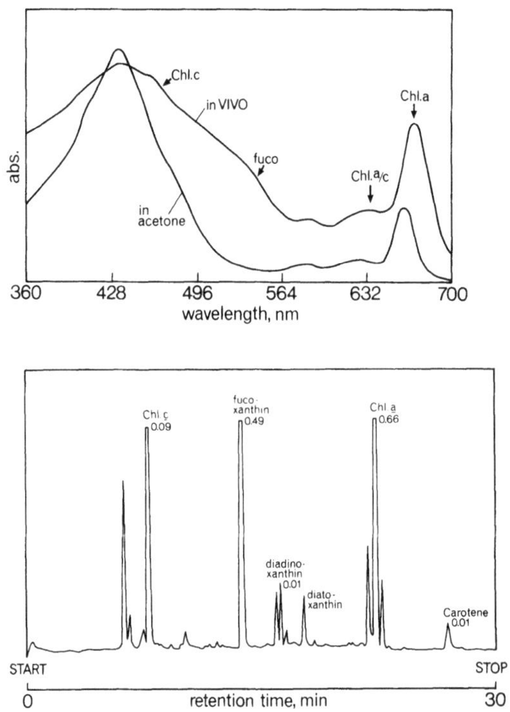

Figure 16. Comparison of absorption spectra: *in vivo* and in acetone extract. Sample of Weddell Sea, November 1986, containing mainly diatoms.

the contribution by detritus, which competes for light with photosynthetic accessory pigments (see Introduction: the fucoxanthins, violaxanthin, Chl *b*, etc.) because of its preferential absorption in the blue range of the spectrum (Yentsch, 1962; Kiefer and SooHoo, 1982; Gieskes and Kraay, 1990). Attempts to estimate photosynthetically absorbed radiation are further confounded by the presence of photosynthetically incompetent pigments (Bigidare *et al.*, 1989a), such as the abundant zeaxanthin of tropical ocean waters.

Quantification of the latter is no longer a problem, since the arrival of HPLC, but systematic assessments of the temporal and spatial importance of detrital absorption have seldom been made in any marine area. We have seen that CZCS satellite images reveal the abundance of detrital material in shallow seas such as the North Sea. In open-ocean waters, detrital presence is certainly less, but also variable, as correlations between chl *a* and Secchi disc visibility are normally low even here (Megard and Berman, 1989). Indeed, detritus, not chl *a* and associated algal pigments, contributes most to absorption of blue light, the spectral window through which the Secchi disc is seen.

The contribution of algal pigments and detritus to total particulate light absorption in marine waters should be mapped systematically on a global scale, now that oceanographic expeditions are not normally made in the framework of international joint efforts such as J.G.O.F.S. and W.O.C.E. Bidigare *et al.* (1989a,b) and Bricaud and Stramski (1990) have started this project by describing methods that are either based on spectral criteria and theoretical considerations (Bricaud and Stramski) or combine HPLC analysis of pigments in samples with a record of *in situ* particulate light absorption using a simple procedure. Bidigare used the concentrations of the various pigments to reconstruct the specific *in situ* light absorption by the whole set of pigments present in the sample, taking into consideration the shifts in maximum absorption of the pigments. These shifts are smallest in the red part of the spectrum (*in vivo* chl *a* peaks at 673 nm, in organic solvents at 660-665 nm) and greater in the blue part. Bidigare *et al.* (1989b) summarize existing knowledge of *in vivo* pigment absorption maxima in their Table 1. Pigment-specific absorption coefficients vary with the degree of pigment packaging, but this effect was found to be of little importance in natural mixed plankton populations by Smith *et al.* (1989). With Bidigare's method, it is possible to obtain the specific detritus absorption at all wavelengths by subtraction of the reconstructed *in vivo* pigment absorption from total *in vivo* particle absorption as measured on glass fiber filters.

I have designed a simpler method, resembling that of Kishino *et al.* (1985), to distinguish the contribution of phytoplankton from that of detritus in total particulate light absorption. The method was conceived on the basis of the experience that *in vivo* light absorption spectra of mixed and unialgal cultures have the same shape as spectra made of acetone extracts - except of course that (see above) the *in vivo* spectrum is always flatter (Duysens, 1956), and its peaks are shifted to longer wavelengths (Fig. 16). Concentration changes are reflected proportionally at all *in vivo*-recorded wavelengths, and wavelength-specific absorptions in the acetone extracts are significantly correlated with absorptions in the *in vivo* spectra when a wavelength shift is taken into consideration (in the red around 10 nm, in the blue up to 40 nm). Normally, an acetone spectrum is composed of the individual absorption spectra of the various pigments as measured by HPLC; these always account for $>80\%$ of the total spectrum. An acetone extract of any natural population is easily made, and its absorption spectrum can then be converted to a reconstructed *in vivo* absorption spectrum by using the conversion of acetone

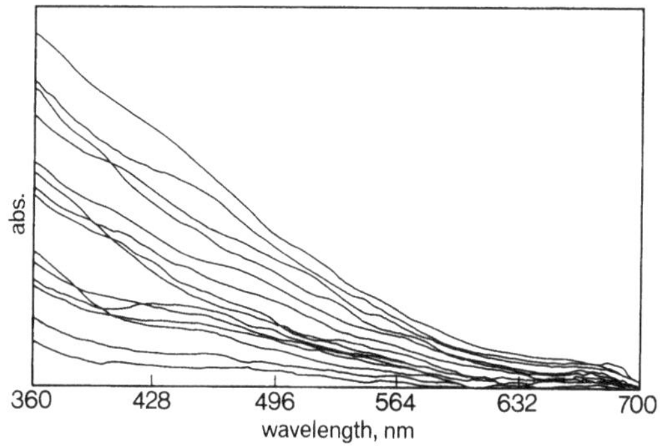

Figure 17. Detritus absorption spectra obtained with the reconstruction precedure (a). A range of spectra in samples with increasing amounts of detritus is also shown (b).

absorption to *in vivo* absorption for all wavelengths between 400 and 700 nm. The difference spectrum (total *in vivo* spectrum as made on a glassfiber filter, minus the reconstructed *in vivo* spectrum) is due to detrital absorption in the sample. In Fig. 17, a number of difference spectra are presented for samples taken in the Weddell Sea in November 1986 (Winter Weddell Sea Project, Polarstern). All have typically the shape of a detritus spectrum (Kiefer and SooHoo, 1982; Bigidare *et al.*, 1989a, 1989b), which supports the value of the method described here. In tropical waters, the difference may show a shoulder between 480 and 580 nm when phycoerythrin (only soluble in water) of PE-rich Cyanobacteria is present.

CONCLUSIONS AND SUMMARY

The stimulus that phytoplankton research received after the arrival of the highly advanced methods of individual particle analysis, such as flow cytometry for application in oceanography is matched by the power of the much simpler technique of algal pigment chromatography for investigations of the structure and function of plankton populations. HPLC is an especially attractive alternative to flow cytometry, image analysis etc., and should become a tool preferred by many oceanographers outside the conglomerate of a few wealthy research centers: the chromatography approach is sophisticated, yet accessible to a large public of scientists because of its technical and financial advantages over the newest laser light-electronic devices. Because of the ease of operation, rapidity, and precision of the measurement of concentrations of the full array of chlorophylls, carotenoids, and even UV-protecting pigments, HPLC can be applied successfully both in monitoring programs to establish phytoplankton succession, in optical oceanography, in ecophysiology, and in food chain and sedimentology studies by marine scientists of all countries of the world.

This is of great importance in an age of large-scale international programs, such as JGOFS, that are born out of the necessity of growing international cooperation: the human influence on the environment is no longer restricted to national boundaries,but has a globe-wide impact. In order to assess the role of phytoplankton in global fluxes of carbon and associated elements between atmosphere, ocean and earth, the activity of all its taxonomic components should be estimated in all ocean regions. A compilation of regional studies to global coverage of data on plankton dynamics is only of value when scientists around the world contribute to the required body of knowledge by producing the best possible results of measurements, with methods that are not less sound than the best ones available, yet are within the limits of their countries' financial budget and technical capacity.

I advocate a globe-wide use of chromatographic methods for phytoplankton pigment analysis to study all aspects of phytoplankton biomass and ecophysiology in ways outlined and discussed in the various sections of this paper. HPLC should not be replaced by the newest alternatives (namely flow cytometry) before HPLC's potential and promises have been fully explored - which until now has only been attempted by a few. They repeatedly found that results are of a quality that can hardly be surpassed by any other method. For all scientists who do have access to the more expensive tools (flow cytometry and also remote sensing of the ocean's phytoplankton with the new generation of satellites), use of conventional methods of phytoplankton production and biomass analysis remains a condition sine qua non for the interpretation of their results. I have presented evidence in this paper that one of these methods, pigment analysis, cannot be omitted in the foreseeable future.

REFERENCES

Abaychi JK, Riley JP (1979) The determination of phytoplankton pigments by high-performance liquid chromatography. Anal Chim Acta 107:1-11

Ackleson S (1989) White waters of the Gulf of Maine. Oceanography 1:18-22

Bazzaz MB (1981) New Chlorophyll *a* and *b* chromophores isolates from a mutant of Zea mays L. Naturwissenschaften 68:94-95

Bazzaz MB, Brereton RG (1982) 4-Vinyl-4-desethyl Chlorophyll *a*: new naturally occurring chlorophyll. FEBS Letters 138:104-108

Berger W, Fischer K, Lai C, Wu G (1987) Ocean productivity and organic carbon flux. Part I. Overview and maps of primary production and export production. Univ. California, San Diego, CA, SIO Ref. 87-30, 69 pp

Bidigare RR, Schofield O, Prézelin BB (1989a) Influence of zeaxanthin on quantum yield of photosynthesis of *Synchococcus* clone WH 7803 (DC2). Mar Ecol Prog Ser 56:177-188

Bidigare RR, Morrow JH, Kiefer DA (1989b) Derivative analysis of spectral absorption by photosynthetic pigments in the Western Sargasso Sea. J Mar Res 47:323-341

Bricaud A, Stramski D (1990) Spectral absorption coefficients of living phytoplankton and non-algal biogenous matter: a comparison between the Peru upwelling area and the Sargasso Sea. Limnol Oceanogr 35:562-582

Brown TE, Richardson FT (1968) The effect of growth environment on the physiology of algae: light intensity. J Phycol 4:38-54
Burger-Wiersma T, Veenhuis M, Korthals HJ, van de Wiel CCM, Mur LR (1986) A new prokaryote containing chlorophylls *a* and *b*. Nature 320:262-264
Burkill PH, Mantoura RFC, Llewellyn CA, Owens NJP (1987) Microzooplankton grazing and selectivity of phytoplankton in coastal waters. Mar Biol 93:581-590
Chisholm SW, Olson RJ, Zettler ER, Waterbury J, Goericke R and Welschmeyer N (1988) A novel free-living prochlorophyte occurs at high cell concentrations in the oceanic euphotic zone. Nature 334:340-343
Cox GC, Hiller RG, Larkum AWD (1985) An unusual cyanophyte, containing phycourobilin and symbiotic with ascidians and sponges. Mar Biol 89:149-163
Czygan FC (1968) Sekundär-Carotinoide in Grünalgen. Arch Mikrobiol 61:81-102 and 62:209-236
Daley RJ, Brown SR (1973) Experimental characterization of lacustrine chlorophyll diagenesis. I. Physiological and environmental effects. Arch Hydrobiol 72:277-304
Daley RJ (1973) Experimental characterization of lacustrine chlorophyll diagenesis. II. Bacterial, viral, and herbivore grazing effects. Arch Hydrobiol 72:409-439
Davies BH (1976) Carotenoids. In: "Chemistry and biochemistry of plant pigments" (Goodwin, T.W., editor). Vol. 2:38-165 Academic Press, London
van Dijken G (1989) Pigmentsamenstelling van Rijn-fytoplankton in de rivier en in de rivierpluim (Noordzee). Dokt.-verslag, Universiteit Groningen:1-39
van Dijken G (1990) Karakterisering van het afwijkende chlorophyll *a* van vrijlevende mariene prochlorofyten. Dokt.- verslag, Universiteit Groningen:1-31
Duysens LNM (1956) The flattening of absorption spectrum of suspensions, as compared to that of solutions. Biochem Biophys Acta 19:2-12
Engelkes M (1985) Derivaatvorming en afbraak van algenpigmenten door copepodenbegrazing gemeten na scheiding met HPLC. Studentenverslag, N.I.O.Z. (Texel, the Netherlands)
Eppley RW, Harrison WG, Chisholm SW, Stewart E (1977) Particulate organic matter in surface waters off Southern California and its relationship to phytoplankton. J mar Res 35:671-696
Fasham MJR, Holligan PM, Pugh PR (1983) The spatial and temporal development of the spring phytoplankton bloom in the Celtic Sea, April 1979. Prog Oceanogr 12:87-145
Faust MA, Norris KH (1985) *In vivo* spectrophotometric analysis of photosynthetic pigments in natural populations of phytoplankton. Limnol Oceanogr 30:1316-1322
Foppen FH (1971) Tables for the identification of carotenoid pigments. Chromatogr Rev 14: 133-298
Foss P, Guillard RRL, Liaaen-Jensen S (1986) Carotenoids from eucaryotic ultraplankton clones (Prasinophyceae). Phytochemistry 25:119-124
Foss P, Lewin RA, Liaaen-Jensen S (1987) The carotenoids of Prochloron sp. (Prochlorophyta). Phycologia 26:142-144
French CS, Brown JS, Allen MB, Elliott RF (1959) The types of chlorophyll *a* in plants. Carnegie Inst. Washington Yr Book 58:327-330
Gieskes WWC, Kraay GW, Tijssen SB (1978) Chlorophylls and their degradation products in the deep pigment maximum layer of the tropical North Atlantic. Neth. J Sea Res 12: 195-204
Gieskes WWC, Kraay GW, Baars MA (1979) Current ^{14}C1 methods for measuring primary production: gross underestimates in oceanic waters. Neth. J Sea Res 13:58-78
Gieskes WWC, Kraay GW (1980) Primary productivity and photoplankton pigment measurements in the northern North Sea during FLEX-76. "Meteor" Forsch.-Erg. A22:105-112
Gieskes WWC, Kraay GW (1983a) Dominance of Cryptophyceae during the phytoplankton spring bloom in the central North Sea detected by HPLC analysis of pigments. Mar Biol 75:179-185
Gieskes WWC, Kraay GW (1983b) Unknown chlorophyll *a* derivatives in the North Sea and the Atlantic Ocean revealed by HPLC analysis. Limnol Oceanogr 28:757-766

Gieskes WWC, Kraay GW (1984) Phytoplankton, its pigments, and primary production at a central North Sea station in May, July and September 1981. Neth J Sea Res 18:51-70
Gieskes WWC, Kraay GW (1986a) Analysis of phytoplankton pigments by HPLC before, during and after mass occurrence of the microflagellate Corymbellus aureus during the spring bloom in the open northern North Sea in 1983. Mar Biol 92:45-52
Gieskes WWC, Kraay GW (1986b) Floristic and physiological differences between the shallow and the deep nanophytoplankton community in the euphotic zone of the open tropical Atlantic revealed by HPLC analysis of pigments. Mar Biol 91:567-576
Gieskes WWC, Elbrächter M (1986) Abundance of nanoplankton-size chlorophyll-containing particles caused by diatom disruption in surface waters of the Southern Ocean (Antarctic Peninsula region). Neth J Sea Res 20:291-303
Gieskes WWC, Kraay GW,Nontji A, Setiapermana D, Sutomo (1988) Monsoonal alternation of a mixed and a layered structure in the phytoplankton of the euphotic zone of the Banda Sea (Indonesia): a mathematical analysis of algal pigment fingerprints. Neth J Sea Res 22:123-137
Gieskes WWC, Kraay GW (1989) Estimating the carbon-specific growth rate of the major algal species groups in eastern Indonesian waters by ^{14}C1 labeling of taxon-specific carotenoids. Deep-Sea Res 36:1127-1139
Gieskes WWC, Kraay GW (1990) Transmission of ultraviolet light in the Weddell Sea: Report of the first measurements made in the Antarctic. BIOMASS Newsletter 12:12-14
Gieskes WWC, Engelkes MM, Kraay GW (1991) Degradation of diatom chlorophyll to colourless, non-fluorescing compounds during copepod grazing. Hydrobiol Bull (in press)
Glover HE, Keller MD, Guillard RRL (1986) Light quality and oceanic ultraphytoplankters. Nature 319:142-143
Glover RS (1967) The continuous plankton recorder survey of the North Atlantic. In: N.B. Marshall: Aspects of marine Zoology Symp Zool Soc London, 19:189-210
Guillard RRL, Murphy LS, Foss P, Liaaen-Jensen S (1985) *Synchococcus* sp. as likely zeaxanthin-dominant ultraplankton in th North Atlantic. Limnol Oceanogr 30:412-414
Hager A, Meyer-Bertenrath T (1967) Beziehungen zwischen Absorptionsspektrum und Konstitution bei Carotenoiden von Algen und höheren Pflanzen. Ber Dtsch Bot Ges Bd 80, H7:426-436
Hager A, Stransky H (1970) Das Carotenoidmuster und die Verbreitung des lichtinduzierten Xanthophyllcyclus in verschiedenen Algenklassen. III Grünalgen Arch Mikrobiol 72: 68-83
Hallegraeff GM (1977) Pigment diversity in freshwater phytoplankton. I. Summer-succession in three Dutch lakes with different trophic characteristics. Int Rev ges Hydrobiol 62: 19-39
Herman AW, Denman KL (1977) Rapid underwater profiling of chlorophyll with an *in situ* fluorometer mounted on a "Batfisch" vehicle. Deep-Sea Res 24:385-397
Hewes CD, Holm-Hansen O (1983) A method for recovering nanoplankton from filters for identification with the microscope: the filter-transfer-freeze (FTF) technique. Limnol Oceanogr 38:389-394
Hoek C, van den (1978) Algen. Einführung in die Phykologie. Thieme Verlag, Stuttgart
Holm-Hansen O, Lorenzen CJ, Holmes RW, Strickland JDH (1965) Fluorometric determination of chlorophyll. J cons perm int Explor Mer 30:3-15
Hooks CE, Bidigare RR, Keller MD, Guillard RRL (1988) Coccoid eukaryotic marine ultraplankters with four different HPLC pigment signatures. J Phycol 24:571-580
Jacobsen TR (1978) A quantitative method for the separation of chlorophylls a and b from phytoplankton pigments by high pressure liquid chromatography. Mar Sci Comm 4:33-47
Jeffrey SW (1961) Paper-chromatographic separation of chlorophylls and carotenoids from marine algae. Biochem J 80:336-342
Jeffrey SW (1968) Quantitative thin-layer chromatography of chlorophylls and carotenoids from marine algae. Biochimica Biophysica Acta, 162, 271-285
Jeffrey SW (1975) Chloroplast pigment patterns in dinoflagellates. Ann Rept 1974-1975, Marine Biochemistry Unit, CSIRO (Australia): 7-9

Jeffrey SW, GF Humphrey (1975) New spectrophotometric equations for determining chlorophylls *a, b, c1* and *c2* in higher plants, algae and natural phytoplankton. Biochem Physiol Pflanz 167:191-194

Jeffrey SW (1976) A report of green algal pigments in the central North Pacific Ocean Mar Biol 37:33-37

Jeffrey SW, Hallegraeff GM (1980) Studies of phytoplankton species and photosynthetic pigments in a warm core eddy of the East Australian Current. II. A note on pigment methodology. Mar Ecol Prog Ser 3:295-301

Jeffrey SW (1981) An improved thin-layer chromatographic technique for marine phytoplankton pigments. Limnol Oceanogr 26:191-197

Jeffrey SW (1989) Chlorophyll *c* pigments and their distribution in the chromophyte algae. In: The Chromophyte algae: Problems and Perspectives (eds Green, Leadbeater and Diver), Syst Ass Spec Vol 38:13-36 Clarendon Press

Jensen A, Sakshaug E (1973) Studies on the phytoplankton ecology of the Trondheim fjord. 3. Chloroplast pigments in relation to abundance and physiological state of the phytoplankton. J exp mar Biol Ecol 11:137-155

Kana TM, Glibert P, Goericke R, Welschmeyer NA (1988) Zeaxanthin and ß-carotene in *Synchococcus* WH 7803 respond differently to irradiance. Limnol Oceanogr 33: 1623-1627

Kiefer D (1973) Fluorescence properties of natural phytoplankton populations. Mar Biol 22:263-269

Kiefer DA, SooHoo JB (1982) Spectral absorption by marine particles of coastal waters of Baja California. Limnol Oceanogr 27:429-459

Kishino M, Takahashi M, Okami N, Ichimura S (1985) Estimation of the spectral absorption coefficients of phytoplankton in the sea. Bull Mar Sci 37:634-642

Klein AWO, Gieskes WWC, Kraay GW (1986) Digestion of chlorophylls and carotenoids by the marine protozoan *Oxyrrhis marina* studied by HPLC analysis of algal pigments. J Plankton Res 8:827-836

Klein B (1988) Variations in pigment content in two benthic diatoms during growth in batch cultures. J exp mar Biol Ecol 115:237-248

Kleppel GS, Pieper RE (1984) Phytoplankton pigments in the gut contents of planktonic copepods from coastal waters off southern California. Mar Biol 78:193-198

Knight R, Mantoura RF (1985) Chlorophyll and carotenoid pigments in Foraminifera and their symbiontic algae: analysis by HPLC. Mar Ecol Prog Ser 23:241-249

Koblentz-Mishke OJ, Volkovinsky VV, Kabanova JG (1970) Plankton primary production of the world ocean. Pp. 183 - 193 in: W. Wooster (ed.): Scientific Exploration of the South Pacific Nat Acad Press, Washington D.C

Kowalenska G, Falkowski L, Hoffmann SK, Szczepaniak LS (1987) Replacement of magnesium by copper (II) in the chlorophyll porphyrin ring of planktonic algae. Acta Physiol Plant 9:43-52

Kowalenska G (1988) Bluegreen algae - the possible source of stable organic copper complexes for the environment. Proc. XVI Conference of Baltic Oceanographers, Kiel

Larsson U et 10 al. (1978) Intercalibration of methods for chlorophyll measurements applied in the Baltic Sea. Merentutkimuslait. Julk./Havsforsk Inst Skr 243:63-76

Leavitt PR, Brown SR (1988) Effects of grazing by *Daphnia* on algal carotenoids: implications for paleolimnology. J Paleolimnol 1:201-213

Lenz J, Fritsche P (1980) The estimation of chlorophyll *a* in water samples: a comparative study on retention in a glass-fibre and a membrane filter and on the reliability of two storage methods. Arch Hydro-biol Beih 14:46-51

Lewis MR, Cullen JJ, Platt TR (1984) Relationships between vertical mixing and photoadaptation of phytoplankton: similarity criteria. Mar Ecol Prog Ser 15:141-149

Lewitus AJ, Broenkow WW (1985) Intermediate depth pigment maxima in oxygen minimum zones. Deep-Sea Res 32:1101-1115

Loftus ME, Carpenter JH (1971) A fluorometric method for determining chlorophylls *a, b* and *c*. J Mar Res 29:319-338

Loftus ME, Seliger HH (1975) Some limitations of the *in vivo* fluorescence technique. Chesapeake Sci 16:79-92

Lopez MDG, Huntley ME, Sykes PF (1988) Pigment destruction by Calanus pacificus: impact on the estimation of water column fluxes. J Plankton Res 10:715-734
Lorenzen CJ (1967a) Vertical distribution of chlorophyll and phaeopigments: Baja California. Deep-Sea Res 14:735-745
Lorenzen CJ(1967b)Determination of chlorophyll and phaeopigments: spectrophotometric equations. Limnol Oceanogr 12:343-346
Lorenzen CJ, Jeffrey SW (1980) Determination of chlorophyll in seawater. UNESCO Technical Papers in marine Science 35:1-19
Lubian LM, Establier R (1980) Taxonomic position of a nannoplanktonic alga on the basis of its pigment composition. ICES, CM 1980 L7:1-9
Mantoura RFC, Jeffrey SW (1983) The rapid determination of algal chlorophyll and carotenoid pigments and their breakdown products in natural waters by reverse-phase high-performance liquid chromatography. Anal Chim Acta 151:297-314
Megard RO, Berman T (1989) Effects of algae on the Secchi transparancy of the southeastern Mediterranean Sea Limnol Oceanogr 34:1640-1655
Moed JR, Hallegraeff GM (1978) Some problems in the estimation of chlorophyll-*a* and phaeopigments from pre- and post-acidification spectrophotometric measurements. Int Rev ges Hydrobiol 63:787-800
Nelson JR (1989) Phytoplankton pigments in macrozooplankton feces: variability in carotenoid alterations. Mar Ecol Prog Ser 52:129-144
Neveux J, Vaulot D, Courties C, Fukai E (1989) Green photosynthetic bacteria associated with the deep chlorophyll maximum in the Sargasso Sea. CR Acad Sci (Paris) III 308:9-14
Nichols PD, Volkman JK, Hallegraeff GM, Blackburn SI (1987) Sterols and fatty acids of the red-tide flagellates *Heterosigma akashiwo* and *Chattonella antiqua* (Raphidophyceae). Phytochemistry 26:2537-2541
Nusch EA (1984) Ergebnisse eines Ringversuches zur Chlorophyll-*a*-Bestimmung im Wasser. Z. Wasser-Abwasser- Forsch. 17:89-94
Oldham PB), Zillioux EJ, Warner IM (1985) Spectral "fingerprinting" of phytoplankton populations by two- dimensional fluorescence and Fourier-transform-based pattern recognition. J mar Res 43:893-906
Olson RJ, Zettler ER, Altabet MA, Dusenberry JA, Chisholm SW (1990) Spatial and temporal distributions of prochlorophyte picoplankton in the North Atlantic Ocean. Deep-Sea Res (in press)
Olson RJ, Zettler ER, Armbrust EV, Chisholm SW (1990) Pigments, size and distribution of *Synchococcus* in the North Atlantic and Pacific oceans. Limnol Oceanogr 35:45-58
Owens TG, Gallagher JC, Alberte RS (1987) Photosynthetic light-harvesting function of violaxanthin in *Nannochloropsis spp.* (Eustigmatophyceae). J Phycol 23:79-85
Pearl HW, Tucker T, Bland PT (1983) Carotenoid enhancement and its role in maintaining bluegreen algal (*Microcystis aeruginosa*) surface blooms. Limnol Oceanogr 28:847-857
Paffenhöfer G-A, Knowles SC (1978) Feeding of marine planktonic copepods on mixed phytoplankton. Mar Biol 48:143-152
Papageorgiou G, Govindjee (1968) Light-induced changes in fluorescence yield of Chlorophyll *a in vivo*. Biophys J 8:1299-1328
Partali V, Liaaen-Jensen S, Slagsvold T, Lifjeld JT (1987) Carotenoids in food chain studies. 2. The food chain of Parus spp. monitored by carotenoid analysis. Comp Biochem Physiol 87B:885-888
Ramus J, Beale SI, Mauzerall D, Howard RL (1976) Changes in photosynthetic pigment concentration in seaweeds as a function of water depth. Mar Biol 37:223-229
Redalje DG, Laws EA (1981) A new method for estimating phytoplankton growth rates and carbon biomass. Mar Biol 62:73-79
Richards FA, Thompson TG (1952) The estimation and characterization of plankton populations by pigment analyses. II. A spectrophotometric method for the estimation of plankton pigments. J mar Res 11:156-172
Riemann B (1978) Carotenoid interference in the spectrophotometric determination of chlorophyll degradation products from natural populations of phytoplankton. Limnol Oceanogr 23:1059-1066

Robinson AL (1979) HPLC: the new King of analytical chemistry? Science (N.Y.) 203:1329-1332

Rott E (1980) Spectrophotometric and chromatographic chlorophyll analysis: comparison of results and discussion of the trichromatographic method. Arch Hydrobiol Beitr 14:37-45

Roy S (1986) Marine chloropigments: chromatographic measurement in phytoplankton and transformations in algal cultures and a spring bloom event in a marine bay. Ph.D. thesis, Dalhousie University

Roy S (1987) High-performace liquid chromatography analysis of chloropigments. Chromatogr 391:19-34

Sartory DP (1985) The determination of algal chlorophyllous pigments by high-performance liquid chromatography and spectrophotometry. Water Res 19:606-610

Schlittenhardt P (1980) Report of comparison of Chlorophyll *a* analyses in some laboratories. ICES, L37:1-12

Schnepf E, Elbrächter M (1988) Cryptophycean-like double membrane-bound chloroplast in the dinoflagellate, *Dinophysis* Ehrenb.: evolutionary, phylogenetic and toxicological implications. Bot Acta 101:196-203

Shibata K, Benson AA, Calvin M (1954) The absorption spectra of suspensions of living microorganisms. Biochim Biophys Acta 15:461-470

Shibata K (1969) Pigments and a UV-absorbing substance in corals and blue-green alga living in the Great Barrier Reef. Plant Cell Physiol 19:325-335

Sieburth JMcN (1988) The nanoalgal peak in the dim oceanic pycnocline: is photosynthesis augmented by microparticulates and their bacterial consortia? Pp. 101-129 in: Biogeochemical cycling and fluxes between the deep euphotic zone and other oceanic realms. National Undersea Research Program, Res Rept 88-1 (C.R. Agegian, Ed.)

Smith RC, Prézelin BB, Bidigare RR, Baker KS (1989) Bio-optical modeling of photosynthetic production in coastal waters. Limnol Oceanogr 34:1524-1544

Sournia A (1982) Is there a shade flora in the marine plankton? J. Plankton Res. 4:391-399

Spitzer D, Wernand M (1979). Photon scalar irradiance meter. Appl Optics 18:1698-1700

Steenbergen CLM, Korthals HJ (1982) Distribution of phototrophic microorganisms in the anaerobic and microaerophilic strate of Lake Vechten (the Netherlands). Pigment analysis and role in primary production. Limno Oceanogr 27:883-895

Stotte W (1988) Afgeleide-analyse van absorptiespectra van mariene algen. Doct versl Mar Bot, Univ Groningen:1-23

Strathmann RR (1967) Estimating the organic carbon content of phytoplankton from cell volume or plasma volume. Limnol Oceanogr 12:411-419

Takahashi M, Hori T (1984) Abundance of picophytoplankton in the subsurface chlorophyll maximum layer in subtropical and tropical waters. Mar Biol 79:177-186

Titlyanova EA, Zvalinskii VI, Yadykin AA, Li BD, Chernova SI (1977) Adaptation of benthic plants to light. II. Resistance to photic destruction of pigment apparatus of the marine green alga *Ulva fenestrata*. Biol Morya 2:3-10

Trees CC, Kennicutt Mahlon II C, Brooks JM (1985) Errors associated with the standard fluorimetric determination of chlorophylls and phaeopigments. Mar Chem 17:1-12

Trüper HG, Yentsch CS (1967) Use of glass fibre filters for the rapid preparation of *in vivo* absorption spectra of photosynthetic bacteria. J Bacteriol 94:1255-1256

Veldhuis MJW, Kraay GW (1990) Vertical distribution and pigment composition of picoplanktonic prochlorophyte in the subtropical North Atlantic: a combined study of HPLC analysis of pigments and flow cytometry. Mar Ecol Prog Ser 68:121-127

Waterbury JB, Watson SW, Guillard RR, Brand LE (1979) Widespread occurrence of a unicellular marine, planktonic cyanobacterium. Nature (Lond) 277:293-294

Weber A, Wettern M (1980) Some remarks on the usefulness of algal carotenoids as chemosystematic markers. In: Pigments in plants. Gustav Fischer Verlag, Stuttgart-New York:104-116

Welschmeyer NA, Lorenzen CJ (1984) Carbon-14 labeling of phytoplankton carbon and chlorophyll *a* carbon: determination of specific growth rates. Limnol Oceanogr 29:135-145

Welschmeyer NA, Lorenzen CJ (1984) Chlorophyll budgets: Zooplankton grazing and phytoplankton growth in a temperate fjord and the Central Pacific Gyres. Limnol Oceanogr 30:1-21

Welschmeyer N, Hoepffner N (1986) Rapid xanthophyll cycling: an *in situ* tracer for mixing in the upper ocean. EOS 67:969

Wood MA (1985) Adaptation of photosynthetic apparatus of marine ultraphytoplankton to natural light fields. Nature 316:253-255

Wright SW, Jeffrey SW (1987) Fucoxanthin pigment markers of marine phytoplankton analysed by HPLC and HPTLC. Mar Ecol Prog Ser 38:259-266

Wyman M, Gregory RPF, Carr NG (1985) Novel role for phycoerythrin in a marine Cyanobacterium, *Synchococcus* Strain DC2. Science (N.Y.) 230:818-820

Yentsch CS (1962) Measurement of visible light absorption by particulate matter in the ocean. Limnol Oceanogr 9:207-217

Zevenboom W (1986) Tracing red-pigmented marine cyanobacteria using *in vivo* absorption maxima. FEMS Microbiol Ecol 38:267-275

Züllig H (1985) Pigmente phototropher Bakterien in Seesedimenten and ihre Bedeutung für die Seenforschung. Schweiz Z Hydrol 47:87-126

Züllig H (1989) Role of carotenoids in lake sediments for reconstructing trophic history during the late Quaternary. J Paleolimnol 2:23-40

ENERGY TRANSFORMATION AND FLUORESCENCE IN PHOTOSYNTHESIS

T. G. Owens

Section of Plant Biology
Cornell University
Ithaca, New York USA
14853-5908

ABSTRACT

The origins of *in vivo* chlorophyll fluorescence in unicellular algae are discussed with regard to the processes that regulate fluorescence yield and interpretation of fluorescence measurements as an indicator of algal primary production and physiological state. Emphasis is placed on limitations in interpretation of field data.

INTRODUCTION

The measurement of *in vivo* chlorophyll (chl) fluorescence has been extensively applied to studies of phytoplankton physiology and growth in both laboratory and field settings. The apparent simplicity of fluorescence measurements, compared with direct measurement of productivity or "physiological state", has contributed significantly to its widespread application. Based on a limited theoretical and experimental background, there is a general perception that measurements of *in vivo* fluorescence can be used to predict important physiological parameters relating to phytoplankton abundance and growth. Although good correlations between fluorescence and biomass or productivity measurements have been demonstrated by several authors, there is reason to use caution in the general application and interpretation of fluorescence data (see Falkowski and Kiefer, 1985 for an excellent review). The principal limitation to interpretation of fluorescence data is the lack of a mechanistic description linking the control of *in vivo* fluorescence yield with the underlying reactions of photosynthesis and coupled processes.

NATO ASI Series, Vol. G 27
Particle Analysis in Oceanography
Edited by S. Demers

It is not my intent that this report serve as an exhaustive review of application and interpretation of *in vivo* fluorescence measurements in phytoplankton studies. Rather, I hope to present a description of our present knowledge concerning the source of *in vivo* fluorescence the physiological processes that regulate the yield of fluorescence, and their relationship to primary photochemistry, electron transport and the dark reactions of photosynthesis. At present, most of this knowledge is derived from studies on higher plant chloroplasts or green algae; the latter only occasionally dominates natural phytoplankton communities. Very little is known about analogous processes in other algal classes. The level of this knowledge is not sufficient, even in green algae, to provide the mechanistic description linking fluorescence with the reactions of photosynthesis. However, this knowledge is sufficient to demonstrate the complexity of processes contributing to the regulation of fluorescence yield and the general ties between these processes and the component reactions of photosynthesis. Most importantly, I hope that this report provides a minimal foundation for understanding fluorescence measurements and sets some significant goals for future research in this area.

IN VITRO FLUORESCENCE

When a photosynthetic pigment in its lowest energy (ground) state absorbs a photon of visible light, the pigment is raised to an electronically excited state. For chl molecules, the major red (560-680 nm) and blue (400-480 nm) absorptions correspond to transitions from the ground state to the four lowest energy excited states (Figure 1). Absorption is restricted to photons whose energy exactly matches the energy difference between the initial (ground) and final (excited) states

$$E_{photon} = \frac{hc}{\lambda} = E_{final} - E_{initial} \tag{1}$$

where h is Planck's constant, c is the speed of light and λ is the wavelength of the photon.

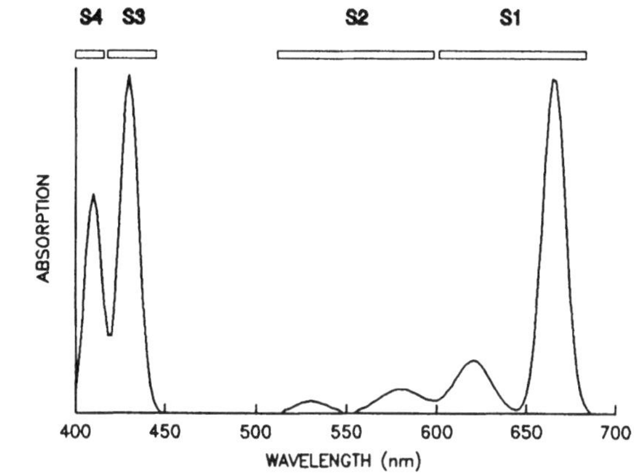

Figure 1. Absorption spectrum of dilute chlorophyll *a* (in diethyl ether) showing the correspondence between the absorption maxima and transitions between the ground state (S0) and the first four excited states (S1-S4).

The formation of the higher energy excited states by photon absorption is followed immediately ($<10^{-12}$ s) by a decay to the first excited state with the corresponding energy difference being lost as heat (Figure 2). All subsequent processes in photosynthesis proceed from the first excited state of chl. In an energetic sense, this loss of energy associated with the absorption of blue light represents a decrease in the overall efficiency of photosynthesis. However, efficiency measurements in photosynthesis are usually determined on a quantum (photon) basis, with the yield of a particular process being normalized to the total number of photons absorbed, independent of the energy (wavelength) of the photon. This formalism is appropriate for photosynthesis since the higher excited states of any photosynthetic pigment decay to the lowest excited state with a quantum yield of unity.

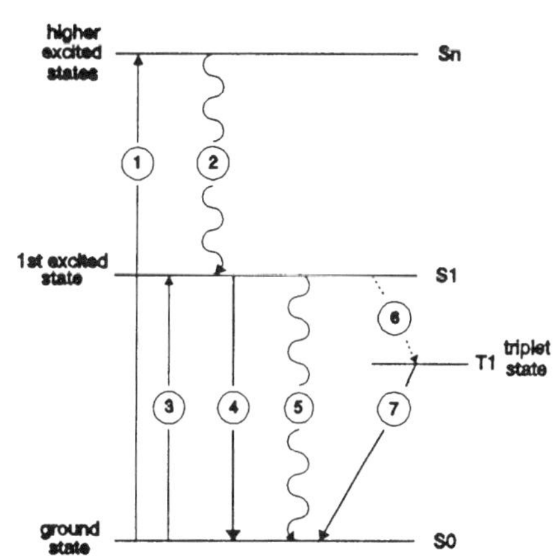

Figure 2. Energy level diagram for dilute chlorophyll *a* showing competing processes for formation and decay of excited electronic states. (1) formation of higher (S2-S4) excited states by photon absorption, (2) rapid non-radiative decay from higher excited states to the first excited state (energy lost as heat), (3) direct formation of first excited state by photon absorption, (4) fluorescence (radiative) decay via emission of a red photon, (5) non-radiative decay to ground state (energy lost as heat), (6) intersystem crossing to form excited triplet state, (7) radiative decay of triplet state (phosphorescence).

The first excited state of chl is metastable, and will rapidly ($<5 \times 10^{-9}$ s) decay back to the ground state by one of several competing processes. For "isolated" chl molecules in a dilute solution, the most important decay processes are fluorescence, thermal emission and decay via triplet excited states (Figure 2). Fluorescence is a radiative decay process in which the energy difference between the first excited state and the ground state is lost through emission of a visible photon; the wavelength of the photon is again determined by the energy gap between the initial and final states. In thermal decay (also called non-radiative decay) the energy of the excited state is dissipated through vibrational modes of the chl without emission of a visible photon. Decay via triplet excited states involves additional changes in the characteristics (electron spin) of the pigment. For a chl molecule in dilute solution, the relative yields of fluorescence, triplet formation and thermal decay are approximately 0.3, 0.65 and 0.05, respectively.

The quantum yield of fluorescence is defined as the number of fluorescence photons emitted divided by the number of photons absorbed. At steady state, this is equivalent to the rate of fluorescent deexcitation divided by the total rate of all processes contributing to decay of the excited state (Clayton, 1980)

$$\phi_f = k_f / (k_f + k_d + k_i) \qquad \textbf{(2)}$$

where k_f, k_d and k_i are the first order rate constants for excited state decay via fluorescence, thermal emission and triplet formation (intersystem crossing), respectively. The decay of a sum of first order reactions is exponential, with the average decay constant (k) and lifetime (τ) given by

$$\tau = 1/k = 1/(k_f + k_d + k_i) \qquad \textbf{(3)}$$

where k is the sum of all processes competing to return the pigment to its ground state. Combining equations 2 and 3 gives the expression

$$\phi_f = k_f \cdot \tau \qquad \textbf{(4)}$$

which shows that the fluorescence quantum yield is proportional to the average lifetime of the excited state. For isolated chl molecules in dilute solution, the lifetime of the first excited state is about 5×10^{-9} s. The rate constants k_f, k_d and k_i are intrinsic properties of each pigment, determined by properties of the molecule itself but also influenced by its local environment. Decreases in the yield of fluorescence (and proportional changes in excited state lifetime) must be attributed to either changes in the pigment environment or to additional processes competing to relax the excited state. In photosynthetic systems, the major process competing for excited state energy is photochemistry. The photochemical reactions of photosynthesis convert chl excited state energy into chemical energy in the form of oxidants and reductants that ultimately drive electron transport and CO_2 fixation.

In quantifying fluorescence emission, it is important to distinguish between fluorescence (quantum) yield, ϕ_f, and fluorescence intensity (F). The fluorescence yield is a constant which relates the rate of light absorption (I_a) to the rate (intensity) of fluorescence emission:

$$F = \phi_f \cdot I_a \tag{5}$$

Thus, for a given state of the sample, the fluorescence intensity F increases linearly with incident light intensity while the fluorescence yield is independent of light intensity. Clearly, fluorescence intensity and quantum yield are not equivalent parameters, although the terms are often (incorrectly) used interchangeably.

The fluorescence emission spectrum of any pigment is quantitatively related to the absorption spectrum for formation of the excited state from which the fluorescence originates. Thus for chls and related pigments, fluorescence arises only from the lowest energy red (620-680 nm) absorption. There is no fluorescence emitted directly from the higher energy absorptions. The absorption bands of photosynthetic pigments are broad, indicating that each electronic state (ground and excited states) has associated with it a distribution of energies. This distribution arises from thermally excited molecular vibrations in the pigment. At physiological temperatures, a large number of these vibrational levels are accessible creating a broad absorption. For both upward (absorption) and downward (fluorescence) transitions between electronic states, the most probable starting point is near the lowest energy vibrational state (Figure 3). However, the most probable end point for either transition is near the middle of the distribution of vibrational states. The energy difference between the absorption and fluorescence transitions is accounted for by thermal emission through the closely spaced vibrational states. This loss of energy accounts for the wavelength difference between the absorption and fluorescence emission maxima (the Stokes shift), which is in the range of 6-8 nm for chl.

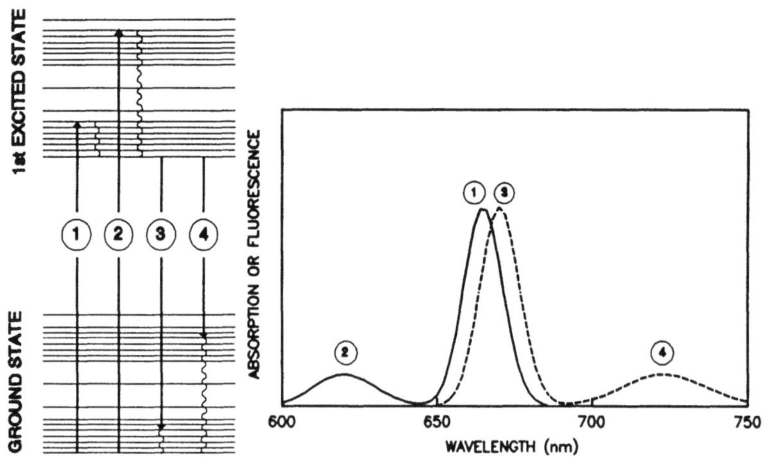

Figure 3. Processes accounting for the absorption and fluorescence emission spectra of chlorophyll *a*. Left: energy level diagram showing vibrational energy subsets of ground and first excited states. Solid lines: transitions involving absorption or emission of a red photon; wavy line: decay via vibrational relaxation (thermal emission). Right: corresponding absorption and fluorescence emission spectra for chlorophyll *a*. Numbers show transitions that correspond to absorption or emission features.

The distribution of vibrational states associated with each electronic state is, in general, not uniform. However, because the positions of the atoms do not change significantly between different electronic states, the distribution of vibrational states is similar in each electronic state. For chls, each electronic state has associated with it two distinct subsets of vibrational energy levels (Figure 3). The main red absorption of chl *a* is due to a transition to the lowest energy vibrational subset in the first excited state while the short wavelength (higher energy) shoulder at 610 nm is due to a transition to the higher energy vibrational subset of the first excited state. Similarly, the main fluorescence emission peak and its long wavelength shoulder are due to downward transitions that end in the lower and higher vibrational subsets of the ground state, respectively (Figure 3). The symmetry of vibrational states in the ground and excited states explains the mirror image relationship between the absorption and emission spectra. Thus for isolated chls, fluorescence emission at wavelengths >700 nm originates from the same electronic state as the main emission band at 680 nm.

A fluorescence excitation spectrum probes those states whose absorptions directly or indirectly lead to the formation of the first excited state of chl. Because the quantum yield of formation of the first excited state from higher energy excited states is essentially unity in chl, the fluorescence excitation spectrum for isolated chl should be identical to the absorption spectrum (when the latter is plotted as fraction of the incident light absorbed). For photosynthetic systems, the fluorescence excitation spectrum probes the absorptions of *all pigments* whose excited state energy is transferred to chl.

IN VIVO ORGANIZATION OF PHOTOSYNTHETIC PIGMENTS

Photosynthetic pigments can be broadly separated into two categories of pigment types depending on their function: antenna and reaction center pigments. There is good evidence to suggest that all functional photosynthetic pigments are bound in specific pigment-protein complexes (Thornber, 1986). The antenna versus reaction center functions of pigments are determined largely by pigment-pigment and pigment-protein interactions in the local binding environment. Differences in these binding environments may also induce significant (5-90 nm) spectral shifts in the lowest energy absorptions of pigments compared to their absorption in organic solvents. The binding environment also provides new avenues for decay of excited state energy among the pigments. Of principal importance are transfer of excited state energy between neighboring antenna pigments and photochemistry on the reaction center pigment. These new processes alter the interpretation of fluorescence yield changes from that of isolated pigments described in the previous section. In addition, there are other structural constraints and physiological processes that alter fluorescence yield *in vivo*, complicating the interpretation of fluorescence and photochemical measurements.

Reaction centers and photochemistry. The reaction center is the pigment in which the energy of absorbed sunlight in the form of an excited electronic state of chl is converted into chemical energy in the form of oxidants and reductants. The pigment and protein composition of the reaction centers appears to be highly conserved among all photosynthetic organisms (Thornber, 1986). The basic pattern of the photochemical reactions, which is also conserved, is schematically shown in Figure 4. Here, P is the reaction center pigment, D is the electron donor to the reaction center, and A is the primary electron acceptor. Absorption or transfer of light energy to P generates the first excited state P^*. P^* is a stronger reductant than P, and a fast (3×10^{-12} s) photochemical reaction generates the charge separated state D P^+ A^-. Now P^+, being a much stronger oxidant than P, removes an electron from D generating the state D^+ P A^-. Subsequent electron transport reactions reduce D^+ and oxidize A^- to regenerate the starting state D P A. Photochemistry is only possible from the state DPA; this is the open state of the reaction center. If either P is oxidized or A is reduced, photochemistry is blocked and the reaction center is said to be closed.

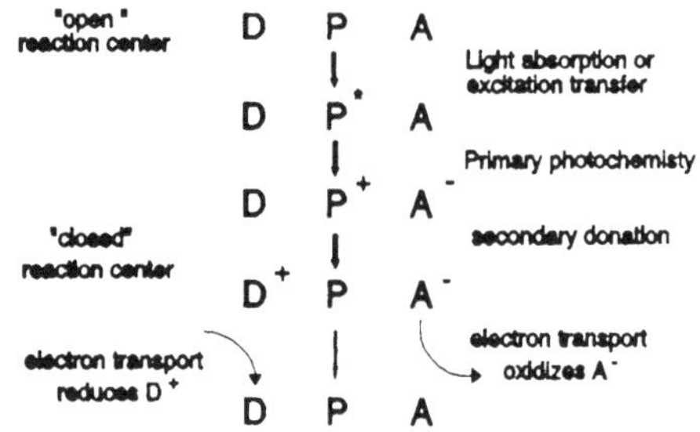

Figure 4. Schematic representation of photochemical and electron transport reactions in a generalized photosynthetic reaction center. P: reaction center pigment (primary electron donor), D: secondary electron donor, A: primary electron acceptor. Photochemistry is possible only in an "open" reaction center where P is reduced and A is oxidized.

In photosynthetic organisms which evolve oxygen, there are two types of reaction center pigments: P680 and P700 (these are chl *a* in special environments, the numbers refer to the wavelength corresponding to the lowest energy absorption of each pigment). The function of P680 and P700 in linear (non-cyclic) photosynthetic electron transport is shown in Figure 5. Photochemistry at P680 produces a weak reductant and a very strong oxidant which removes electrons from water (producing oxygen). At P700, photochemistry produces a

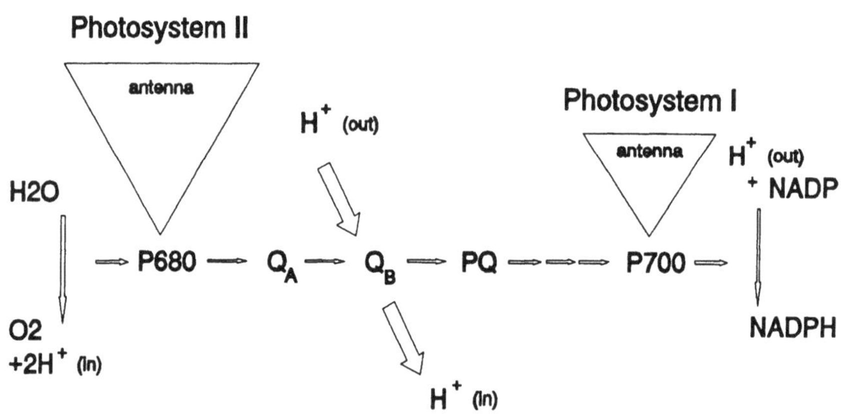

Figure 5. Pattern of photosynthetic electron transport in O_2-evolving plants and algae. Reaction centers P680 and P700 function in series to move electrons from water to $NADP^+$. Q_A and Q_B: primary and secondary quinone acceptors of PS II, PQ: plastoquinone pool; (in) and (out) refer to inside and outside to the thylakoid membrane.

weak oxidant and a strong reductant that reduces $NADP^+$.

Electron transport between the weak reductant produced by P680 and the weak oxidant produced by P700 is coupled to the vectorial transport of protons into the interior of the space enclosed by the photosynthetic (thylakoid) membrane, and ultimately to ATP synthesis. The photochemical reactions are thus directly or indirectly coupled to all major reactions producing or utilizing NADPH or ATP (Foyer *et al.*, 1990). Principal among these are the reactions of electron transport and CO_2 fixation. Since photochemistry and fluorescence are both competing for excited state energy, variations in any of the coupled reactions may potentially show up as a change in the yield of fluorescence.

The fact that linear electron transport from water to NADP requires that P700 and P680 function in series does not mean that P700 and P680 occur in a 1:1 stoichiometry in thylakoid membranes. In green algae and higher plants, the P680 and P700 reaction center complexes are widely separated in the membrane, connected only by pools of mobile electron carriers (Ort, 1986). P680:P700 ratios may vary between species and with environmental conditions (Falkowski *et al.*, 1981).

Antenna pigments and excitation transfer. Although a reaction center pigment is fully capable of direct absorption of light energy, its physical size limits the rate at which it can absorb light in a typical terrestrial or aquatic environment. Even in maximum sunlight, the rate of photochemical turnover of the reaction centers without antennae would be insufficient to support a phototrophic metabolism. Nature has overcome this problem by coupling the absorption of hundreds of light-harvesting antenna pigments to each reaction center. This coupling occurs in the form of highly efficient excited state energy transfer among the antenna pigments. (The mobile excited state is often called an excitation or exciton.) This coupling of antenna pigments effectively increases the area through which the reaction center can absorb light energy and results in a proportional increase in the rate of photochemical turnover. Each type of reaction center has associated with it a specific antenna of defined size and composition. P700, its associated electron transport components and antenna form an aggregate called photosystem I (PS I), while the corresponding aggregate containing P680 is called PS II.

Unlike the reaction center pigments which are always chl *a*, photosynthetic antenna pigments may be chl *a*, *b* or *c*, phycobilins, or a variety of carotenoids. These pigments are bound in several classes of pigment-protein complexes (Anderson and Barrett, 1986). The diversity of light-harvesting pigments is especially pronounced in the algae, where pigment composition is often a major characteristic separating algal classes. In all cases, chl *a* is thought to be the central component of the antenna, with all excitation energy ultimately being transferred to antenna chl *a* prior to transfer to the reaction center (Figure 6). Because of the energetics of the excitation transfer reactions, transfers from other types of antenna pigment to chl *a* are essentially irreversible; ie, back transfer from chl *a* does not occur. For the accessory chls *b* and *c* and the carotenoids, it is thought that energy initially absorbed by one of these pigments is immediately transferred to chl *a* in the same pigment-protein complex without intermediate transfers involving other pigment types (Trautman *et al.*, 1990). All subsequent transfers in the antenna occur between chl *a* pigments. In contrast, algae with phycobilisome antennas have long-range excitation transfer among the bilins, eventually reaching chl *a* only in a small antenna aggregate near the reaction centers (Bryant, 1986).

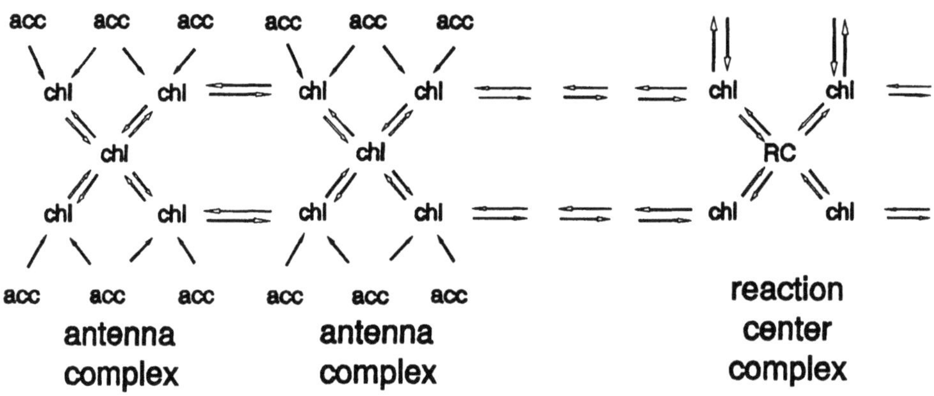

Figure 6. Schematic representation of excited state energy transfer dynamics in chlorophyll *a*-based antenna systems. The figure is not intended to model any specific antenna system, but rather to emphasize the relative motion of the excited state among accessory pigments (acc), chlorophyll *a* (chl) and the reaction center (RC).

The diversity of antenna pigments also increases the range of wavelengths that can be utilized in photosynthesis. This is especially important in aquatic environments where the red and blue light absorbed directly by chl *a* are the wavelengths of light that are most rapidly attenuated with increasing water depth (Kirk, 1983). The absorptions of the accessory chls, carotenoids, and phycobilins in the green region of the spectrum is central to the success of aquatic algae. Finally, antenna complexes serve several regulatory functions which optimize the overall efficiency of photosynthesis under conditions of varying light intensity and spectral quality (Fork and Satoh, 1986). As we shall see, these regulatory functions can also be detected as changes in chl fluorescence.

Source of in vivo fluorescence. Equations 2-4 describe the factors that determine the quantum yield of fluorescence for individual isolated pigments in a dilute solution. Photosynthetic systems are considerably more complex because of the large and heterogeneous aggregate of coupled pigments and because of the additional processes that are competing for excitation energy. However, both the rate constant and lifetime formulations of fluorescence yield can be useful when appropriately applied to the large antenna systems. A useful starting point is a comparison of the contributions of PS I and PS II to total fluorescence emission. It has long been known that nearly all of the room temperature fluorescence from O_2-evolving organisms is derived from PS II pigments. Recent excitation lifetime measurements on PS I and PS II from green algae and higher plant chloroplasts provide a straightforward explanation for this observation: the lifetime of excitations in PS I is about 70-90 x 10^{-12} s, a factor of five shorter than that measured in PS II (Holzwarth, 1987). In addition, there are about twice as many PS II as PS I pigments, with the net result that PS I contributes only about 5% of the total steady state fluorescence. Thus, the direct contribution of PS I to fluorescence at physiological temperatures can be essentially ignored. (At cryogenic temperatures, fluorescence emission is dominated by PS I emission at 710-740 nm; Butler, 1978).

The relative contribution of accessory pigments to total fluorescence is also an important consideration. The ability to detect emission from accessory pigments could be useful in determining the taxonomic characteristics of a sample by detection of emission from pigments unique to a particular group of algae. Unfortunately, most of the accessory pigments contribute very little to the total fluorescence of a sample under physiological conditions. For accessory chls and carotenoids, the rate constant for excitation transfer to chl *a* is thought to be on the order of 10^{12} s^{-1} . Adding this term into the denominator of equation 2 gives a fluorescence yield of about 0.02%. Thus, even if the accessory chls accounted for 50% of the cellular pigment, they would not contribute significantly to total fluorescence emission.

The exceptional case in emission from accessory pigments occurs in phycobilisome antennae. The phycobilisome is a large aggregate of phycobilin pigments. The lifetime of the excited state in the phycobilisome may be comparable to or greater than its lifetime in the chl *a* antenna (Bryant, 1986) with the result that significant emission from the bilins can be detected in many cases.

The final consideration in the source of *in vivo* fluorescence is the relative importance of antenna chl *a* versus P680. The reaction center pigment is frequently referred to as a "trap" for excitation energy in the antenna. In a sense this is correct because the major process leading to loss of excitations in PS II is photochemistry in the reaction center. Efficient photochemistry is the main factor that limits the lifetime of the excited state in any antenna system. However, the concept of a trap is a misnomer because once the excitation reach the reaction center, it is not immediately destined to be used in photochemistry. Rather, the excitation may migrate back into the antenna only to return to the reaction center many more times prior to being utilized in photochemistry. In PS II, the excited state is essentially in a dynamic equilibrium between all antenna chl *a* including the reaction center (Schatz *et al.*, 1988). Because there are hundreds of antenna chl *a* per reaction center, the total emission is dominated by the antenna chl *a*. Thus, with the exception of algae with phycobilisome antennae, the ultimate source of most fluorescence from photosynthetic samples is the chl *a* pigments in the PS II antenna.

PROCESSES AFFECTING FLUORESCENCE YIELD *IN VIVO*

When a dark-adapted photosynthetic sample is suddenly exposed to continuous (constant intensity) light, the yield of fluorescence passes through a series of maxima and minima before reaching a steady-state level (Lavorel and Etienne, 1977). This induction phase of chl fluorescence (the Kautsky effect, Figure 7) is characterized by an instantaneous rise to the initial (O or F_o) level followed by a complex rise to a maximum P and a decline to a minimum S. These fast changes are completed in a few seconds and may be followed by a slower (minutes) rise to a new maximum M and finally a decline to the steady-state level T. The time scale of these changes and the relative amplitudes of the minima and maxima will depend on the illumination intensity, on the species, and on the previous illumination history of the sample. However, the general shape of the induction curve is seen in most organisms. The changes in yield during induction are the result of several regulatory processes in photosynthesis (Sivak and Walker, 1985) and have been widely used to evaluate the relationship between photosynthetic physiology and fluorescence yield.

Source of variable fluorescence: photochemical quenching. In photosynthetic systems, the dominant process utilizing excited-state energy is photochemistry. However, the possibility of photochemistry depends on the open or closed state of the PS II reaction center. In practice, electron transport from secondary donors to $P680^+$ is much faster than from the primary electron acceptor to subsequent acceptors, such that a closed reaction center most frequently occurs when the primary electron acceptor is reduced (Diner, 1986). Thus, the yield of fluorescence will change depending on the fraction (α) of reaction centers in the open state according to the equation:

$$\phi_f = k_f / (k_f + k_d + k_i + \alpha \cdot k_p) \qquad (6)$$

This dependence of fluorescence yield on the redox state of an electron transport component was originally proposed by Duysens and Sweers in their Q-hypothesis (Lavorel and Etienne, 1977). They suggested the presence of a specific fluorescence quencher Q. In its oxidized state, Q is a strong quencher of excitation energy and thus of fluorescence, but when Q is reduced, its quenching is blocked. Subsequent experiments identified the quencher Q as the primary quinone acceptor of PS II (also called Q_A, Figure 5). When Q_A is oxidized, photochemistry is possible and fluorescence is low; when Q_A is reduced, the reaction center is closed and fluorescence is high. This effect largely accounts for the fluorescence rise from F_o (all traps open in dark adapted sample) to F_p (some fraction of traps closed) during fluorescence induction (Figure 7). If fluorescence induction is measured on a sample which has been treated with an inhibitor such as DCMU (inhibits electron transport after Q_A), the fluorescence rapidly rises to a constant, maximum level F_m (or F_{max})(Figure 7). The difference between F_{max} and F_o is called the maximum variable fluorescence, or F_v. Since the rate-limiting step of photosynthetic electron transport is the oxidation of reduced plastoquinone, and Q_A is in rapid redox equilibrium with plastoquinone, some variable fluorescence rise is observed even at low light in the absence of inhibitors. This dependence of fluorescence yield on the photochemical state of the PS II reaction center has recently been termed photochemical fluorescence quenching (q_P). In the presence of DCMU, the rise in variable fluorescence represents the transition from a state of maximal photochemical quenching at F_o (q_P = 1) to minimal quenching at F_{max} (q_P = 0). (For a review of standardized fluorescence nomenclature see van Kooten and Snel, 1990).

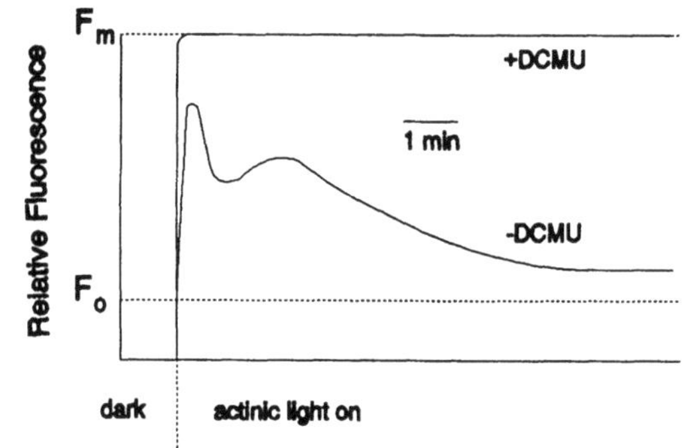

Figure 7. Typical patterns of fluorescence induction following sudden illumination of a dark-adapted photosynthetic sample. Representative traces are shown for samples with or without the inhibitor DCMU, which blocks electron transport between Q_A and Q_B of PS II.

Other quenching processes: non-photochemical quenching. Although the fluorescence yield changes between O and P are almost entirely attributable to photochemical quenching, subsequent fluorescence changes between P and T have contributions from several other processes that are independent of the redox state of Q_A. These processes are cumulatively termed non-photochemical fluorescence quenching (q_N). Under normal conditions, the largest contribution to q_N is due to quenching associated with accumulation of a ΔpH across the thylakoid membrane. This "energy-dependent" quenching, or q_E, has been the subject of intense investigation in green algae and higher plant chloroplasts (Demmig-Adams, 1990). Several types of studies suggest that as q_E increases, there is a corresponding decrease in PS II fluorescence and in the quantum yield of photochemistry in open PS II reaction centers (Weis and Berry, 1987). This quenching of excitation energy occurs in the PS II antenna with the lost energy being dissipated as heat (by a mechanism that differs from that of k_d). The observed increase of q_E with increasing light intensity suggests that this quenching affords some protection to the photosynthetic apparatus against photoinhibition.

The mechanism of q_E-type quenching is also under active investigation. Clearly there are at least two components of q_E. The largest component, especially at moderate to high light intensity, is highly correlated with the light-dependent conversion of violaxanthin to zeaxanthin (xanthophyll cycle) in higher plant chloroplasts (Demmig-Adams, 1990). A similar cycling of carotenoids between diadinoxanthin and diatoxanthin has been identified in some chromophyte algae (Stransky and Hager, 1970), although there have been no quantitative reports correlating the changes in carotenoid composition with fluorescence quenching. The smaller component(s) is independent of the xanthophyll cycle, saturates at low light intensity and may involve quenching in the PS II antenna or reaction center. The reaction(s) that accounts for this quenching of excited states must compete with the fast (efficient) transfer of excitations between neighboring antenna pigments. Under these conditions, collisional interactions between antenna chls and quenching molecules (xanthophylls or otherwise) within the antenna pigment-protein complexes are too slow to compete effectively with the fast (0.2-5 x 10^{-12} s) excitation transfer reactions. Thus, quenching must be of the static type (Lakowicz, 1986), in which the antenna chl and the quencher are bound in a complex at the time the excitation arrives at the chl. The actual mechanism of excited-state quenching by carotenoids remains under investigation (Demmig-Adams, 1990).

The term "fluorescence quenching" is in some ways misleading because the physiological processes that result in changes in the fluorescence yield do not in fact have any direct effect on k_f, the rate constant for fluorescence decay. Rather, those processes that are quantified in measurements of q_P and q_N affect the rate constants of processes that compete with fluorescence for excited states in PS II. Equation 6 shows that the fluorescence yield decreases (q_P increases) with increases in the fraction of reaction centers in which photochemistry (k_P) competes with k_f, k_d and k_i for excited states in the antenna. Similarly, decreases in the fluorescence yield associated with q_N can be quantified by introducing a new rate constant k_q into the denominator of equation 6:

$$\phi_f = k_f / (k_f + k_d + k_i + \alpha \cdot k_p + k_q) \qquad (7)$$

Here, k_q represents the cumulative rate constant for all non-photochemical processes (other than fluorescence, thermal emission, or triplet formation) that compete for excited states in PS II. Thus for both q_P and q_N, the fluorescence yield decreases as the result of excited-state "quenching" reactions competing for excited-states in PS II. Thus, an increase in k_q associated with q_E or any other component of q_N will result in a decrease in the yield of photochemistry.

Photoinhibition. Exposure of photosynthetic organisms to moderate or high intensity light eventually results in a loss of electron transport capacity which has been broadly termed photoinhibition (Neale, 1987). The intensity and duration of illumination required to produce a given amount of photoinhibition varies between species and with the growth conditions to which the organism is adapted. Photoinhibitory damage may be reversible or irreversible, the latter generally requiring higher intensity/longer duration exposure. Under physiological conditions, reversible photoinhibition is much more common. Damage appears to be localized in the PS II reaction center complex, resulting in a loss of photochemical charge separation (Kyle and Ohad, 1986). Recovery requires synthesis and insertion of new proteins into the complex. A common characteristic for reversible photoinhibition of PS II is a loss of variable fluorescence. Under a fixed set of measuring conditions, the extent of variable fluorescence measured in the presence of DCMU declines in proportion to the fraction of PS II reaction centers that are photoinhibited (Bjorkman and Demmig, 1987). The decrease in variable fluorescence accompanying the loss of reaction center photochemistry suggests that photoinhibition induces the formation of a new quenching reaction in PS II. A satisfactory mechanism explaining these observations is not yet known. As would be expected, photoinhibition also results in a decreased capacity for electron transport and quantum yield of photochemistry in PS II (Bjorkman and Demmig, 1987). Fluorescence quenching due to photoinhibitory damage (q_I) occurs at both saturating and limiting light intensities in approximately a dose dependent manner. Recovery from photoinhibition shows a similar dependence, requiring tens of minutes to many hours.

Quenching related to changes in antenna size. The lifetime of an excitation in the PS II (and PS I) antennae is determined by the efficiency of photochemistry in the reaction center and by the size of the antenna (Owens *et al.*, 1987; Holzwarth, 1987). The contribution of antenna size can be understood as follows: it is known that an excited state is essentially delocalized among the antenna pigments during its lifetime, spending no more time on the reaction center than on any other pigment. Because fluorescence yield and excited state lifetime are linearly related (equation 4), the fluorescence yield increases with increasing antenna size because the lifetime of the excited state increases as it migrates over a larger number of antenna pigments.

Changes in antenna size occur on two physiological time scales: as long-term (days) adaptation to average growth conditions and short-term (minutes) response to changes in light intensity and spectral quality (Ley, 1980). The latter response is called a light state transition and is generally characterized by changes in the functional antenna sizes of PS II and/or PS I to maintain optimal photosynthetic efficiency. In green algae or higher plants, if light affecting the turnover of PS I is limiting the rate of electron transport, antenna complexes coupled to PS II are reversibly dissociated, decreasing the PS II antenna size and rate of PS II turnover (helping to balance the distribution of light energy between PS I and PS II) (Fork and Satoh, 1986). The dissociated complexes may become coupled to PS I or may simply dissipate energy as heat. In either case, excited state energy is being dissipated by processes that compete with PS II fluorescence. The time scale of light state transitions is on the order of 1 to 10 minutes. Quenching of fluorescence due to light state transitions (q_T) has been identified in higher plants, and several algal classes. Although mechanisms of light state transitions may differ greatly between classes (Biggins and Bruce, 1989), their contribution to total q_N is probably ubiquitous.

Excited state annihilation. In general, the yield of fluorescence emitted by a sample is linearly related to the intensity of light exciting the fluorescence (equation 5). However, at very high light intensities, there is a probability that two or more excited states may exist in the antenna of a reaction center at the same time. Under these conditions, an annihilation process may occur between two excited states where the energy of one or both excitations is lost as heat (Campillo and Shapiro, 1978). The threshold for the occurrence of multiple excitations occurs only at very high, non-physiological light intensities (focused laser illumination, for example). Annihilation reactions at high light intensity decrease fluorescence yield in a non-photochemical manner that is independent of the biological reactions regulating fluorescence yield.

INTERPRETATION OF FLUORESCENCE DATA

Present limitations. The description of processes that regulate the yield of chl fluorescence *in vivo* presented in the previous sections is largely derived from studies on higher plant chloroplasts or green algal suspensions. The interpretation of fluorescence signals from other types of algae (i.e.: diatoms, dinoflagellates, etc.) based on the green algal/higher plant data, even when measured under identical conditions, is certainly open to criticism. There are presently several major obstacles to broad interpretation of fluorescence yield data.

The mechanism of photochemical fluorescence quenching, that is, the regulation of fluorescence yield by the redox state of the primary quinone acceptor of PS II, is probably identical among all oxygen-evolving plants and algae. Although fluorescence quenching studies have not been applied systematically to all algal classes, it is widely accepted that the proteins, pigments, and electron transport components of the PS II (and PS I) reaction center complex have been highly conserved throughout evolution (Thornber, 1975; Bryant, 1986). Since the components involved in regulating q_P are all localized in the PS II reaction center, their function in regulating q_P is probably also conserved. Similarly, the mechanism of photoinhibitory damage to PS II may be conserved among the algae. Studies in several algal classes have demonstrated a loss of variable fluorescence coincident with a decline in photosynthetic capacity (Neale, 1987).

In contrast, there is no reason to expect that any of the processes contributing to non-photochemical fluorescence quenching (with the exception of photoinhibition) are conserved among the algae. The reaction(s) of non-photochemical quenching is thought to occur in the light-harvesting antenna complexes, and these complexes exhibit considerable diversity among the algal classes (Owens, 1988; Anderson and Barrett, 1986). This is not to say that formation of a trans-thylakoid ΔpH, regulation of functional antenna size through light state transitions, and photoinhibition do not occur in all algal classes. Nor should we expect that non-photochemical quenching will be absent in other algal classes. Dealing with variable light intensities is a common problem for all photosynthetic organisms. Because the non-photochemical quenching processes are components of an overall photosynthetic process that has evolved in a common environment, it is likely that non-photochemical quenching occurs in some form in all algal classes. However, they may differ in their mechanistic contributions to fluorescence quenching.

Regulation of q_E provides an example of the potential problem. In green algae and higher plants, the major component of q_E requires that at least three conditions be met: the presence of a proton gradient across the thylakoid membrane, the light-dependent conversion of violaxanthin to zeaxanthin, and proper redox-poising of an unidentified, antimycin-sensitive electron transport component (Demmig-Adams, 1990). Interconversion of violaxanthin and zeaxanthin via the xanthophyll cycle has not been identified in any other algal class. Violaxanthin and zeaxanthin are absent in several algal classes, particularly the blue-green algae. It is certainly possible that other carotenoids such as diadinoxanthin and diatoxanthin may be capable of functioning in quenching of chl excited states. Clearly, a broader understanding of non-photochemical quenching processes is a prerequisite for using fluorescence as a diagnostic tool in most algal classes.

There is a more general problem in interpreting fluorescence measurements from any photosynthetic sample. Our present knowledge of processes regulating fluorescence yield is based on relatively few experiments involving a limited number of species grown under a particular set of conditions. One cannot neglect the possibility that the quantitative relationship between a measured photosynthetic parameter and fluorescence yield may vary with growth conditions, or that variation in growth conditions may produce a range of effects between species. For example, the relationship between ΔpH and q_E has been examined in higher plants. There is strong evidence suggesting that this relationship varies between species and especially with stress conditions during growth (Laasch, 1987). Under these conditions, even if all other factors remain constant, it would be difficult to unambiguously relate fluorescence changes to the size of the pH gradient across the thylakoid membrane.

Fluorescence as a predictor of photosynthetic rates. There is a widespread interest among the biological oceanographic community in the possibility that *in vivo* fluorescence measurements can be used to predict *in situ* rates of photosynthesis. This hypothesis is largely based on the assumption that parameters quantifying the rise in variable fluorescence from F_o to F_{max}, measured in the presence of the inhibitor DCMU, are proportional to the amount of PS II present and are therefore related to photosynthetic capacity (Malkin *et al.*, 1981). Similar measurements have utilized enhancement of steady-state *in vivo* fluorescence upon addition of DCMU as a measure of photosynthetic rate (Roy and Legendre, 1979; Vincent, 1980). Laboratory experiments have generally demonstrated a reasonable correlation between variable fluorescence parameters and photosynthetic capacity measured by ^{14}C incorporation

(Samuelsson and Oquist, 1977; Bates and Platt, 1984). Analogous field experiments have met with varied success (Cullen and Renger, 1979; Bates and Platt, 1985; Keller, 1987; Heath, 1988). In general, good correlations were found when samples were restricted to well defined algal populations; correlations determined in different populations showed larger variability. Even in the best correlations, broad confidence limits on regression data provide for largely qualitative rate predictions, especially in low biomass samples.

There are several factors that suggest caution in correlating fluorescence parameters and photosynthetic capacity. Of principal importance is the fact that the variable fluorescence rise is proportional only to the PS II pool size; estimation of the maximum capacity for electron transport through PS II requires knowledge of the maximal photochemical turnover rate of PS II. PS II turnover rate has been shown to vary between species and with growth conditions (Falkowski *et al.*, 1981; Boardman, 1977). There is also an implicit assumption that PS II electron transport capacity limits the rate of photosynthesis at light saturation. Recent data from higher plants contradicts this assumption. Stitt (1986) has demonstrated a 50-100% excess of electron transport capacity at light saturation and assigned the rate limiting step to the dark reactions: translocation of inorganic phosphate from the cytoplasm to the chloroplast (Sharkey *et al.*, 1986). The limitation of photosynthesis by inorganic phosphate translocation was also concluded by Sivak and Walker (1986). In a green alga, Sukenik *et al.*. (1987) showed that light saturated rates correlated with the amount of ribulose biphosphate carboxylase, the primary CO_2 fixing enzyme of the dark reactions. These results suggest that the mechanistic basis for expecting a relationship between photosynthetic capacity and variable fluorescence should be reinvestigated in the algae.

Experimentally, samples are routinely dark adapted for some length of time prior to measurement of variable fluorescence (Bates, 1985). This dark adaptation period is required so that all non-photochemical fluorescence quenching processes are completely relaxed ($q_N = 0$) and that Q_A is completely oxidized ($q_P = 1$) and permits the accurate measurement of both F_o and F_{max}. In practice, the time required for complete relaxation of all components of q_N may be quite variable and dependent on species and physiological state of the sample (Owens, 1986; Walters and Horton, 1990). In addition, the assumption that Q_A is completely oxidized after dark adaptation is known to be incorrect in some algal classes (Owens, 1986;

Bennoun, 1982). A dark reaction known as "chlororespiration" has been shown to maintain Q_A and the plastoquinone pool in a semi-reduced state even after long dark adaptation periods (Bennoun, 1982). However, brief illumination of the sample with low intensity, far red (700-710 nm) light just prior to the onset of actinic illumination may oxidize Q_A by forcing turnover of PS I alone (Owens, 1986). The incomplete relaxation of q_N and partial reduction of Q_A following dark adaptation will lead to an underestimation of fluorescence parameters associated with the variable fluorescence rise and are likely to increase the error in the correlation between regression of fluorescence and photosynthetic capacity.

One potentially promising technique for predicting rates of photosynthesis from measurements of q_P and q_N has been introduced by Weis and Berry (1987). This technique assumes that (1-q_P) is a good estimate of the fraction of open PS II reaction centers and that q_N correlates with the yield of photochemistry in open PS II traps. The technique is calibrated by measuring q_P, q_N and photosynthetic rate under standard conditions (usually by varying light intensity) and determining appropriate regression parameters. This procedure has been successfully applied to several studies in higher plants and at least one algal system (Holmes *et al.*, 1989). The universal applicability of this technique relies on a constant relationship between q_N and the yield of photochemistry in open PS II traps under all conditions. That is, will the relationship between q_N, q_P and photosynthetic rate be the same for samples measured under optimal conditions and in samples differing in the degree of nutrient or light stress or in the previous light history? This important question remains to be answered.

Fluorescence as an indicator of algal biomass. The quantum yield of fluorescence is defined as the number of fluorescence photons emitted by a sample divided by the number of photons absorbed. This suggests that in dilute algal suspensions, the *in vivo* fluorescence emission intensity is proportional to the number of photons absorbed by the sample, the proportionality constant being the fluorescence quantum yield. Since absorption (in dilute samples) is largely determined by the amount of pigment present, *in vivo* fluorescence should be an indicator of algal biomass. This relationship probably represents the most highly utilized aspect of chl fluorescence (Lorenzen, 1966). However, the influence of environmental conditions on fluorescence quenching processes would suggest that natural variability in the fluorescence quantum yield would negate, or at least limit, the ability of fluorescence to predict biomass.

Indeed, there is strong evidence that *in vivo* fluorescence per unit chl *a* is highly variable (Loftus and Seliger, 1975; Falkowski and Kiefer, 1985; Keller, 1987). Even if the samples are carefully adapted to the state where $q_P = 1$ and $q_N = 0$ (proper F_o), there is no evidence in green algae or higher plant chloroplasts that F_o fluorescence correlates with chl *a* (TG Owens and CS Ting, unpublished results). *In vivo* fluorescence should therefore be regarded only as a qualitative indicator of algal biomass.

Fluorescence as an indicator of "physiological state". The number of factors that affect the yield of chl fluorescence *in vivo* outlined above might suggest that we should expect a poor quantitative relationship between the yield of fluorescence and photosynthesis under varying environmental conditions. However, Walker and co-workers have elegantly demonstrated (in higher plants) that *relative* changes in fluorescence yield are tightly coupled to photosynthetic rate under conditions where photosynthetic rates are changing rapidly (changes in light intensity or CO_2 availability) (Sivak and Walker, 1985). Subsequent work by a number of investigators in higher plants and algae suggests that the measurement of photochemical and non-photochemical fluorescence quenching can provide at least qualitative information on the effects of environmental conditions on the basic reactions of photosynthesis (ie: physiological state). For example, Holmes *et al.* (1989) have used the relationship between fluorescence quenching and photosynthetic rate described by Weis and Berry (1987) to correctly predict the diversion of reductant from CO_2 to nitrogen assimilation under conditions of transient nitrogen uptake. During nitrogen uptake, changes in the fluorescence quenching are due to direct effects of altered demand for ATP and NADPH as well as to changes in functional antenna size of PS II (light state transitions)(Turpin and Bruce, 1990). Miller *et al.* (1988) have similarly demonstrated diversion of ATP from CO_2 fixation to inorganic carbon uptake.

Under controlled laboratory conditions, a number of studies have shown that the quantum yield of photosynthesis is affected by both light and nutrient availability (Welschmeyer and Lorenzen, 1981; Bjorkman and Demmig, 1987; Cleveland and Perry, 1989; Kolber *et al.*, 1988; Herzig and Falkowski, 1989). Equations 6 and 7 suggest that decreases in the yield of photochemistry must be accompanied by decreased fluorescence yield and are likely to be attributed to increased non-photochemical quenching in PS II. In all laboratory studies, changes in fluorescence yield in response to growth conditions are determined in a controlled environment and can be unambiguously assigned to specific physiological responses of the organism. The success of this type of experiment depends critically on the presence of control samples, to which the responses of the treated samples can be compared. A lack of suitable controls is one of the main factors which limits the interpretation of fluorescence techniques in field samples. Considering the number of environmental factors that potentially influence photosynthesis, and the complex coupling of these reactions to fluorescence yield, an unambiguous assignment of field fluorescence measurements to specific environmental influences and physiological responses is nearly impossible without additional information.

This is not to say that field measurements of *in vivo* or *in situ* chlorophyll fluorescence are useless. There are excellent examples of studies that have used correlations between fluorescence measurements and other physical and biological variables, such as natural gradients in light or nutrient availability, to assess phytoplankton responses (Vincent, 1980; Kolber *et al.*, 1989; Therriault *et al.*, 1989; Cleveland *et al.*, 1989; Yentsch and Campbell, 1991; Olson *et al.*, 1989a). Powerful new fluorescence techniques, such as the "double-flash" technique used by Falkowski and co-workers (see below), also appear to be quite useful in evaluating many aspects of photosynthetic physiology. One additional technique, which remains to be explored, involves the use of field samples as controls against which aliquots of the same sample briefly incubated with added nutrients, metabolic inhibitors, or under various light conditions, can be compared. Under these conditions, one could potentially evaluate the bulk properties of a field sample in terms of the relative importance of multiple environmental factors that affect photosynthetic physiology.

Fluorescence excitation and emission spectra. A fluorescence emission spectrum probes the spectral properties of those pigments that directly emit fluorescence. Since the origin of fluorescence emission in nearly all photosynthetic algae and higher plants is the chl *a* antenna of PS II, emission spectra are of little use in distinguishing between the photosynthetic properties of samples. The exception is the phycobilin-containing algae, where a significant fraction of total fluorescence emission originates from the phycobilisome antenna (Bryant, 1986). This factor has been utilized to evaluate the abundance and distribution of marine *Synechococcus* species (Olson *et al.*, 1989a).

In contrast, fluorescence excitation spectra probe the spectral properties of those pigments that are capable of transferring absorbed light energy to chl *a*. Because of the diversity of light-harvesting antenna pigments among the algal classes (Anderson and Barrett, 1986), fluorescence excitation spectra can be of taxonomic value in evaluating differences between field samples (Yentsch and Yentsch, 1979; Hinton *et al.*, 1989). In addition, fluorescence excitation spectra have been used to evaluate photoadaptation (Neori *et al.*, 1984) and to identify light-harvesting function of pigments that are unique to specific groups or species of algae (Owens *et al.*, 1986).

Modeling of photosynthetic light reactions. When evaluating the complex processes of the photosynthetic light reactions, it is useful to have a kinetic model which parameterizes the component reactions which are competing for excited state energy. Equations 6 and 7, which express the quantum yields of fluorescence and photochemistry in terms of rate constants for processes competing for excited state energy in the antenna, grossly oversimplify the complexity of the photosynthetic light reactions. They are explicitly correct only when each of the rate constants is an "effective" rate constant integrated over all pigments in the antenna. For example, k_p in equations 6 and 7 represents the rate at which excited states are quenched by photochemistry from the entire antenna, not just on the reaction center pigment. The most widely accepted models describing photochemistry, fluorescence and thermal emission in PS II have been the bipartite and tripartite models of Butler (Butler, 1978). These models have been generally successful in predicting changes in the yield of fluorescence and photochemistry under many physiological conditions. Recently, Holzwarth and co-workers have presented an alternative model based on ultrafast time-resolved spectroscopy in PS II (Schatz *et al.* 1988). None of these models is completely successful at explaining all phenomena related to the light reactions of photosynthesis. Thus, although Butler's model remains useful in many situations, the limitations of this or any model should be recognized in predicting or evaluating physiological phenomena.

TECHNIQUES FOR MEASURING IN VIVO FLUORESCENCE

Measurement of chl fluorescence requires external illumination for two different types of processes: to generate excited states in the antenna whose decay via fluorescence is directly detected (the measuring or excitation function), and to drive the photosynthetic reactions or to poise the sample in some known physiological state (the actinic function). Techniques for measuring chl fluorescence can be divided into three categories depending on the type of light source(s) used to provide these functions: (1) a single, continuous light source which acts both as a measuring and actinic source, (2) two or more sources using one low intensity, modulated source to excite fluorescence and separate, non-modulated source(s) to provide actinic illumination, (3) pulsed light sources. Each of these general techniques has inherent benefits and disadvantages which will be discussed briefly. In general, each of these techniques can be applied to dilute suspensions of phytoplankton or to samples in which the cells have been concentrated on a filter substrate. The minimum concentration of chl required for analysis is highly instrument-dependent. Any manipulation of samples prior to measurement (concentration, filtration or storage) must be considered as potentially altering the sample from its *in situ* characteristics.

Instruments using a single light source as both exciting and actinic illumination are most commonly used for measurement of fluorescence induction. In general, the sample is dark adapted to bring $q_P = 1$ and $q_N = 0$. The sample is then suddenly exposed to continuous illumination (usually using a shutter) and the time-dependent changes in fluorescence emission are monitored (Figure 7). Illumination both excites the fluorescence and drives the induction process. Estimation of the components of fluorescence quenching can be accomplished in this system by taking advantage of the different rates at which the various quenching processes relax when electron transport is inhibited by the rapid (1 s) addition of DCMU (Walters and Horton, 1990). The single source technique has the benefits of being inexpensive, requiring relatively unsophisticated apparatus and has been widely used in a number of different applications. The system suffers from considerable inflexibility in measurements, primarily the inability to provide internal standards (for normalizing fluorescence between samples and to quantify the components of fluorescence quenching) without poisoning the sample with DCMU. In addition, any change in source intensity will be detected as changes in fluorescence emission independent of the quenching state of the sample.

Measurement of fluorescence using a modulated excitation source and continuous actinic source(s) overcomes the experimental limitations in the single source technique. Use of the modulated excitation has two advantages. First, its intensity may be kept low enough that the measuring source alone produces no actinic effects (such that the modulated source alone measures only F_o in a dark adapted sample). Use of a lock-in amplifier in the detection circuitry provides for sufficient signal to noise even at the low excitation intensity. Second, the modulated fluorescence signal is essentially independent of any direct effects of all non-modulated sources illuminating the sample. This means that an instantaneous change in the intensity of a non-modulated source will have no affect on the instantaneous fluorescence yield. However, the non-modulated sources will have an indirect, actinic effect by altering the state of the photochemical reactions in the sample. Thus the modulated fluorescence signal measures a true, relative fluorescence yield. The modulated fluorescence technique and its advantages are described in detail by Schreiber (1983, 1986).

An additional advantage of the modulated fluorescence measurements is that it provides for a non-destructive technique for separating photochemical and non-photochemical fluorescence quenching (Schreiber *et al.*, 1986). In this procedure, brief (0.1-2 s) illumination with high intensity light is used to induce a transient, complete reduction of Q_A, bringing q_P to 0. Any fluorescence quenching remaining when $q_P = 0$ is by definition q_N (Figure 8). This technique of "light doubling" was introduced by Bradbury and Baker (1984). The combination of modulated fluorescence detection and light doubling also allows for direct comparison of fluorescence measurements on widely differing samples. In a true, dark adapted sample (q_P = 1 and q_N = 0), the modulated measuring beam alone provides an accurate measure of F_o. Superimposing a saturating light pulse to transiently reduce Q_A defines the F_{max} level (q_P = 0, q_N = 0; Figure 8). These fluorescence levels provide limits between which all other signals can be normalized (relative fluorescence yield with $F_o = 0$ and $F_{max} = 1$), permitting direct comparison of steady-state fluorescence levels and quenching parameters between samples. The disadvantage of the modulated fluorescence technique is its cost (compared to the single source technique). Several commercial instruments are available and are seeing widespread application in higher plants, but in general they lack sensitivity for work with dilute algal samples. With care, they will work on samples containing 1 μg chl *a*/ml, and improvements in design could produce another factor of 10 increase in sensitivity.

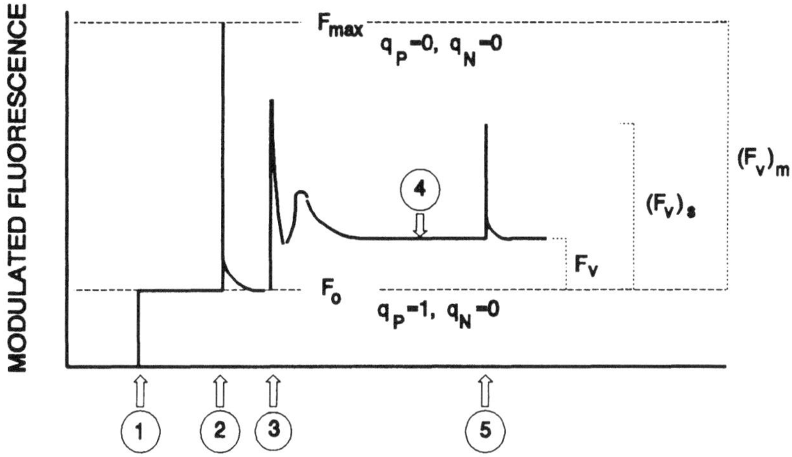

Figure 8. Determination of fluorescence parameters using the modulated fluorescence and "light-doubling" technique. Prior to (1), the sample is brought to the dark-adapted state ($q_P=1$, $q_N=0$). (1) sample is illuminated with low intensity modulated source, defining F_o level; (2) sample is briefly illuminated with a saturating light to completely reduce Q_A, defining the F_{max} level; (3) continuous background illumination initiates a transition to steady-state photosynthesis; (4) determination of steady-state fluorescence yield in presence of continuous background illumination; (5) brief saturating illumination to transiently reduce Q_A. The parameters q_P and Q_N are defined by the following equations:

$$q_P = \frac{(F_v)_s - (F_v)}{(F_v)_s} \qquad q_N = \frac{(F_v)_m - (F_v)_s}{(F_v)_m}$$

The final method for measuring fluorescence parameters utilizes short (μs) pulses of light generated by flashlamps of pulsed laser sources. The light pulses are short enough that a maximum of one photochemical turnover occurs at each reaction center during the duration of the pulse ("single-turnover" pulses). Here, the fluorescence induced by a weak probe (measuring) flash is measured following a higher intensity pump (actinic) flash (hence the name double-flash or pump-probe technique). The intensity of the actinic flash is adjusted to provide complete or partial reduction of Q_A. In addition, continuous background light may be used to poise the sample in various states. The technique is similar to modulated fluorescence detection in that the fluorescence yield of the probe flash is relative; it is insensitive to the direct effects of continuous actinic illumination. The application of the pump-probe technique to phytoplankton studies (field and laboratory) has been pioneered by Falkowski and co-workers (Falkowski *et al.*, 1986; 1988). Depending on the time delay between the pump and probe flashes, this technique can provide direct measures of PS II antenna sizes, the rate of electron transport away from Q_A, or the fraction of PS II traps closed by continuous actinic illumination. It can also be used to quantify q_P and q_N directly. As with the modulated technique, its main disadvantages are cost and complexity of the apparatus.

The modulated fluorescence and double-flash techniques have the significant advantage that they can easily be combined with simultaneous measurements of O_2 evolution, a direct measure of photosynthetic rate. The combination of fluorescence and photosynthetic rate data derived from samples whose physiological state has been manipulated by light intensity, stress conditions, metabolic inhibitors, etc., provides the most promising protocols for understanding the processes that control photosynthesis and how best to use fluorescence measurements to predict *in vivo* or *in situ* rates and physiological state.

FLUORESCENCE MEASUREMENTS IN INDIVIDUAL CELL ANALYSES

The measurement and interpretation of chl fluorescence data described in the previous sections rely on the assumption of a homogeneous sample, that the average fluorescence properties measured simultaneously on a large number of cells is representative of the properties of each individual cell. Unfortunately, natural phytoplankton populations are rarely dominated by a single species, thus the heterogeneous nature of the sample must be considered (Yentsch and Campbell, 1991). Within a heterogeneous population, individual species may exhibit different growth rates, varying degrees of adaptation to the physical/chemical/biological environment at the time of sampling, and may have growth rates limited by different environmental factors. Thus the fluorescence properties of the entire sample may have little relationship to average photosynthetic rate of physiological properties of the sample. More importantly, measurement of bulk sample fluorescence may mask critical information of individual components of the population.

With the introduction of individual cell analysis into phytoplankton studies using flow cytometric techniques, one may address the problem of heterogeneous natural samples (Yentsch and Campbell, 1991; Olson *et al.*, 1985; Legendre and Yentsch, 1989). *In vivo* fluorescence from chls and phycobilins are the major components of phytoplankton "autofluorescence" and have become a routine component of the multiparameter flow cytometric analyses of laboratory and field phytoplankton samples (Cullen *et al.*, 1988). Using flow cytometry, separation of photosynthetic cells from heterotrophic organisms and detrital materials is routinely achieved. Flow cytometry has the ability to distinguish between individual species or groups of related species within a single sample (Olson *et al.*, 1989b). The ability to detect both chl and phycoerythrin fluorescence with flow cytometry has been instrumental in demonstrating the ubiquitous distribution of marine *Synechococcus* species and their contribution to algal biomass and productivity (Olson *et al.*, 1988, 1989a). Fluorescence and particle size data were also central to the recent identification of free-living marine prochlorophytes (Chisholm *et al.*, 1988).

In addition to these taxonomic considerations, fluorescence analyses from flow cytometry measurements have also been used to evaluate the photosynthetic physiology of algal populations in laboratory and field studies. Sosik *et al.* (1989) investigated the relationship between flow cytometric fluorescence signals and pigment composition between species and with adaptation to various light conditions. Perry and Porter (1989) demonstrated the relationship between fluorescence per cell measured by flow cytometry and the effective cellular absorption cross-section. Several field studies have shown changes in fluorescence per cell that appear to be linked to light adaptation (Olson *et al.*, 1989a; Demers *et al.*, 1989; Campbell *et al.*, 1989).

All of the fluorescence techniques described in previous sections, which may be used for estimating photosynthetic capacity or physiological state, rely on quantifying relative fluorescence yields by scaling of fluorescence data either between measurements on a single sample or between samples. With dilute algal suspensions or algae concentrated on filters, fluorescence scaling presents minor but correctable problems in instruments with a fixed excitation and detection geometry. Internal normalization to F_o and F_m fluorescence levels using the light doubling and modulated excitation techniques provides the most unambiguous scaling of data. Unfortunately, these scaling procedures are not easily applied to flow cytometric analyses in which individual cells are entrained in a rapidly flowing medium. Thus there is considerable uncertainty as to the nature and interpretation of fluorescence signals measured in flow cytometry (Neale *et al.*, 1989). Several problems must be addressed.

One problem has to do with differences in the optical properties of individual cells (Iturriaga and Siegel, 1989). On the assumption that each particle is uniformly illuminated within the flow cell and that fluorescence detection is equally uniform (independent of the transverse position of the particle in the flow), then the amount of fluorescence (F) detected from each particle will be a function of three physical parameters: I_a, the total light absorbed by functional photosynthetic pigments in the particle, G_s, the fraction of fluorescence emission that is reabsorbed by pigments within the cell, and G_a, the fraction of fluorescence emission that is scattered off of the optical detection axis. In addition, there is the parameter of interest, ϕ_f, the true fluorescence yield as determined by biological fluorescence quenching at the instant before the particle entered the flow cell.

$$F = I_a \cdot (1 - G_s) \cdot (1 - G_a) \cdot \phi_f \tag{8}$$

Each of the physical parameters depends upon the total amount of pigment and its distribution within the particle, the size and shape of the particle, and the light scattering properties (internal and external) of the particle. These factors may differ significantly, even between cells in synchronous, unialgal cultures, as suggested by the approximately factor of 2 variation in fluorescence counts at any given cell size (Campbell and Yentsch, 1989).

Despite this large variability within a homogeneous population, changes in chl fluorescence resulting from cellular adaptation to environmental conditions can be resolved (Campbell *et al.*, 1989; Olson *et al.*, 1989a; Demers *et al.*, 1989). The critical question remaining to be answered is: what is the relative fluorescence yield that is being measured in flow cytometry? This question was addressed by Neale *et al.* (1989), who concluded that an EPICS V system measures relative fluorescence yields that are intermediate between F_o and F_m. The authors also concluded that the relative fluorescence measured would vary with excitation irradiance and mean residence time of cells in the illuminated volume of the flow cytometer.

Since flow cytometry is essentially a single excitation source measurement (as described in the previous section), one must consider both the excitation and actinic effects of the source illumination. Equation 8 shows that for any given fluorescence yield, the intensity of fluorescence emission (F) is determined by the rate of light absorption (I_a). I_a is in turn determined by the incident photon flux (determined by laser power and focused spot size) and the absorption properties of the cell. The actinic effects of the laser illumination depend on the residence time of the cell in the illuminated volume and the rate of light absorption by the antenna of each PS II reaction center. For a typical residence time of 1-20 μs, the major actinic effect should be due to photochemical quenching (electron transport from P680 to Q_A); all other reactions of electron transport and non-photochemical quenching occur on time scales slower than 20 μs. Essentially, the actinic effect determines how much ϕ_f will increase as the cell traverses the illuminated volume due to closure of PS II traps. However, at focused laser intensities, the possibility of excited-state annihilations cannot be neglected. Clearly, the actual ϕ_f will be instrument-dependent and direct comparisons between different instruments may be difficult.

Recently, a modification of the double-flash method has been applied successfully to suspensions of higher plant chloroplasts and green algae to determine the rate of electron transport away from Q_A (Petit *et al.*, 1989). Here, the pump beam from one laser (closing some fraction of PS II reaction centers) is positioned at a fixed distance upstream from the probe (measuring) beam. By varying the flow rate, the delay time between the pump and probe beams can be varied; fluorescence yield from the probe beam as a function of delay time gives the kinetics of Q_A^- oxidation. Additional modifications would permit the quantification of q_P and q_N. Thus, with continued evolution of multiparametric analyses in flow cytometry, it is likely that the properties of individual cells can be quantified sufficiently to allow direct measurement of relative fluorescence yield.

CONCLUSIONS

Emission of chl fluorescence is a common characteristic of all photosynthetic organisms. Measurement of *in vivo* fluorescence from phytoplankton samples, using a variety of instruments, is sensitive and relatively straightforward. Based on both theoretical foundation and correlative experiments, there is evidence that *in vivo* chl fluorescence can be useful in predicting several important parameters including algal biomass, rates of photosynthesis, and information on physiological state. This is of obvious importance to both laboratory and field studies; fluorescence analyses require smaller samples and are considerably less time consuming than direct measurement of these parameters.

In the past 20 years, there has been considerable progress in understanding the source of *in vivo* fluorescence and of the processes that regulate the yield of fluorescence. Our present challenge is to use this knowledge to provide a better foundation for interpreting *in vivo* fluorescence data. Clearly the quantum yield of fluorescence is variable, under the control of many physiological processes. Our knowledge of these processes is sufficient to define the complexity of the problem (relating yield to rates or physiological state) and to provide quantitative information only under narrowly defined experimental conditions. Our lack of knowledge on variations in the processes controlling fluorescence yield between species (particularly between algal classes) and depending on previous growth history of the sample are the major obstacles to a more general and quantitative interpretation of fluorescence data. Without question, this should be a major component of future research in this area.

Using several techniques, it is possible to quantify the individual components of photochemical and non-photochemical fluorescence quenching in any algal sample. With proper scaling of the fluorescence, it is possible to directly compare measurements between physically different samples and to assess the effects of experimental treatments (light, nutrient addition, etc). With knowledge of the processes controlling fluorescence yield in hand, it should be possible to use the components of fluorescence quenching to evaluate any process that has a direct or indirect effect on photosynthesis. At present, application of these types of techniques to flow cytometry is limited by the inability to scale fluorescence data between individual cells due to differences in the properties of the cells. Further improvement in multiparameter analyses may permit these properties to be quantified sufficiently to permit direct comparison of fluorescence yield between cells.

REFERENCES

Anderson JM, Barrett J (1986) Light-harvesting pigment-protein complexes of algae. *In* Staehelin AL, Arntzen CJ, eds. Encyclopedia of Plant Physiology, New Series vol 19. Springer-Verlag, Berlin pp. 269-285

Bates SS (1985) Sample conditioning for measurement of fluorescence induction of chlorophyll *a* in marine phytoplankton. J Plankton Res 7:703-714

Bates SS, Platt, T (1984) Fluorescence induction as a measure of photosynthetic capacity: response of *Thalassiosira pseudonana* (Bacillariophyceae) and *Dunaliella tertiolecta* (Chlorophyceae). Mar Ecol Prog Ser 18:67-77

Bates SS, Platt T (1985) Fluorescence induction of chlorophyll *a* in the Sargasso Sea and on the Grand Banks: correlation with photosynthetic capacity. Mar Ecol Prog Ser 27:29-38

Bennoun P (1982) Evidence for a respiratory chain in the chloroplast. Proc Nat Acad Sci USA 79:4352-4356

Biggins J, Bruce D (1989) Regulation of excitation energy transfer in organisms containing phycobilins. Photosynth Res 20:1-34

Bjorkman O, Demmig B (1987) Photon yield of O_2 evolution and chlorophyll fluorescence characteristics at 77K among vascular plants of diverse origins. Planta 170:489-504

Boardman NK (1977) Comparative photosynthesis of sun and shade plants. Ann Rev Plant Physiol 28:355-377

Bradbury M, Baker NR (1984) A quantitative determination of photochemical and non-photochemical quenching during the slow phase of the chlorophyll fluorescence induction curve of bean leaves. Biochim Biophys Acta 765:275-281

Bryant DA (1986) The cyanobacterial photosynthetic apparatus: comparison of those of higher plants and photosynthetic bacteria. Can Bull Fish Aquatic Sci 214:423-500.

Butler WL (1978) Energy distribution in the photochemical apparatus of photosynthesis. Ann Rev Plant Physiol 29:345-378

Campbell JW, Yentsch CM (1989) Variance within homogeneous phytoplankton populations: analysis of clonal cultures. Cytometry 10:596-604

Campbell JW, Yentsch CM, Cucci TL (1989) Variance within homogeneous phytoplankton populations, III: analysis of natural populations. Cytomerty 10:605-611

Campillo AJ, Shapiro SL (1978) Picosecond fluorescence studies of exciton migration and annihilation in photosynthetic systems. A review. Photochem Photobiol 28:975-1013

Chisholm SW, Olson RJ, Zettler ER, Goericke R, Waterbury JB, Welschmeyer NA (1988) A novel free-living prochlorophyte abundant in the oceanic euphotic zone. Nature 334:340-343

Clayton RK (1980) Photosynthesis: Physical Mechanisms and Chemical Patterns. Cambridge University Press, Cambridge, London, New York

Cleveland JS, Perry MJ (1987) Quantum yield, relative specific absorption and fluorescence in nitrogen-limited *Chaetoceros gracilis*. Mar Biol 94:489-497

Cleveland JS, Bidigare RR, Perry MJ (1989) Maximum quantum yield of photosynthesis in the northwestern Sargasso Sea. J Mar Res 47:869-886.

Cullen JJ, Renger EH (1979) Continuous measurement of the DCMU-enhanced fluorescence response of natural phytoplankton populations. Mar Biol 53:13-20

Cullen JJ, Yentsch CM, Cucci, TL, MacIntyre HL (1988) Autofluorescence and other optical properties as tools in biological oceanography. Proc SPIE 925:149-156

Demers S, Davis K, Cucci TL (1989) A flow cytometric approach to assessing the environmental and physiological status of phytoplankton. Cytometry 10:644-652

Demmig B, Winter K, Kruger A, Czygan F-C (1987) Photoinhibition and zeaxanthin formation in intact leaves. A possible role of the xanthophyll cycle in the dissipation of excess light energy. Plant Physiol 84:218-224

Demmig-Adams B (1990) Carotenoids and photoprotection in plants: a role for the xanthophyll zeaxanthin. Biochim Biophys Acta 1020:1-24

Diner B (1986) The reaction center of photosystem II. *In* Staehelin AL, Arntzen CJ, eds. Encyclopedia of Plant Physiology, New Series vol 19. Springer-Verlag, Berlin pp. 422-436

Falkowski PG, Kiefer DA (1985) Chlorophyll *a* fluorescence in phytoplankton: relationship to photosynthesis and biomass. J Plankton Res 7:715-731

Falkowski PG, Owens TG, Ley AC, Mauzerall DC (1981) Effects of growth irradiance levels on the ratio of reaction centers in two species of marine algae. Plant Physiol 68:969-973

Falkowski PG, Wyman K, Ley A, Mauzerall D (1986) Relationship of steady state photosynthesis to fluorescence in eucaryotic algae. Biochim Biophys Acta 849:183-192

Falkowski PG, Kolber Z, Fujita Y (1988) Effect of redox state on the dynamics of photosystem II during steady state photosynthesis in eucaryotic algae. Biochim Biophys Acta 933:432-443

Fork DC, Satoh K (1986) The control by state transitions of the distribution of excitation energy in photosynthesis. Ann Rev Plant Physiol 37:335-261

Foyer C, Furbank R, Harbinson J, Horton P (1990) The mechanism contributing to photosynthetic control of electron transport by carbon assimilation in leaves. Photosynth Res 25:83-100

Iturriaga R, Siegel DA (1989) Microphotometric characterization of phytoplankton and detrital absorption properties in the Sargasso Sea. Limnol Oceanogr 34:1706-1726

Heath MR (1988) Interpretation of *in vivo* fluorescence and cell division rates of natural phytoplankton using a cell cycle model. J Plankton Res 10:1251-1272

Herzig R, Falkowski PG (1989) Nitrogen limitation of *Isochrysis ganbana* 1. Photosynthetic energy conversion and growth efficiencies. J Phycol 25:462-471

Hinton J, Rigs E, Jaworski G (1989) Algal identification using *in vivo* fluorescence spectra. J Plankton Res 11:65-74

Holmes JJ, Weger HG, Turpin DH (1989) Chlorophyll *a* fluorescence predicts total photosynthetic electron flow to CO_2 or NO_3^-/NO_2^- under transient conditions. Plant Physiol 91:331-337

Holzwarth AR (1987) A model for the functional antenna organization and energy distribution in the photosynthetic apparatus of higher plants and green algae. In: Biggins J (ed) Progress in Photosynthesis Research, vol I. Nijhoff, Dordrecht, p53

Keller AA (1987) Mesocosm studies of DCMU-enhanced fluorescence as a measure of phytoplankton photosynthesis. Mar Biol 96:107-114

Kirk JTO (1983) Light and Photosynthesis in Aquatic Ecosystems. Cambridge University Press, Cambridge, London, New York

Kolber Z, Wyman KD, Falkowski PG (1990) Natural variability of photosynthetic energy conversion efficiency: a filed study in the Gulf of Maine. Limnol Oceanogr 35:72-79

Kolber Z, Zehr J, Falkowski PG (1988) Effects of growth irradiance and nitrogen limitation on photosynthetic energy conversion in photosystem II. Plant Physiol 88:923-929

Kyle DJ, Ohad I (1986) The mechanism of photoinhibition in higher plants and green algae. *In* Staehelin AL, Arntzen CJ, eds. Encyclopedia of Plant Physiology, New Series vol 19. Springer-Verlag, Berlin pp. 468-475

Laasch H (1987) Non-photochemical quenching of chlorophyll *a* fluorescence in isolated chloroplasts under conditions of stressed photosynthesis. Planta 171:220-226

Ley AC (1980) The distribution of light energy for algal photosynthesis. *In* Falkowski PG, ed., Primary Productivity in the Sea. Plenum, New York, pp.59-82

Lakowicz JR (1986) Principles of Fluorescence Spectroscopy, 3rd edn. Plenum, New York

Lavorel J, Etienne A-L (1977) In vivo chlorophyll fluorescence. *In* Barber J, ed. Primary Processes of Photosynthesis, Elsevier, Amsterdam pp.203-268

Legendre L, Yentsch CM (1989) Overview of flow cytometry and image analysis in biological oceanography and limnology. Cytometry 10:501-510

Loftus ME, Seliger HH (1975) Some limitation of the *in vivo* fluorescence technique. Chesapeake Sci 16:79-92

Lorenzen CJ (1966) A method for continuous measurement of *in vivo* chlorophyll concentration. Deep-Sea Res 13:223-227

Malkin S, Armond PA, Mooney HC, Fork DC (1981) Photosystem II photosynthetic unit sizes from fluorescence induction in leaves. Plant Physiol 67:570-579

Miller AG, Espie GS, Canvin DT (1988) Chlorophyll a fluorescence yield as a monitor of both active CO_2 and HCO_3^- transport by the cyanobacterium *Synechococcus* UTEX 625. Plant Physiol 86:655-658

Neale PJ (1987) Algal photoinhibition and photosynthesis in the aquatic environment. *In* Kyle D et al, eds. Photoinhibition, Elsevier, Amsterdam

Neale PJ, Cullen JJ, Yentsch CM (1989) Bio-optical inferences from chlorophyll *a* fluorescence: What kind of fluorescence is measured in flow cytometry? Limnol Oceanogr 34:1739-1748

Olson RJ, Vaulot D, Chisholm SW (1985) Marine phytoplankton distributions measured using shipboard flow cytometry. Deep-Sea Res 32:1173-1180

Olson RJ, Chisholm SW, Zettler ER, Armbrust EV (1988) Analysis of *Synechococcus* pigment types in the sea using single and dual beam flow cytometry. Deep-Sea Res 35:425-440

Olson RJ, Chisholm SW, Zettler ER, Armbrust EV (1989a) Pigments, size and distribution of *Synechococcus* in the North Atlantic and Pacific Oceans. Limnol Oceanogr 35:45-58

Olson RJ, Zettler ER, Anderson OK (1989b) Discrimination of eukaryotic phytoplankton cell types from light scatter and autofluorescence properties measured by flow cytometry. Cytometry 10:636-643

Ort DR (1986) Energy transduction in oxygenic photosynthesis: an overview of structure and mechanism. *In* Staehelin AL, Arntzen CJ, eds. Encyclopedia of Plant Physiology, New Series vol 19. Springer-Verlag, Berlin pp. 143-196

Owens TG (1986) Photosystem II heterogeneity in the marine diatom *Phaeodactylum tricornutum*. Photochem Photobiol 43:535-544

Owens TG (1988) Light-harvesting antenna systems in the chlorophyll a/c-containing algae. *In* Stevens SE and Bryant DA, eds. Light-Energy Transduction in Photosynthesis: Higher Plant and Bacterial Models. American Society of Plant Physiologists, Bethesda, pp. 122-136)

Owens TG, Gallagher JC, Alberte RS (1987) Photosynthetic light-harvesting function of violaxanthin in *Nannochloropsis* spp. (Eustigmatophyceae). J Phycol 23:79-85

Owens TG, Webb SP, Mets L, Alberte RS, Fleming GR (1987) Antenna size dependence of fluorescence decay in the core antenna of photosystem I. Experimental estimates of charge separation and energy transfer rates. Proc Nat Acad Sci USA 84:1532-1536

Perry MJ, Porter SM (1989) Determination of the cross-section absorption coefficient of individual phytoplankton cells b analytical flow cytometry. Limnol Oceanogr 34:1727-1738

Petit PX, Diolez P, de Kouchkovsky Y (1989) Flow cytometric analysis of energy transducing organelles: mitochondria and chloroplasts. In: Yen A (ed) Flow Cytometry: Advanced Research and Clinical Applications, vol. I. CRC Press, Boca Raton, p 271

Roy S, Legendre L (1979) DCMU-enhanced fluorescence as an indicator of photosynthetic activity in phytoplankton. Mar Biol 55:93-101

Samuelsson G, Oquist G (1977) a method for studying photosynthetic capacity of unicellular algae based on *in vivo* chlorophyll fluorescence. Physiol Plant 40:315-319

Schatz GH, Brock H, Holzwarth AR (1988) A kinetic and energetic model for the primary processes in photosystem II. Biophys J 54:397-405

Schreiber U (1983) Chlorophyll fluorescence yield changes as a tool in plant physiology. I. The measuring system. Photosynth Res 4:261-272

Schreiber U (1986) Detection of rapid induction kinetics with a new type of high frequency modulated chlorophyll fluorometer. Photosynth Res 9:261-272

Schreiber U, Schliwa U, Bilger W (1986) Continuous recording of photochemical and non-photochemical chlorophyll fluorescence quenching with a new type of modulation fluorometer. Photosynth Res 10:51-62

Sharkey TD, Stitt M, Heineke D, Gerhardt R, Raschke K, Heldt HW (1986) Limitation of photosynthesis by carbon metabolism. II: O_2-insensitive CO_2 uptake results from limitation of triose phosphate utilization. Plant Physiol 81:1123-1129

Sivak MN, Walker DA (1985) Chlorophyll *a* fluorescence: can it shed light on fundamental questions in photosynthetic carbon metabolism? Plant Cell Environ 8:439-448

Sivak MN, Walker DA (1986) Photosynthesis in vivo can be limited by phosphate supply. New Phytol 102:499-512

Sosik HM, Chisholm SW, Olson RJ (1989) Chlorophyll fluorescence from single cells: interpretation of flow cytometric signals. Limnol Oceanorg 34:1749-1761

Stransky H, Hager A (1970) The carotenoid pattern and the occurrence of the light-induced xanthophyll cycle in various classes of marine algae. Archiv Mikrobiol 73:315-323

Stitt M (1986) Limitations of carbon metabolism. I. Evidence for excess electron transport capacity in leaves carrying out photosynthesis in saturating light. Plant Physiol 81:1114-1122

Sukenik A, Bennett J, Falkowski PG (1987) Light-saturated photosynthesis - limitation by electron transport or carbon fixation? Biochim Biophys Acta 891:205-215

Therriault J-C, Booth D, Legendre L, Demers S (1990) Phytoplankton photoadaptation to vertical excursion as estimated by an *in vivo* fluorescence ratio. Mar Ecol Prog Ser 60:97-111

Thornber JP (1986) Biochemical characterization and structure of pigment-proteins of photosynthetic organisms. *In* Staehelin AL, Arntzen CJ, eds. Encyclopedia of Plant Physiology, New Series vol 19. Springer-Verlag, Berlin pp. 98-142

Trautman JK, Shreve AP, Owens TG, Albrecht AC (1990) Femtosecond dynamics of carotenoid to chlorophyll energy transfer in thylakoid membrane preparations from *Phaeodactylum tricornutum* and *Nannochloropsis* sp.. Chem Phys Lett 166:369-374.

Turpin DH, Bruce D (1990) Regulation of photosynthetic light harvesting by nitrogen assimilation in the green alga *Selenastrum minutum*. FEBS Lett 263:99-103

van Kooten O, Snel JFH (1990) The use of chlorophyll fluorescence nomenclature in plant stress physiology. Photosynth Res 25:147-150

Vincent WF (1980) Mechanisms of rapid photosynthetic adaptation in natural phytoplankton communities: changes in photochemical capacity as measured by DCMU-induced chlorophyll fluorescence. J Phycol 16:568-577

Walters RG, Horton P (1990) The use of light pulses to investigate the relaxation in the dark of chlorophyll fluorescence quenching in barley leaves. In: Baltscheffsky M (ed) Current Research in Photosynthesis, vol. I, Kluwer, Dordrecht, p631

Weis E, Berry JA (1987) Quantum efficiency of photosystem II in relation to 'energy'-dependent quenching of chlorophyll fluorescence. Biochim Biophys Acta 894:198-208

Welschmeyer NA, Lorenzen CJ (1981) Chlorophyll-specific photosynthesis and quantum efficiency at subsaturating light intensities. J Phycol 17:283-293

Yentsch CM, Campbell JW (1991) Phytoplankton growth: perspectives gained by flow cytometry. J Plankton Res 13 (supplement):83-108

Yentsch CS, Yentsch CM (1979) Fluorescence spectral signatures: the characterization of phytoplankton populations by the use of excitation and emission spectra. J Mar Res 37:471-483

Marine optics and light scattering

OPTICS OF MARINE PARTICLES AND MARINE OPTICS

A. MOREL

Laboratoire de Physique et Chimie Marines
Université Pierre et Marie Curie et CNRS
BP 08
F 06230 Villefranche sur Mer
France

INTRODUCTION

Optical oceanography was initially developed as a specific, somewhat isolated part of physical oceanography. More or less independently, a few marine biologists and ecologists were concerned with optical data and measurements, particularly concerning the study of oceanic primary production (Steele and Menzel, 1962) and the estimation of the part of radiant energy that can be absorbed within the algal compartment compared to that absorbed by detritus (Yentsch, 1962; 1963; Riley, 1965). A paper by Yentsch and Yentsch (1984) *Emergence of optical instrumentation for measuring biological properties*, described the new panorama resulting from the introduction of powerful optical techniques in the 1980s. The techniques which appear most promising paradoxically address the biological problems via opposite scales: the microscopic (cell by cell) scale, typical of flow cytometry, and the global scale, typical of satellite borne ocean color sensor. The interest of biologists with regard to optics has been stimulated by the capacity of these new tools.

The measurements of the photosynthetic carbon fixation at sea were accelerated by the introduction of the ^{14}C method in the early 1960s. The need for adequate measurements of available radiant energy was acknowledged and created an increasingly active participation of optical oceanographers in the field of primary production studies. An efficient cooperation was organized through the SCOR-Working Group 15, set up by the Scientific Committee on Oceanic Research (1963) and chaired by J.E. Tyler (1966). Thanks to this cooperation in particular, physicists realized that in the open ocean, the biological activity in general, and

NATO ASI Series, Vol. G 27
Particle Analysis in Oceanography
Edited by S. Demers

specifically the abundance of phytoplankton, were the source of most of the variations they had detected in the optical properties of oceanic waters. This is now well established and substantiated by useful relationships between bulk optical and biological parameters, empirically derived from numerous field experiments (see *e.g.* Morel and Prieur, 1977; Smith and Baker, 1978; Gordon *et al.*, 1988; Morel, 1988) as now used in the interpretation of ocean color data.

Analytical Optics was developed to increase the understanding and to separately assess the optical influences of each particular component of the oceanic system. Among the various particles present in this system, priority was naturally given to algal cells. Their photosynthetic performance relies on their absorptive capabilities and they also can be isolated and grown in axenic cultures for detailed experiments. In the fundamental study of photosynthesis, physiologists and biophysicists were often far advanced in this bio-optical research (see e.g. Rabinowitch and Govindjee, 1969) and marine biologists subsequently benefited from this experience. Optical (*in vitro*) techniques were developed for the study of absorption, scattering and fluorescence by various phytoplankters in relation to their pigment content and composition. Due to these spectrophometric measurements, the knowledge of the optical properties at the level of individual cells is considerable in phototrophic and heterotrophic organisms. With the introduction of flow-cytometric methods, a new technique became available with unprecedented capabilities in terms of the number of processed particles and of detectable parameters. Cell analysis and discrimination rest on their optical properties, eapecially light scattering in different directions and auto- or induced fluorescence. A meaningful interpretation of the optical signatures as well as improvements in this technique still require a continuous effort in optical research.

The present paper is organized as follows: the physical characteristics of a cell or of a population of particles (refractive index, typical size, or size distribution) which rule their optical behavior are examined and the necessary optical parameters are introduced (Part 2). In Part 3, the relevant theories to be operated when interpreting or trying to predict the optical properties of individual cells or particles are briefly reviewed (a more detailed presentation was previously given in Morel and Bricaud, 1986). Their applicability is demonstrated by using some examples selected from among the phototrophic and heterotrophic organisms. The general agreement between theoretical and experimental results concerning both absorption and total scattering is analyzed. Information about the angular scattering pattern and its magnitude is provided for various particles, with particular emphasis on the possibilities offered by theory in the quantitative interpretation of scattering signals as detected in flow cytometry.

The last part deviates slightly from the main topic, to the extent that it does not strictly coincide with the title of the Advance Study Institute (*Individual Cell and Particle Analysis in Oceanography*). In this last discussion, the cell-by-cell level or the microscopic scale are abandoned and the optical oceanography viewpoint is adopted. In principle, the formation of the bulk optical properties of oceanic waters could be explained from the knowledge of the optical role and contribution of each category of materials, as supposedly provided by the *analytical optics* method. The solution of this problem is as yet out of reach: all types of particles are not presently studied or even identified, their numerical abundance is not sufficiently documented and so on. Therefore, this tentative discussion, which does not claim to be exhaustive, mainly aims at putting forward some working hypotheses for future research.

RELEVANT PHYSICAL CHARACTERISTICS OF PARTICLES

Among the physical characteristics of individual particles or of a population of particles, some are of direct interest when dealing with the optical properties. The bulk index of refraction and absorption capability of the material forming a single cell or particle is, along with its size, the parameter which primarily governs its optical properties. Heterogeneities and internal and external structures, as well as actual shape, are additional factors also to be taken into account for detailed studies, requiring much more complex theoretical support and computational effort. To the extent that marine particles are randomly oriented when their optical properties are to be measured or predicted, or when they are producing the in-water radiative field, the simplified approach of considering that they behave as quasi-spherical and homogeneous bodies is an acceptable approximation, which at least allows the main features to be understood and predicted.

When dealing with the natural environment and the variety of suspended living and non-living material, another characteristic at the level of the entire population is the size distribution. The description of the number of particles as a function of their size is directly involved in the formation of the optical properties of a water body which contains such a population. The present knowledge concerning these successive points is briefly reviewed below.

Index of refraction of marine particles

After having impinged upon a homogeneous substance, the propagation of electromagnetic waves is affected in two ways: *i*) the phase velocity is modified in relation to the real part of the index of refraction, denoted n_s and *ii*) the field strength and thus the energy transported by the waves are progressively damped by absorption, which is described by the imaginary part of the index of refraction, denoted n'_s. The corresponding definitions are as follows:

$$n_s = v_o / v_s \tag{1}$$

$$n'_s = a_s \lambda / 4\pi \tag{2}$$

where v_o and v_s are the light speed in vacuo and in the substance respectively, λ is the wavelength, and a_s is the absorption coefficient of the substance. To the extent that a marine particle can be seen as a sufficiently homogeneous body in such a way that a bulk refractive index can be assigned, this index is in general a complex number, $n_s - in'_s$. The scattering process by a particle can only occur if the real part of its index departs from that of the surrounding medium (water in our case) and, in effect, the pertinent parameter is the *relative* complex index, m, defined as

$$m = n - in' = \frac{n_s}{n_w} - i\frac{n'_s}{n_w} \tag{3}$$

where n_w is the refractive index of water, (about 1.34 for sea water), which is assumed to be a real number in the visible part of the spectrum. Actually, its imaginary part, n'_w, is negligible; it is at most 4.10^{-8}, *i.e.* at the red end of the spectrum (700 nm).

A brief review of the known or expected n and n' values for particulates in oceanic waters is given below. A detailed study has been provided by Aas (1981).

Minerogenic particles originating from continental margins and river discharges or reaching the open ocean through eolian transport, are relatively well known with respect to their composition and thus their refractive index. Quartz particles have n values around 1.15, diminishing when the amorphous phase replaces the cristallized state. Aluminosilicate and iron hydroxyde particles have n values ranging from 1.10 to 1.18; various clays studied by Ahn (1990) exhibit n values in the range 1.10 - 1.14 and n' values in the range 4 to 8 10^{-3} (at the most, in the blue part of the spectrum).

Other mineral particles, biologically produced, originate from the tests of diatoms (amorphous silica or opale) and cocolithophorids (calcite and aragonite plates). Their relative indices are in the ranges 1.05 - 1.06 and 1.19 - 1.21, respectively, with no notable absorption.

At least in the upper parts of the open ocean, the predominant particles are of biogenous origin and organic in composition. They include all living organisms (heterotrophs and phototrophs, uni- or pluricellular) and all kinds of detritus, generated along the prey - predator trophic chain and through their natural decay; they are permanently modified by the physical or chemical actions exerted by the sea water medium. They have in common bodies with high water conten (50 - 80 % of the water volume, perhaps a bit less for some compact detritus and very small bacteria; Norland *et al.*, 1987; Simon and Azam, 1989). Therefore their index cannot dramatically depart from that of water. Besides water, the other constituent, (the "dry" organic matter) is composed of mixtures in variable proportions of carbohydrates, proteins and lipids. These compounds have similar influence in governing the index increment with respect to that of water, in such a manner that their relative proportions are not crucial in determining the final bulk index of the particle. For this reason, it has been possible to show that n can be simply related to the organic carbon concentration within the cell material, denoted C_{ic} (Kg C m^{-3}), according to (Morel and Ahn, 1990)

$$n = 1 + 0.233 \cdot 10^{-3} C_{ic} \tag{4}$$

Note that this relationship does not apply to species with hard external walls. It has been verified for free-living bacteria (C_{ic} = 180 - 275 kg m^{-3}), for heterotrophic flagellates (150 - 170 kg m^{-3}), and for a cyanobacterium (140 - 220 kg m^{-3}) (see Stramski and Morel, 1989). Its general applicability in the case of other phototrophic organisms remains to be verified. Mostly in response to the variable water content rather than to the organic composition, phytoplanktonic organisms exhibit real relative index in the range 1.01 - 1.09 (Spinrad and Brown, 1986; Ackleson and Spinrad, 1988; Bricaud *et al.*, 1988; and references therein). The lowest values, which have been distinctly observed for diatoms, account for the presence of vacuoles. Gas vacuolate cyanobacteria form a special category and also behave as very "soft" particles, with low n values of the order of 1.01 (Dubelaar *et al.*, 1987).

Very likely, the most colored and absorbing particles in the open ocean are algal cells. With the additive effects of chlorophyll and carotenoids in the blue part of the spectrum (430 - 450 nm), the imaginary (relative) part of the index, n', could be as high as 0.015 for strongly pigmented cells (*Hymenomonas elongata*; Ahn, 1990). Nonetheless, n' rarely reaches 0.01 at these wavelengths, (see Table II.2 in Bricaud, 1989), a value still high and comparable to (or actually higher than) that of a common "black" glass. In the red absorption peak of the sole chlorophyll, the n' values are obviously slightly lower (1 to 6 10^{-3}, Bricaud, 1989); they may be enhanced by the presence of phycocyanin (as in *Synechocystis*; see in Stramski and Morel, 1989) or other phycobilin pigments.

Almost colorless heterotrophic bacteria and other larger protists have very low n' values (10^{-4} - 10^{-3}), except in the vicinity of the (d) Soret band of cytochrome (around 412 nm), where n' peaks at 0.8 to 1.6 10^{-3} for free-living bacteria and at 0.6 to 0.9 10^{-3} for ciliates and flagellates (Morel and Ahn, 1990; 1991).

In summary, marine particles (living, detritic, or even minerogenic) are never very refractive bodies: n is close to 1, and n', even for the most pigmented algal cells, remains close to 0. Therefore, their complex relative index of refraction fullfills the condition below

$$m - 1 << 1$$

This is of interest because the theoretical approach can be considerably simplified in such a case.

Another point is worth noting and refers to the links which may exist between the real and the imaginary parts of the index. Inasmuch as absorbing substances in living organisms are more or less correlated to the total carbon content, n' and n could be related. Such a "chemical" link cannot be tight because the pigment-to-carbon ratio is not stable; it exists only as a first approximation, useful for interspecific algae comparison. A "physical" and exact relationship results from the anomalous dispersion phenomenon. The Ketteler - Helmoltz theory predicts that, as soon as n' departs from zero (inside an absorption band), n is affected, being depressed and then enhanced on both sides of the absorption band (see for instance Fig. 13 and comments in Morel and Bricaud, 1987). For algal cells with several pigments and the resulting intricate absorption bands, the $n'(\lambda)$ spectrum exhibits complicated patterns. In response, the $n(\lambda)$ spectrum is also featured around a "mean" value related, as said before, to the carbon content. This is part of the complexity when modeling the optical properties of phytoplankters (see Bricaud and Morel, 1986; Stramski *et al.*, 1988).

Size distribution of marine particles

It has long been thought that the population of particles, even if widely distributed with respect to size, is nevertheless more or less centered around an "average" size. Many studies were devoted to determining this supposedly existing average size and its possible change with location and depth. From microscopic observations, however, Gordon (1970) concluded that the size frequency distribution is continuous and monotonic, with an almost exponential increase of the number of particles for decreasing sizes. Thus the concept of an average size appeared to be meaningless, as necessarily "close to the smallest observable size" (Sheldon *et al.*, 1972). Meanwhile, electronic sizing and counting techniques (Coulter counter) provided a new insight into the size spectrum over a wide range of diameters, between 3 and 100 μm (Sheldon and Parsons, 1966). The steeply ascending slope toward the smaller particles proved to be the general rule, even though some relative maxima or dominance of certain sizes, due to the occurence of particular planktonic species, were often found to be superimposed over the general trend.

When the number of particles within a size increment, $d\,N/d\,d$, is plotted as a function of the mean size inside this increment (note that d is the diameter of a sphere with the same volume as the particle), the size distribution of marine particles appears to be well approximated by a power function:

$$dN/d\ d \text{-} Kd^{-j} \tag{5}$$

(Bader, 1970), which is similar to the equation introduced by Junge (1963) to describe the aerosol size distribution (*K* and *j* are positive constants). In addition, the most common exponent value (*j*) turned out to be around 4, as in the atmosphere, for the 1 to 100 μm size domain studied (Bader, 1970; Brun-Cottan, 1971). If, instead of numbers, the volumes of particles inside each size increment are now considered, the above expression becomes

$$d\ V/d\ d \text{-} K'd^{(-j+3)}$$

or

$$d\ V/d\ \ln(d) \text{-} K'd^{(-j+4)} \tag{6}$$

This representation using volumes instead of numbers was that adopted by Sheldon and Parsons (1967). They plotted the particle concentration, (expressed as ppm by volume and with a linear scale) *versus* the equivalent sphere diameter, *d* (log-scale). Independently from Bader, Sheldon *et al.* (1972), analyzing systematic data gathered in the Atlantic and Pacific Oceans, concluded that within the 1-128 μm (or even 1-1000 μm) range, similar amounts of particulate material occur in logarithmically equal size intervals. Strictly equal amounts are observed if *j* is assigned a value of 4 in eq. 6; the graphical representation of this equation reduces to a horizontal straight line. In practice, there are often some "bumps" in actual volume distributions due to peculiar dominances. A slightly decreasing slope toward larger sizes often appears (meaning that *j* would be slightly above 4). The same tendency of an approximately constant biomass concentration (zooplankton, micronekton, fishes...) in logarithmically equal size ranges seems to pervade along the food chain in the direction of larger organisms (Sheldon *et al.*, 1972), as well as in the direction of smaller ones, such as picoplankton and bacterioplankton (Rassoulzadegan and Sheldon, 1986).

The existence of a continuous size spectrum, characterized by an exponent in the vicinity of 4, is an important result for the prediction of the optical properties of such an assemblage of particles, with interesting consequences examined later on. It is acknowledged that local and temporary deviations from this somewhat ideal distribution may occur. As an operative hypothesis, however, this size spectrum can provide a useful tool.

OPTICAL PROPERTIES OF INDIVIDUAL CELLS OR PARTICLES

Some definitions

The interaction between incident radiations and a particle considered as an isolated body has two aspects. The first one, already described when defining the complex refractive index, corresponds to the phenomenon of absorption and affects the amount of energy carried by the radiation, whereas the second process deals with the direction of propagation of the radiation and is called scattering (or "elastic" scattering, as there is no loss of energy). Absorption results mostly or exclusively in heating and providing energy for chemical reactions (such as photosynthesis), or possibly for re-emission of radiation at other generally larger wavelengths (Raman "inelastic" scattering and fluorescence emission). A proper definition of absorption thus implies that the wavelength be fixed.

If we assume that the incident electromagnetic radiation propagates as a plane wave, it is possible to define the geometrical cross section, S_g, of any particle as being the projected ("shadow") area of this particle on the plane of the wave. The above processes (absorption and scattering) are conveniently described by two dimensionless "efficiency factors", Q_a and Q_b, respectively defined as the ratios of the radiative energy absorbed within, and the energy scattered by, the particle, to the energy impinging on its geometrical cross section. The sum, $(Q_a + Q_b) = Q_c$, defines the efficiency factor for attenuation, which expresses the global removal of energy from the incident plane wave (note that the scattered energy is diverted from its initial direction, but not lost for the medium). The products $S_g\ Q_a$ and $S_g\ Q_b$ (dimension L^2) are the absorption and scattering cross sections of the particle, denoted S_a and S_b, respectively.

According to the direction of the incident light, an irregularly shaped particle has various geometrical sections S_g; if this particle is convex and randomly oriented in the light field, its geometrical cross section is on average one-fourth of its total surface area ("this theorem, apparently, is one of those which gets discovered and rediscovered many times", van de Hulst, 1957, p 111). The efficiency factors Q_a and Q_b of irregularly shaped particles and thus S_a and S_b also change with the orientation of the particle in the light field. In a diffuse light field, or even in a directional field if particles are randomly oriented (e.g. algae in a cuvette during classical spectrophotometric measurements), "average" cross sections are determined. In cytometry (very short duration) or in microspectrophotometric techniques (organisms fixed in position), orientation - dependent cross sections are obtained (Iturriaga *et al.*, 1988). In these cases, even a perfectly monodisperse population of particles (same size and even same shape within the entire population) will give rise to variable signals in scattering, in absorption, and even in fluorescence (primarily dependent on absorption capabilities) as a consequence of this shape-effect.

This effect obviously vanishes for spherical particles, as it also vanishes for irregular particles, as far as macroscopic techniques allow many particles with various orientations to be simultaneously examined. The first step in a theoretical approach consists of considering spherical particles.

Under the provisos that these spherical particles are i) larger than the wave length and ii) that their relative index m remains close to 1 (a condition satisfied as seen before), the anomalous diffraction theory developed by van de Hulst (1957) can apply. It allows the efficiency factors Q_a and Q_b to be predicted as a function of dimensionless parameters ρ and ρ' defined as

$$\rho = 2\,\alpha\,(n-1) \qquad (7)$$

$$\rho' = 4\,\alpha\,n' \qquad (8)$$

where

$$\alpha = \pi d/\lambda_w = \pi d\,n_w/\lambda_o \qquad (9)$$

α is the size parameter, relative to the wavelength in the surrounding medium (λ_w in water, λ_o in vacuo), and ρ and ρ' combine the relative size α with the real part or the imaginary part of the relative refraction index. By combining Eq. 8 with Eq. 2, it is clear that ρ' can also be defined as

$$\rho' = a_s d \tag{8'}$$

the product of the absorption coefficient of the substance forming the particle with its diameter.

Absorption cross section

In the frame of the van de Hulst approximation, the efficiency factor for absorption is only a function of ρ'

$$Q_a = 1 + 2[\exp(-\rho')]\,\rho'^{-1} + 2[\exp(-\rho') - 1]\,\rho'^{-2} \tag{10}$$

This function asymptotically approaches 1 when ρ' approaches ∞ ,meaning that in such a case, the particle is able to absorb all the energy impinging on its geometrical cross section and then is a black body. The behavior of this function (see Fig. 1) is at the origin of the so-called *package effect* (Kirk, 1975 a, b), or *discreteness effect* (Morel and Bricaud, 1981). This effect is experienced by most of the phytoplankters, which often are rather strongly absorbing bodies. The practical consequence is that the actual absorption capability of an algal cell is not simply proportional to its pigment content, as easily seen in Fig. 1. For instance, if we assume that a certain cell has an initial ρ' value equal to 2 before doubling its pigment content, ρ' becomes 4 after the doubling, whereas the absorption capability, depicted by Q_a (which goes from about 0.7 to 0.9), is only increased by about 30 %. The ratio of actual absorption to pigment concentration is no longer a constant, as it depends on the initial ρ' value (hence Beer's Law does not apply in this case).

The "flattening" of the absorption spectrum is another aspect of the same phenomenon. If, by changing the wavelength, a_s is found to be doubled (and thus ρ' doubled), Q_a in general is not doubled and the amplitude of the absorption peak for a discrete absorbing substance is not doubled as it would be in a continuous medium (a "solution" of the same substance).

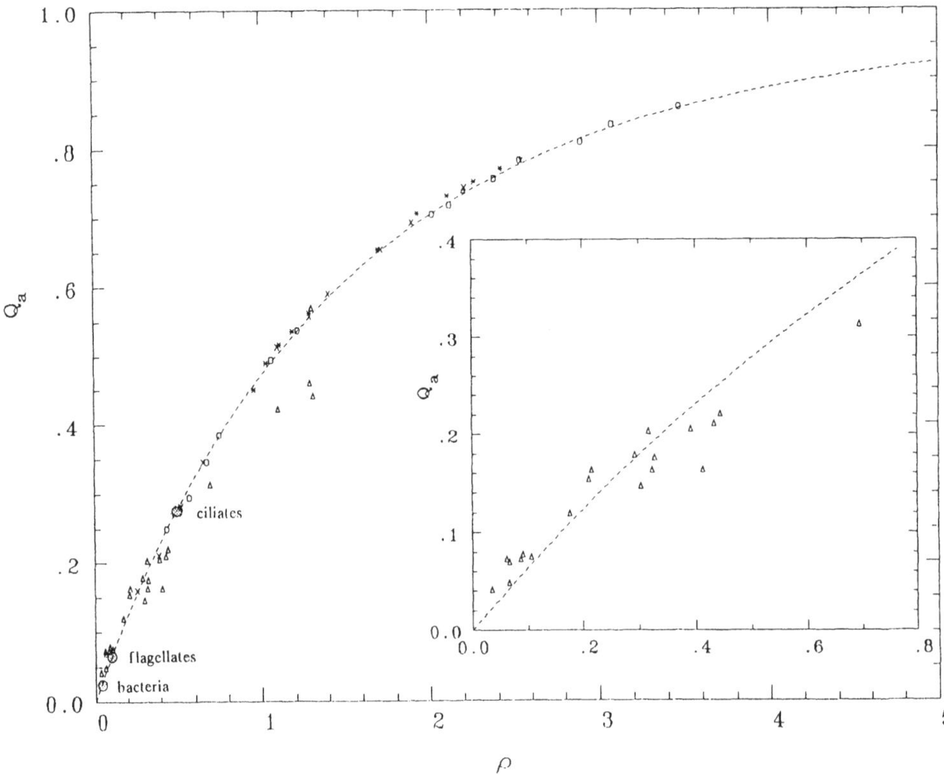

Figure 1. The theoretical variations of Q_a, the efficiency factor for absorption (dashed curves), as a function of the dimensionless parameter ρ',. The triangles are experimental determinations of Q_a (at 675 nm) for various algae (Morel and Bricaud, 1986; Ahn, 1990); other symbols are for determinations of 3 algal species studied by Sosik (1988). The values for heterotrophic organisms, as indicated, come from Morel and Ahn (1990, 1991). The inset is an enlargment of the initial part of the curve.

As shown in Fig. 1, the Q_a values determined for various phytoplankters at l = 675 nm nicely follow the theoretical predictions. This wavelength is chosen because the predominant absorbing pigment in this part of the spectrum is almost exclusively Chlorophyll *a*. Thus ρ' (Eq. 8') can be computed with the modal diameter of the algal population and with a_s obtained through

$$a_s - C_{i\,Chl}\, a^*_{Chl}(675)$$

where $C_{i\,Chl}$ is the intracellular Chl *a* concentration and a^* is the Chl-specific absorption coefficient, here taken equal to 20.7 m^2 (g Chl $a)^{-1}$, as in Bricaud *et al.*, (1983).

To the extent that the initial part of the Q_a curve can be regarded as almost linear, the package effect no longer exists. When expanded, Eq. 10 becomes

$$Q_A = (2/3)\,\rho' - (1/4)\,\rho'^2 \qquad \textbf{(10')}$$

and can be reduced to its first term (within an accuracy of 10 %) as long as ρ' does not exceed 0.26. This value delimits the quasi-linear domain with no package effect. This linear approximation is relevant not only for very small algal cells, but also for bacteria or heterotrophic flagellates (even at their absorption maximum, 415 nm). Some larger species, like diatoms, may nevertheless belong to this domain, because of their very low pigment per cell content (see Table 1 and Fig. 11 in Morel and Bricaud, 1986). Conversely small species with high $C_{i\,Chl}$ values can experience an important package effect, in particular those species able to develop a dense pigmentation in response to low radiative level, (see *e.g. Isochrysis galbana*, Table 1 in the above reference). The absorptive properties of tiny recently discovered marine prochlorophytes (Chisholm *et al.*, 1988; Vaulot *et al.*, 1990) remain to be studied.

Scattering cross section

The approximation of the van de Hulst anomalous diffraction also provides the efficiency factor for scattering, Q_b, as a function of ρ (Eq. 7), when the (spherical) particle is non-absorbing, or as a function of ρ and ρ' when absorption occurs within the particle. Considering the first case (ρ' = 0 and thence Q_a = 0), Q_b is expressed as

$$Q_b = 2 - (4/\rho)\sin\rho + (4/\rho^2)(1-\cos\rho) \qquad \textbf{(11)}$$

graphically represented in Fig. 2 (note that $Q_b = Q_c$ in this case). The successive maxima and minima are progressively reduced in amplitude and the limiting value is 2 (when ρ approaches ∞). For such large particles, twice the amount of energy geometrically

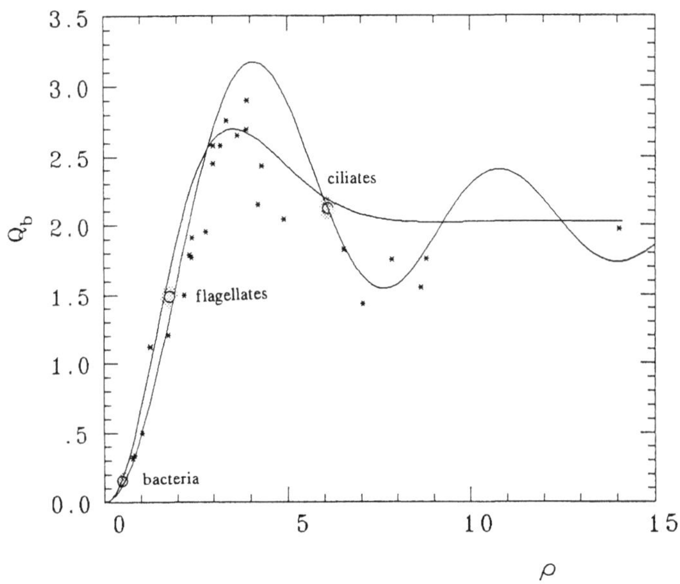

Figure 2. The theoretical variations of Q_b , the efficiency factor for scattering by non absorbing spheres (solid curve with marked oscillations) as a function of the dimensionless parameter ρ. The smoothed curve is for an averaged $\bar{Q}_b$ to be applied for population with a log - normal size distribution. The crosses are the $\bar{Q}_b$ values (at $\lambda \sim 580$ nm) determined for various phytoplankters grown in culture (see Table 1 in Morel and Bricaud, 1986); additional data for algal cells come from Ahn (1990). The circles indicate the $\bar{Q}_b$ values (at $\lambda \sim 550$ nm) determined for free living marine bacteria, heterotrophic flagellates, and naked ciliates, (Morel and Ahn, 1990; 1991).

intercepted by the particle is removed by scattering. This so-called "extinction paradox" originates from the addition in the scattered flux of the diffracted light around the particle and of the transmitted light through the particle which is deflected from its original direction by the reflection and refraction processes. Favorable (unfavorable) interferences between diffracted and transmitted radiations are the causes for the sequence of maxima (minima) observed in the Q_b curve of Fig. 2. It is easily conceivable that when absorption occurs inside the particle, the transmitted light will be reduced, therefore the possibilities of interferences and then of pronounced extrema will also be reduced.

For an absorbing particle, Q_b differs from the predictions of Eq. 11 and the anomalous diffraction approximation leads to an expression (see van de Hulst, 1957) which involves both ρ and ρ', or ρ and tan ξ , with tan $\xi = (1/2)\, \rho'/\rho$

$$Q_C = 2 - 4\exp(-\rho\tan\xi)\left[\frac{\cos\xi}{\rho}\sin(\rho-\xi) + \left(\frac{\cos\xi}{\rho}\right)^2\cos(\rho-2\xi)\right] + 4\left(\frac{\cos\xi}{\rho}\right)^2\cos 2\xi \qquad (12)$$

from which Q_a (Eq. 10) has to be subtracted to obtain Q_b. Note that the limiting value for Q_c (when ρ approaches ∞) remains equal to 2 as before. Given that the limit for Q_a is 1 for large particles, the limit for Q_b is necessarily 1 (the scattered radiations reduce to the diffracted part). It is easy to see that Eq. 12 degenerates into Eq. 11 when tan ξ becomes zero. Q_c (ρ) and Q_b (ρ) curves corresponding to realistic n' values (tan ξ values) which are encountered when dealing with algal cells have been shown elsewhere (Fig. 3 in Morel and Bricaud, 1986). The Q_b values as well as their oscillations are considerably reduced for the most absorbing phytoplankters, when n' (or tan ξ) departs markedly from zero.

The experimental determination of Q_b is generally performed with populations of particles (*e.g.* a monospecific culture) inside which the sizes are not uniform. Only a "mean" efficiency factor $\bar{Q}_b$, valid for a "mean" particle typical of the envisaged population can be arrived at. The effect of such a polydispersion with respect to sizes can be theoretically predicted. It also results in smoothing the oscillations of the $\bar{Q}_b$ curve, when plotted against the $\bar{\rho}$ value which corresponds to the mean particle (see Figure 2, where this effect is simulated).

Actual values of Q_b for phytoplanktonic cells, determined at about 570-590 nm (*i.e.* at the wavelength of minimal absorption), are shown in Fig. 2; results are added for heterotrophic organisms, namely naked ciliates, flagellates, and free bacteria, at λ = 550 nm (where absorption is negligible). There exists a general agreement between the span of measured values and that allowed by theory. Beside the inevitable inaccuracies attached to such experiments, and also the fact that even at the selected wavelengths a residual absorption for algae causes some departure from the curves (established for strictly non-absorbing cells), other deviations of second order originate from the non-sphericity and the non-homogenous constitution of the particles under examination.

This last question has been studied in detail by Aas (1984) (see also Asano and Sato, 1980). Equations similar to Eq. 10, 11, 12 are given in Aas, under the same conditions as for anomalous diffraction approximation (i.e. large particle and relative index close to 1) to predict the Q's factors for randomly oriented cylinders and disks, and also for two-layered spheres, with absorbing material concentrated within the inner core or in the outer shell. The Q_c, Q_a curves for such particles have in common with those for spheres the same limiting values and a similar sequence of progressively diminishing maxima and minima. With respect to curves for spheres, their positions are not directly comparable, to the extent that the "ρ" parameter, unambiguously defined in the case of sphere, can be defined in different ways when dealing with different shapes of particles. A representative ρ parameter for such non-spherical particles could be computed by introducing a "size" defined as being the ratio of the volume to the mean geometrical cross section, V/S_g, (Aas, 1984). To our knowledge, experimental verifications of these theoretical findings for cylindrically shaped particles remain to be effected. Experiments with various powders, crushed minerals, and other irregularly shaped particles have confirmed the expected disappearance of oscillations in the Q_b curve (see also papers in "*Light scattering by irregularly shaped particles*" edited by D.W. Schuerman, Plenum, 1980, where emphasis is put on aerosol particles with transferable results in the case of marine particles, and also theoretical papers in "*Electromagnetic scattering*" edited by M. Kerker, 1963).

The comparison made between theory and experiments in Figure 2 is restricted to a single wavelength. With given index and size (or actually a range of sizes for a population of living organisms) and when changing the wavelength, a large portion of the ρ domain can be explored. In effect, ρ is multiplied by the ratio 750 / 400 when λ goes from 400 to 750 nm, as a consequence of Eq. 7 and 9. Therefore, the comparison between theory and experiments is possible over an extended ρ-range (Fig. 3), and this is actually the basis on which the possibility of inferring the real part of the refractive index of biological particles rests (see Bricaud and Morel, 1986 and discussion therein). Nearly non-absorbing bacteria exhibit a regularly ascending Q_b (λ) curve, from red to blue, which corresponds to the initial part of the theoretical Q_b curve in Fig. 2. Their actual Q_b values fix the possible range of variations in ρ and hence, (their size being known), the compatible n-value. The same reasoning cannot apply when naked ciliates are considered. With their rather big sizes (13 - 18 μm), they enter into the ρ domain where Q_b is equal to 2 and thus remains constant whatever the wavelength. The refractive index for ciliates consequently remains undetermined. Conversely, heterotrophic flagellates and the cyanobacterium *Synechocystis* belong again to the initial

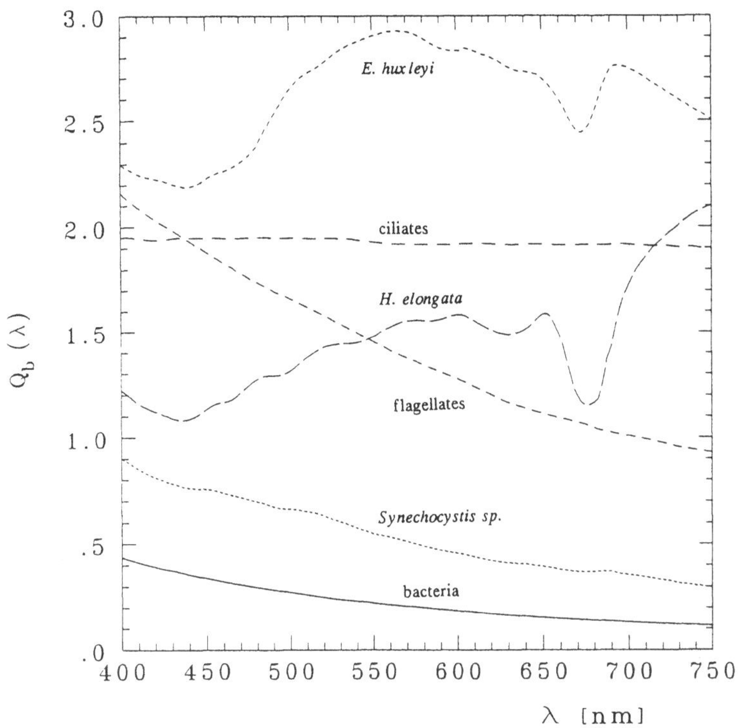

Figure 3. Spectral variations of Q_b within the 400-750 nm range of various phototrophic and heterotrophic organisms as experimentally determined (Morel and Ahn, 1990, 1991).

domain, before the first maximum of Q_b is reached, whereas the spectrum for *Emiliana huxleyi* is centered on this maximum and thus can exhibit very high Q_b values. For these organisms *n* can thus be deduced. For algal cells, the effect of pigments and of related broad absorptions (which peak at about 440 and 675 nm), is clearly seen as minima in the Q_b (λ) spectrum (see curves for *E. huxleyi* and for *H. elongata*). This second species exhibits a Q_b (λ) curve characterized by an ascending slope, from blue to red. Such a particular trend orginates from the combination of the size (11.3 μm) and index, which entails that the ρ domain involved is just beyond the first maximum of the Q_b (ρ) curve, *i.e.* in its descending part.

To the extent that the presence of absorption bands ($n'! = 0$) reacts upon the real part of the index *n*, the parameter ρ is not exclusively ruled by the size, and the above explanations (assuming *n* constant) are only a first approach. Full computations making use of Eq. 12 and involving the Ketteler - Helmoltz theory are required to account for all aspects of the optics of these pigmented bodies. With this aim, a model has been developed (Bricaud and Morel, 1986; Stramski *et al.*, 1988) where the input parameters are the size distribution function, the actual Q_a (λ) and Q_c (λ) spectra (as derived from experiments), and its outputs, beside the understanding of all phases of the phenomenon, are the n' (λ), n (λ) and Q_b (λ) spectra (the spectral efficiency factor for backscattering, Q_{bb} (λ) is also derived).

The tiny cells have some interesting scattering properties which result from the behavior of the theoretical Q_b (λ) curve in its initial ascending part (Fig. 2). This portion closely resembles a parabola, as Eq. 11. When expanded, it becomes

$$Q_b(\rho)-(1/2)\rho^2-(1/36)\rho^4+.... \qquad (11')$$

and can reduce to its first term (to within 5 %), provided that ρ remains below 0.95. This condition is fully satisfied by free-living bacteria and almost by *Synechocystis* (ρ varies from 0.65 to 1.2 when λ varies from 750 to 400 nm). For ρ values in this range, Q_b (at given λ) is therefore proportional to the square of the size and the scattering cross section (S_g Q_b) becomes proportional to d^4. For a given (small) size and with ρ inversely related to λ (Eq. 7 and 9), Q_b will vary as λ^{-2}. Both these properties have been perfectly confirmed by experimental determinations carried out with bacterioplankton of changing size (0.55 to 1 μm, Morel and Ahn, 1990). *Synechosystis* (1.15 μm) is also not far from exhibiting a Q_b (λ) spectrum with a λ^{-2} dependency (Stramski and Morel, 1990).

The question of even smaller particles, *i.e.* smaller than the wavelength, in such a manner that the van de Hulst approximation begins to fail, is not a purely academic one (as we shall see later on). Such particles belong to a transition domain toward the pure Rayleigh scattering. For a Rayleigh dipole, Q_b is proportional to α^4 (*i.e.* to d^4) or equivalently inversely proportional to λ^4. Therefore the scattering cross section is varying as d^6. Computations show that the departure from the ρ^2 dependency as predicted by Eq. 11' takes place when ρ falls below about 0.3, at least for particles having a relative index smaller than 1.06 (see Fig. II. 11 and discussion in Morel, 1973; and also Figure 5 later). Detrital particles, and very small microbes with sizes above 0.3 μm still will behave like "common" bacteria and are characterized by a ρ^2 dependency. Smaller particles, viruses, colloidal particles, and big molecules, with their nature and abundance largely unknown, are definitely relevant to the Rayleigh scattering and characterized by a ρ^4 dependency.

Angular scattering pattern

The description of the angular distribution of the scattered light from a particle involves the volume scattering function $\beta(\theta)$ (units m^{-1} sr^{-1}) defined as

$$\beta(\theta) = dI(\theta)/E\,dv$$

where $dI(\theta)$ is the scattered radiant intensity at a polar angle θ (with respect to the initial direction of propagation, corresponding to $\theta = 0$), E is the incident irradiance, and dv a volume element. The absence of any azimuthal dependency is justified in as much as spherical, or even non-spherical particles with random orientation, are under consideration.

The integral over the whole space ($\omega = 4\pi$) of the above equation provides the (total) scattering coefficient b with

$$b = \int_0^{4\pi} \beta(\theta)\, d\Omega = 2\pi \int_0^{\pi} \beta(\theta) \sin\theta \, d\theta \qquad \textbf{(13)}$$

A dimensionless scattering function (or phase function) is simply defined as the ratio

$$\tilde{\beta}(\theta) = \beta(\theta)/b \qquad \textbf{(13')}$$

the integral of which is 1 (over 4π). For a single particle (or N similar particles) in a given volume, b is equal to S_b (or NS_b), the scattering cross section (m^2), when divided by this volume (m^3).

Spherical particles of any size are relevant to Mie theory, which allows the scattering pattern to be accurately predicted (modifications exist for non-spherical and also for non-homogeneous spherical particles). It is out of the scope of the present paper to enter into the details of this theory (see van de Hulst, 1957). Only results useful for what will follow are succinctly presented.

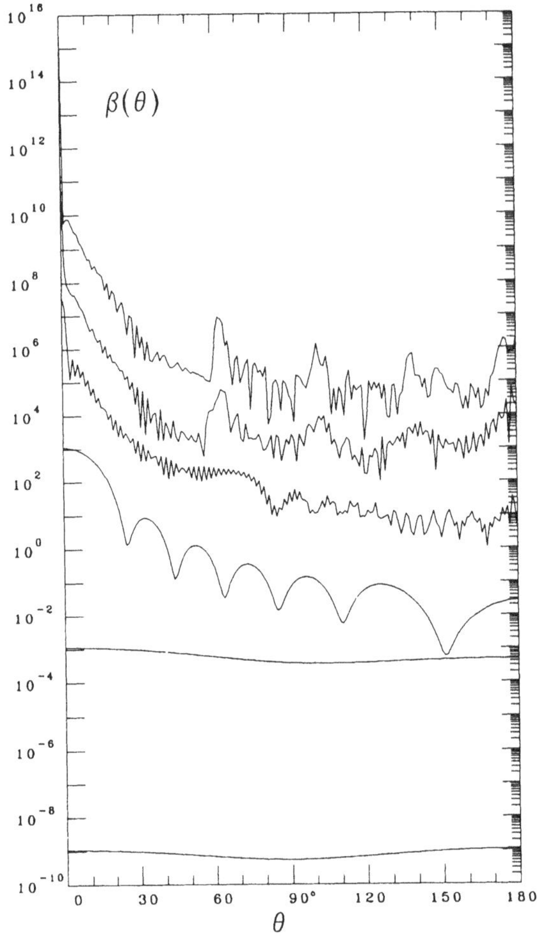

Figure 4. Volume scattering functions, $\beta(\theta)$, computed through Mie theory when the index of refraction is real ($n = 1.05$) and the size parameter α is given successive values of 0.1, 1, 10, 100, 1000 and 10000 (curves from bottom to top). The numerical values of $\beta(\theta)$ (dimension m^{-1} sr^{-1}) are internally consistent, albeit arbitrary (because the wavelength is not fixed).

In Figure 4 shows the volume scattering functions (VSF) computed through Mie theory for a single spherical particle with a real relative index $n = 1.05$ (no absorption) and with increasing values given to the size parameter α (namely 0.1, 1, 10, 100, 1000, 10 000). If the wavelength is assumed to be 421 nm, it comes $\alpha = 10$ d, so that the corresponding sizes range from 0.01 to 100 μm. Two important features emerge from this figure: *i*) the shape of the VSF is symmetrical with respect to the scattering angle q = 90° only in the Rayleigh domain (when $\alpha = 0.1$ and $\rho = 0.01$), and departs from symmetry by the progressive

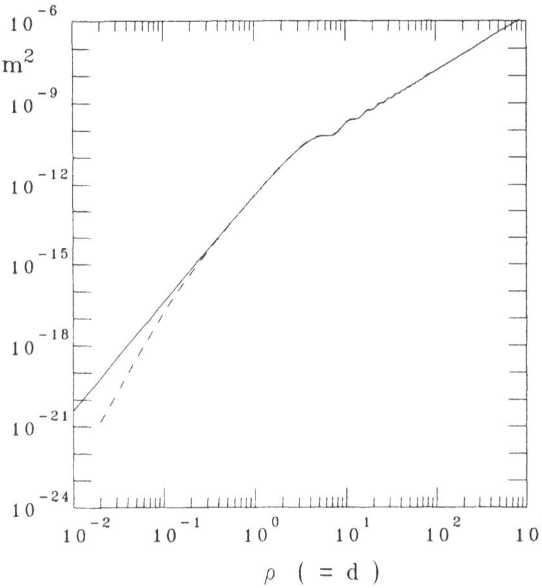

Figure 5. Scattering cross section (S_g Q_b) of spherical particles as a function of the parameter ρ. (Note that ρ = d, the diameter, when n = 1.05 and λ = 421 nm). The dashed portion (Rayleigh domain) replaces the solid curve for small ρ-values (see text).

development of an intense forward lobe (at small θ) when α increases. For very large particles ($\alpha > 1000$), the further modification essentially consists of a narrowing of this peak centered on $\theta = 0$; *ii*) the magnitude of the scattered intensities increases with size. When integrated over the θ angular domain (Eq. 13), β (θ) provides the scattering cross section, S_b (Figure 5), which is a measure of the total amount of scattered energy. Between the two first VSF (α = 0.1 and 1) the change in S_b is about α^6 (or d^6), typical of the Rayleigh domain (ρ from 0.01 to 0.1) and thus different from that obtained by extrapolation of Eq. 11 toward small ρ values, as already mentioned. In the frame of the above approximation (Eq. 11'), S_b is varying as α^4 when α goes from 0.1 to 1 (ρ from 0.01 to 0.1) and for bigger particles ($\alpha > 100$) the stability of Q_b (~ 2) leads to a α^2 dependency for S_b.

Because of the changing shape of the VSF, the amount of energy scattered within a particular angular domain cannot follow what is predicted for the total amount, apart from the general trend. As a supplementary comment to Fig. 4, the oscillations superimposed on the general

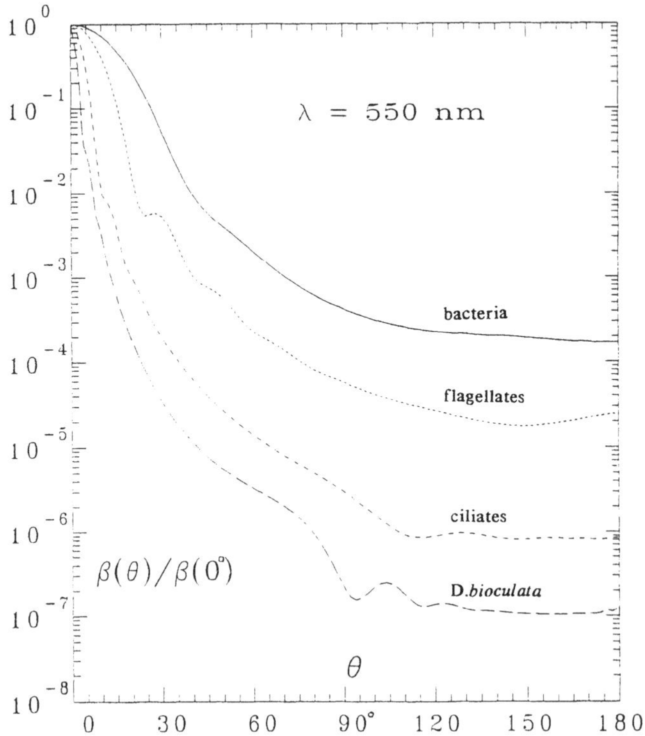

Figure 6. Volume scattering function (normalized at $\theta = 0°$ and for $\lambda = 550$ nm) computed for various organisms by using their refractive index and size distribution as experimentally determined (see text).

pattern (the number of which increases with increasing size) could only be observed for a single perfectly spherical particle (or for a collection of spherical particles, if their size is perfectly constant). They are quickly smoothed, as soon as sizes are not strictly uniform within the population (and shapes not spherical). As examples, are shown the VSF computed through Mie theory for several organisms when are used, as input parameters, the actual size distribution functions determined with a size analyzer (100 channels) and the refractive indices computed as previously explained (Ahn, unpublished).

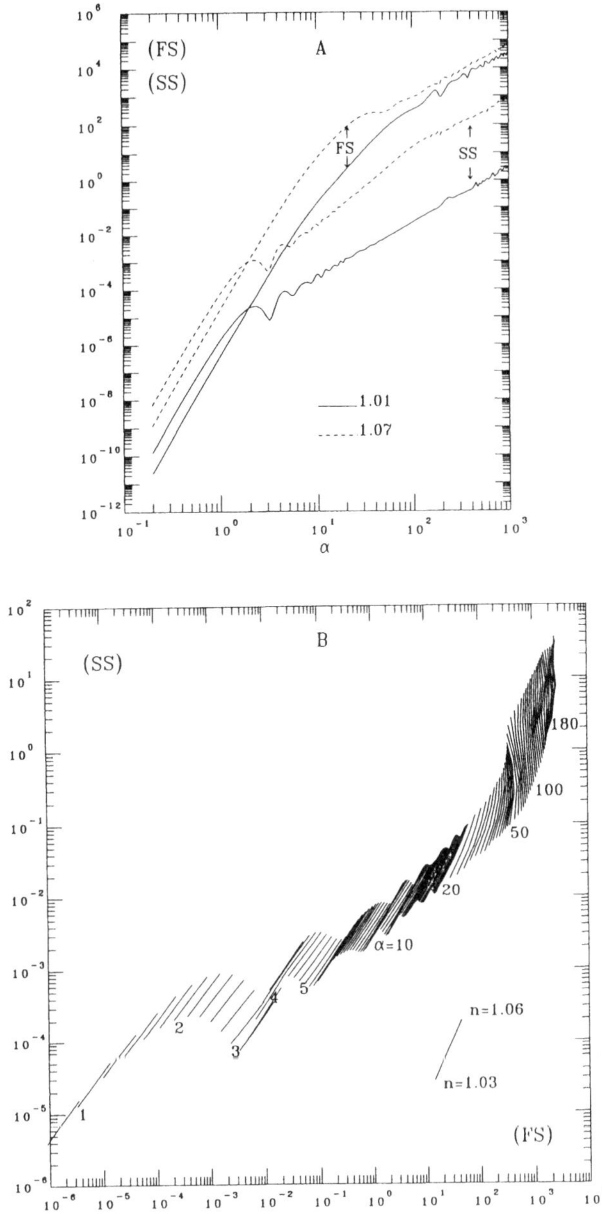

Figure 7. A) Side Scatter (SS) and near-forward scatter (FS) for spherical particles with increasing size parameter α (abcissae) and for two values of the refraction index (real, 1.01 and 1.07 respectively). The units for SS and FS are m^{-1} sr^{-1} (but see remark in the legend of Fig. 4); B) SS plotted *versus* FS for various α values, as indicated. Each rod corresponds to a given α-value and a varying index of refraction (between 1.03 and 1.06). Note that the diameter d (in μm) is numerically equal to α if λ is supposed to be 0.421 μm.

Even with relatively narrow size distribution for each of these populations (see Morel and Ahn, 1991), there are only a few remnants of oscillations. Also figure 4 exemplifies the large span of the lateral scattering (around $\theta = 90$ °) with respect to the forward lobe. This property is largely used in flow cytometry techniques to identify and characterize various particles according to their scattering capabilities. In this perspective, Eq. 13 has been integrated between the angular limits 2-18° and 72-108°, respectively, in view of simulating the near forward-scatter and the side-scatter detectors of a flow cytometer. Before integrations, the individual VSF were computed for increasingly discrete α-values (from 2 to 1000) and various values given to the index of refraction n (real, between 1.01 and 1.09). The results are shown in two different ways: in Fig. 7a the side scatter (SS) and forward scatter (FS) are separately plotted as a function of the size parameter and for only two n - values (1.01 and 1.07); in Fig. 7b, SS is plotted as a function of FS, and each "rod" in this representation corresponds to a single α - value with varying n value, between 1.03 (bottom end) and 1.06 (top end).

For very small α - values (say below 1.5), FS and SS are both increasing as α^6 (extended Rayleigh domain) and both are equally depending on n through $(n^2-1 / n^2 + 2)^2$. In this domain (left hand side of Fig. 7b, where the slope is 1), there is an ambiguity between the effects of the size and of the index upon the SS and FS values (note these particles are excessively small, in the range of 0.1 - 0.2 μm). When α exceeds 2, SS starts to experience oscillations (see Fig. 7a) around a general trend which corresponds to an α^2 (*i.e.* d^2) dependency, whereas FS continues to increase according to α^6. In this domain, which extends up to $\alpha = 20$ (d $\simeq$ 2-3 mm) or more (depending on the index), FS is the signal which is the most sensitive to the size (through approximately d^6, progressively becoming d^4). In principle, separate knowledge of SS and FS, provided that they are transcribed in absolute units (*via* calibrating beads), allow the size and index to be separately determined in many occasions. The restriction comes from the "folds" which are clearly seen in Fig. 7b. The enlargments of such folds show, however, that a couple of SS-FS values leads to a plausible domain of size and index which remains rather restricted (under the proviso of sphericity; see also Spinrad and Brown, 1986 and Ackleson and Spinrad, 1988). With reduced oscillations in SS for greater α, the situation tends to simplify. As soon as α exceeds 50 or 60 (d > 6-7 mm), FS as well as SS become proportional to α^2. In Fig. 7a it can be noticed that within this domain of larger α, the distance between the nearly parallel curves for SS and FS is governed by the refractive index (decreasing distance for increasing n). Whatever α, the ratio FS/SS takes a fixed value which is typical of n; the "rods" in Fig. 7b tend to be vertically oriented. In this size range, FS could be seen as a measure of the geometrical cross section, whereas its associated SS value could mainly indicate the refractive index value.

Any precise statement about the significance of SS and FS signals is always simplistic. The log-log representation used in Fig. 7a and b also conceals and smooths some more complicated patterns. A more systematic resort to theory and computations is certainly possible and able to resolve many ambiguities.

PARTICLES AND BULK OPTICAL PROPERTIES OF A WATER BODY

The local inherent optical properties of an absorbing - scattering medium, namely: (1) its absorption coefficient, a, (2) its scattering coefficient, b, and (3) its volume scattering function, $\beta(\theta)$, are strictly additive properties with respect to each component contributing to the formation of the bulk coefficients (see e.g. Preisendorfer, 1961). It is therefore logical to examine to what extent the optical properties of a water body can be explained in light of the present knowledge concerning the particles it contains, at least those presently identified.

Besides particles, water itself and dissolved materials, there are also optical components not considered in the scope of the present discussion. It would suffice to recall that the optical properties (b; $\beta(\theta)$; a) of optically pure sea water (water molecules and ions) are relatively well established and theoretically understood (Morel, 1974; Smith and Baker, 1981). Their spectral variations are also known, even though the absorption coefficient at its minimum (in the blue part of the spectrum) is still difficult to ascertain. The effect of dissolved organic compounds is also relatively well documented. They are introduced into the sea by land drainage (and can be important in coastal zones influenced by rivers), or they can be locally produced through the biological activity in the open ocean where they have a long residence time. This dissolved organic matter globally acts as a "blue absorber" and, from an optical viewpoint, it is identified as the "yellow substance". Many studies after that of Kalle (1938) have concluded that the absorption spectrum exhibits a regular exponential increase from the visible toward UV, with a rather constant coefficient (see Equation 17). In oceanic waters, far from terrigenous influence, the endogenous yellow substance has a weak, albeit not negligible influence (Bricaud *et al.*, 1981). Some questions remain about its true abundance, because its optical influence is easily blended with that of filterable organisms (small bacteria or viruses, for instance, or tiny "colored" debris also absorbing in the blue part of the spectrum). Notwithstanding the restriction, arising from the imprecise limit between dissolved and particulate matter, the optical influence of all other commensurable particles is to be studied. Their role in contributing to the absorption and scattering properties of oceanic waters will be successively examined below; for the scattering and backscattering coefficients, conclusions will be summarized, since details can be found elsewhere (Morel and Ahn, 1991).

Absorption coefficient of oceanic waters

Analytical approach

Among the colored particles, there is no doubt that phytoplanktonic cells play the major role. This is evident by the green shift of oceanic waters with increasing algal biomass, finally resulting in the dark green color in eutrophic situations contrasting with the deep blue color of oligotrophic waters. The general shape of absorption spectra by phytoplanktonic cells accounts for this shift. The complexity of the real-world, however, renders any precise prediction of the actual absorption by mixtures of various algae occuring in the natural environment difficult.

The inter- or even the intraspecific variability of algal absorption is the first difficulty to cope with when trying to predict the influence of this compartment. The adoption of an "average" absorption spectrum obviously cannot account for all populations, which have various pigment compositions and various package effects. Another cause of uncertainty is from that algae in sea water are inseparable from detritus, more or less colored, and from heterotrophic protists bearing respiratory pigments also absorbing in the violet part of the spectrum. Because of their relative abundance, bacteria are likely the more important absorbers among these organisms. In oceanic waters, the absorption coefficient due to all these particulate materials could ideally be written as

$$a = a_a + a_h + a_d \tag{14}$$

with

$$a_a = a^* \langle Chl \rangle \tag{15}$$

where the subscripts *a, h* and *d* stand for algae, heterotrophs and detritus (mineral terrigenous particles being excluded), and where a_a is then expressed as the product of the Chlorophyll concentration, $\langle Chl \rangle$, with the Chl-specific absorption coefficient of algae, a^* (the λ - dependency is omitted everywhere). Various approaches have been proposed, consisting of merging chlorophyll and pheophytin concentrations and thus merging the effect of algae with some of their derivatives, or alternatively, approaches consisting of relating a_d with the Pheo *a* concentration (see e.g. Kiefer and Soohoo, 1982; Soohoo and Kiefer, 1982; Mitchell 1987). The practicality of the above analytical method is impeded by: *i*) the considerable interspecific variations in a^* (λ) (*e.g.* Bricaud *et al.*, 1983; 1988), *ii*) the difficulties in properly defining (and separately estimating) the detritus absorption, and finally *iii*) the still insufficient knowledge of the heterotrophic compartment.

Fortunately, progress in the analytical approach can reasonably be expected, thanks to methods recently developed in the analysis of the biogenous particle absorption. An experimental and direct method (Kishino *et al.*, 1985), consisting of measuring the absorption spectra of particles retained on a glass-fiber filter before and after methanol extraction, allows the amount of extractable pigments to be estimated by difference and the algal and non-algal contributions to be estimated. Their respective influences are not unambiguously separated by this technique (pheopigments in debris are also extracted and bleached cell materials are seen as detritus - see discussion in Bricaud and Stramski, 1990). The single-cell microspectrophotometric method developed by Iturriaga *et al.* (1988) is a very direct one: any kind of particle (apart from the smallest ones) can be identified and then its Q_a factor spectrally determined. Extensive measurements combined with enumeration, in principle, could give access to the three terms in Eq. 14.

High performance liquid chromatography (HPLC) permits the separation and quantification of the extractable pigments and the subsequent reconstruction of a hypothetical absorption spectrum by the whole algal population (Bidigare *et al.*, 1987). The important unknown in such a reconstruction is the actual packaging which affects these pigments when they are within the cells. To the extent that Chl *a* fluorescence (at 685 nm) and absorption cross section (within the λ - domain corresponding to excitation) are tightly linked (Perry and Porter, 1989; Sosik, 1988; Sosik *et al.*, 1989), the flow cytometric technique, via the fluorescence signal, could also provide valuable information about the algal absorption (accounting for the packaging effect). The advantage of such a method lies in that a great number of cells can be analyzed and thus a significant figure could be obtained.

Indirect methods, as alternatives to experimental ones, have also been recently proposed (Morrow *et al.*, 1989; Bidigare *et al.*, 1989; Bricaud and Stramski, 1990). Though they are based upon different assumptions, they essentially rest on spectral criteria and typical slopes of the absorption spectrum. The actual absorption spectra determined for the total particulate matter collected on a cleared filter can be numerically decomposed and divided into the contribution by living algal cells and by the non-algal materials (probably mainly detrital). These methods (also that of Kishino *et al.*, 1985) have a flaw which affects the absolute values of the absorption coefficient when measured with the cleared filter technique: the pathlength amplification of the measured absorption (the "β factor", Butler, 1962) is poorly understood. Its prediction remains questionable, as the configuration of joined and multilayered particles settled on a filter itself acts as a diffusing plate.

In spite of drawbacks in any of the above methods, substantial improvements on the knowledge concerning algae and detritus are to be expected from their systematic use and their mutual complementarity. Concerning the heterotrophic organisms, recent determinations of their absorbing capabilities (Morel and Ahn, 1990; 1991; Stramski, 1990) open the possibility of assessing their optical role in oceanic waters.

Empirical approach

Mainly for the purpose of modeling (see Gordon, 1987 or Morel, 1991), empirical and global equations (derived from field data) can be adopted to predict the absorption coefficient of oceanic waters in the upper layers as a function of a single variable, the Chlorophyll *a* +

Pheophytin *a* concentration, denoted (C). The spectral values of the absorption coefficient, $a(\lambda)$, can be expressed according to

$$a(\lambda) = [a_w(\lambda) + 0.06A(\lambda)(C)^{0.65}][1 + 0.2Y(\lambda)] \quad (16)$$

with

$$Y(\lambda) = \exp - [0.014(\lambda - 440)] \quad (17)$$

where $a_w(\lambda)$ is the spectral absorption coefficient for pure water, $A(\lambda)$ the relative coefficients governing the spectral shape of absorption by natural mixed populations of algae ($A = 1$ when $\lambda = 440$ nm), and where $Y(\lambda)$ stands for (oceanic) yellow substance absorption, depending on λ according to Eq. 17 (Bricaud *et al.* ,1981). The first equation and the set of the $A(\lambda)$ values were proposed by Prieur and Sathyendranath (1981) and result from a statistical analysis performed by using *in situ* data of $K(\lambda)$, the spectral attenuation coefficient for downwelling irradiance after they have been transformed into $a(\lambda)$ values (through the inversion of a simple radiative transfer model tailored for the surface layer).

The non-linear character of Eq. 16 with respect to (C) accounts for the fact that absorption by bio-detritus and heterotrophs is not dissociated from that by the sole living algal cells. The exponent below 1 indicates that this unidentified absorbing material plays an increasing role as the algal concentration diminishes. An obvious, albeit inevitable, weakness of such an approach is to assume a constant spectral shape (the A values) for the entire biogenous compartment, while there is every chance that the shape is not stable, considering the changing proportions between living algae and non-algal particles. This is clearly evident when the spectral values of the "chlorophyll-specific attenuation for downwelling irradiance", $k_c(\lambda) = d\,K(\lambda)\,/\,d(C)$, are examined as a function of C. These $k_c(\lambda)$ coefficients increase when the pigment concentration decreases, and in addition, the shape of the $k_c(\lambda)$ spectra changes according to the pigment content. They resemble that of pure phytoplankton absorption in eutrophic waters and are comparable to that of detritus absorption (including heterotrophs) in very oligotrophic waters (Morel, 1988). Such spectral changes would mean that the exponent (constant in Eq. 16) is in effect variable with λ. Other parametrizations for $K(\lambda)$ as a function of C are also possible to account for this effect (Baker and Smith, 1982).

It is worth recalling that $a(\lambda)$ is similar to, but differs from, $K(\lambda)$; nevertheless the general spectral trends exhibited by both these coefficients are essentially the same.

Yellow substance absorption (possibly filterable particle absorption) appears as a complementary term in Eq. 16. It partly compensates for the implicit ignorance of the detrital contribution, since it intervenes as a multiplied "correction", applied even on the pure water absorption a_w (becoming a dominant term at low C concentration). With the steep absorption increase toward the short wavelengths (as implied by Eq. 17), this corrective term actually modifies the shape of the global $a(\lambda)$ spectrum.

Agreement between analytical and empirical approaches

In order to comparing these approaches, the absorption values to be attributed to each component in Eq. 14 have to be specified. This is possible for algae and for heterotrophic protists. The sum of both these contributions has to be at least compatible with the empirical expression (Eq. 16, 17). The differences could be tentatively interpreted as the effect of the ignored compartment (detritus and other unidentified substances).

The chlorophyll-specific absorption spectrum for living algae is computed as being the average of 14 spectra determined for algae grown in culture (Morel, 1988; see also discussion in Bricaud, 1989). Two particular a^* values, 40 and 45 m^2 (g Chl a)$^{-1}$ for λ = 415 and 440 nm, respectively, are selected and a_a at these wavelengths will be linearly related to the chlorophyll concentration according to Eq. 15 (see Fig. 8a).

The heterotrophic organisms are weakly but significantly absorbing at λ = 415 nm. Among them, free living bacteria are relevant to the approximation given in Eq. 10'

$$Q_a \approx (2/3) a_s d$$

then

$$S_a \approx Q_a S_g \approx (\pi/6) a_a d^3 \quad \textbf{(18)}$$

A population of microbes with a numerical abundance N/V (cell per m^{-3}) will give rise to an absorption coefficient $a_m = (N/V)\ S_a$. Typical marine bacteria have sizes between 0.4 and 0.7 μm (for the present computations, a mean size of 0.55 μm is adopted), and the absorption coefficient of the cellular material, a_s, (at λ = 415 nm) is about 0.45 10^5 m^{-1} (Morel and Ahn, 1990). The absorption cross section of such a bacterium amounts to 3.9 10^{-15} m^2 at its maximum, 415 nm.

In the photic zone, bacterial numbers (m^{-3}) and Chl *a* concentrations (in mg m^{-3}) seem to be significantly correlated through a non linear expression proposed by Cole *et al.* (1988) and adopted here, written as

$$(N/V) = 0.91\ 10^{12} (Chl) 0.52 \qquad \textbf{(19)}$$

Therefore, the partial absorption due to the presence of bacteria can be established as a function of Chl *a* concentration (Fig. 8a) according to $a_m = (N/V)\ S_a$, leading to

$$a_m = 3.57\ 10^{-3}\ (Chl)^{0.52}$$

To go further and account for bigger protists, some hypotheses are needed. The absorption cross sections for heterotrophic flagellates (mean size 3.4 μm) and naked ciliates (mean size 13.5 μm) have been determined (Morel and Ahn, 1991), and the absorption coefficients of their cellular material (at λ = 415 nm) have been deduced. They do not markedly differ from that of bacteria and the above a_s value can be kept for further computations. The contribution of these protists to absorption can be estimated, provided that their numerical abundance is known or assumed. Abundances of flagellates and microbes have been studied in the field (Sieburth, 1983; Geider, 1988); flagellate and aloricate ciliate biomasses have also been compared (Sherr *et al.*, 1986). This information, however, remains relatively scarce and generalization is presently risky. A working hypothesis (and in effect an upper limit) could consist of considering that, within the chain bacteria - flagellates - ciliates, equal bio-volumes occur in logarithmically equal size intervals (as in Sheldon *et al.*, 1972) and thus that the whole heterotrophic population is continuously distributed according to a Junge power function (with j = 4; see Eq. 5). Very likely such a hypothesis will lead to an overestimate of the heterotrophic compartment. Even if the Sheldon rule is verified between bacterioplankton, nanoplankton and then microplankton, the later categories include not only heterotrophic but also phototrophic organisms with comparable biomasses (see discussion in Beers *et al.*, 1982). With this reservation in mind, the absorption coefficient for the bacterial compartment alone or for the whole heterotrophic compartment, a_h, can be computed by integrating

$$a = \frac{\pi}{4} \int_{d_{min}}^{d_{max}} Q_a(d)\, d^2 (K d^{-4})\, d(d) \qquad \textbf{(20)}$$

where d_{min} is arbitrarily set at 0.1 μm and d_{max} successively is equal to 0.7 μm (bacteria only) or ∞ (all heterotrophs); note that K is an arbitrary constant and Q_a is obtained through Eq. 10, with $a_s = 4.5\ 10^5\ m^{-1}$. With respect to absorption originating solely from the bacterioplankton, the absorption is only increased by 25 % when all bigger heterotrophic organisms are added ($a_h = 1.25\ a_m$), in spite of the fact that this increase is likely overestimated.

Fig. 8a and 8b show the emperical a- values which result from Eq. 16 for the wavelengths considered (415 and 440 nm) and as a function of the chlorophyll concentration (upper solid curves). The horizontal lines correspond to the a_w values at the two wavelengths. The a_a variations (with a slope of 1) and the a_m and a_h variations at only $\lambda = 415$ nm (with a slope of 0.52), computed as explained above, are displayed on the same figure. Note that a_m or a_h would be halved if λ shifts from 415 to 440 nm (not shown on Fig. 8a); for greater wavelengths, absorptions by heterotrophs are negligible.

The "measured" (empirical) and "reconstructed" (analytical) absorption values over the whole range of Chl concentration are roughly the same, (Fig. 8b), the reconstructed values being generally below the empirical ones, except in the eutrophic waters where the converse holds true. In these waters the algal absorption is probably overestimated because the mean a^* values adopted in the reconstruction are too high. There are many reasons to think that in nutrient-rich waters there is a predominance of rather big algal cells, with high packaging effect and consequently reduced a^* values. Recent evidence is shown by measurements in the Peruvian upwelling, where on average a^* (440) is about 38 m^2 (gChl)$^{-1}$, while it is 77 m^2 (g Chl)$^{-1}$ in the Sargasso Sea (Bricaud and Stramski, 1990). It is worth noting that in these waters, the absorption by heterotrophs appears to be negligible (Fig. 8a), even if the assumptions concerning their numerical abundances are drastically revised.

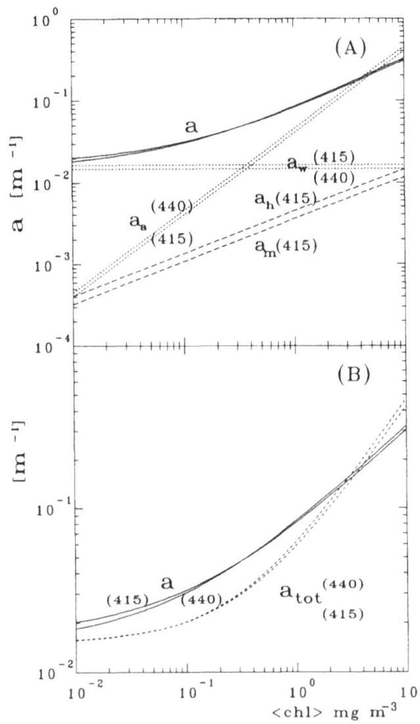

Figure 8. A) The absorption coefficients originating from algae, a_a, microbes, a_m, and all heterotrophes a_h, as a function of the chlorophyllous pigment concentration, $<Chl>$. The constant values for absorption by pure water (at λ = 415 and 440 nm, as indicated), a_w, are represented by the horizontal lines. Note that for a_m and a_h only the values at 415 nm are shown. The solid curves (for 415 and 440) come from the empirical Eq. 16 giving the bulk absorption coefficient α; B) the same empirical a-curves are compared to $a_{tot} = a_h + a_a + a_w$ for the two wavelengths in question.

In the middle range of common oceanic waters, with Chl concentrations around 0.1-0.3 mg Chl m^{-3}, heterotrophs would play a relatively weak role, even at 415 nm (1/5 or 1/10 of that of algae). Absorption, essentially by algae, appear to be nearly below the empirical bulk values resulting from Eq. 16. The missing part could be explained by the absorbing effect of biogenic detritus; many observations using the Kishino *et al.* (1985) method tend to support this explanation.

In ultra-oligotrophic situations, heterotrophic and phototrophic organisms would be of similar importance in forming the absorption coefficient at 415 nm, (but not at 440 nm, where absorption by bacteria is divided by two). The sum ($a_a + a_h$) again remains below the bulk empirical absorption value and leaves room for other absorbers, supposedly detritus. It is also possible that absorption by algae (at 415 and 440 nm) has been underestimated (by 50-70 % ?) in these waters, where small species are dominant (cf. remark concerning a^* in Sargasso Sea).

The above absorption budget is tentative and rests heavily on the validity of the Cole expression. Refinements are possible as soon as more information becomes available with to the rapid development of analytical approaches. The main conclusions, however, concerning the relative role of the compartments and the dominance of algae in the absorption process will probably be globally maintained.

Scattering coefficient of oceanic waters

By virtue of the additivity principle, equations similar to Eq. 14 and 15 can be written for b, the scattering coefficient resulting from the presence of various particulate materials in suspension:

$$b = b_a + b_h + b_d \tag{21}$$

and

$$b_a = b^* \langle Chl \rangle \tag{22}$$

where b^* is the Chl-specific scattering of algae and the subscripts (a, h, d) have the same meaning as before. By following the same scheme as for absorption, it must first be determined what information is available to complete the analytical expression 21, and then examine if the cumulated contributions of the various components are compatible with the global and empirical expressions, as previously derived from field data. This study has been presented in detail elsewhere (Morel and Ahn, 1991) and only the main results are given below.

Analytical approach

The scattering cross sections of various living organisms (photo- and heterotrophic) have been determined. Examples already provided (Fig. 2, 3) demonstrate that Q_b may experience wide inter-specific variations leading to still wider variations in the scattering cross section, $S_g\ Q_b$ (from about 0.2 for bacteria to more than 1000 μm^2 for diatoms). The highly variable character of the "chlorophyll-specific" coefficient b^* is more important when calculating Eq. 22 (see *e.g.* Bricaud *et al.*, 1988). In the spectral domain 570-590 μm (where b^* is less affected by the absorption bands), algal cells tend to exhibit lower b^* values ($< 0.2\ m^2\ mg\ Chl^{-1}$) when they are big (> 5-$10\ \mu m$), whereas higher b^* values (> 0.2 and up to $0.6\ m^2\ mg\ Chl^{-1}$) would occur in the size range 1 to 5 μm (Morel, 1987). Note that a rigorous size-dependency could be predicted if the refractive index increment, (n-1), was sufficiently constant and if C_{iChl}, the intracellular chlorophyll concentration, was also constant. The second condition, in particular, is far from being satisfied, as C_{iChl} may exhibit large inter- and even intra-specific changes, from 1 to 15 Kg m^{-3} (see Table 1 in Morel and Bricaud, 1986). Therefore, the general trends for b^*, as described above (and adopted later on), are to be considered only as possible hypotheses.

The weakly absorbing heterotrophic organisms act mainly as scattering objects with predictable properties under the proviso that a value can be assigned to the refractive index of the cellular material. The term b_h can thereafter be computed if the numerical abundance and typical sizes of heterotrophs are known (or assumed). In summary, in spite of an increasing knowledge relative to the scattering properties at the level of individual cells (phototrophs and heterotrophs), analytical methods are still tentative because field data about cell densities are rather scarce.

Empirical approach

The scattering coefficient b (at $\lambda = 550$ nm) has been measured within the upper layers of the open ocean and directly related to C, the Chlorophyll + Pheophytin concentration, without any further identification or analysis of the particles contributing to its formation. The empirical relationship (Gordon and Morel, 1983), based on the analysis of about 500 coupled observations (in the world ocean and after the turbid - Case 2 waters have been excluded) is

$$b = 0.30(\pm 0.15)\langle C\rangle^{0.62} \tag{23a}$$

where the coefficient 0.30 is a mean value which acknowledges that a certain flexibility occurs in such a "natural" law, yet limited to oceanic waters (the upper value for this coefficient (0.45) corresponds to the transition between Case 1 and turbid-Case 2 waters). By noting that Eq. 23a can be transformed into

$$\frac{b}{\langle Chl\rangle} = 0.30\langle C\rangle^{-0.38} \tag{23b}$$

it appears that $b/\langle C\rangle$, (formally equivalent to b^* in Eq. 22), depends on the biological state of the water and increases with decreasing C concentration. For instance, when C goes from 0.02 to 20 mg m^{-3}, b varies from 0.027 to 2 m^{-1}, therefore $b/\langle Chl\rangle$ decreases from 1.33 to 0.10 m^2 $(mg)^{-1}$.

The non-linear character of Eq. 23 with respect to the pigment concentration originates from the fact that all suspended material (living + detrital), is responsible for the formation of b; b has been found to be roughly proportional to the total particulate organic carbon (Morel, 1988). The exponent lower than 1 in Eq. 23a indicates that the non-algal compartment has, in comparison to the algal compartment, an increasing importance when C diminishes. The Chl-specific scattering coefficient b^* is a lower limit for $b/\langle Chl\rangle$, attainable only in the absence of detritus. The compatibility of these two parameters has been discussed elswhere (Morel, 1987). For diminishing phytoplankton biomass, the increasing role of biodetritus (and of small, not easily-observable organisms) results in an increase of $b/\langle Chl\rangle$.

Another important and relatively old experimental finding deals with the volume scattering function. It was found that the phase function for particles is surprisingly stable in various marine situations. This conclusion was first reached in 1968 (Jerlov) and has always been found to be true, even if the shape of the VSF was found slightly variable in the backward directions (where measurements admittedly are more questionable). The phase function for marine particles as determined by Petzold (1972) is often used for modeling purposes (Kirk, 1981; Gordon, 1989) as being of ubiquitous significance. Interestingly, the shape of this phase function is theoretically accounted for (via Mie theory), provided that the population of scattering particles follows a Junge Law with an exponent value of 4 and if the mean relative refractive index is close to 1.05 (Morel, 1973). Fluctuations around these values are rather limited when correlation between experimental and computed VSF is sought. Incorporation of a small minerogenic fraction (with $n = 1.15$) in order to more precisely explain the backscattering lobe has also been suggested (Brown and Gordon, 1973; Zaneveld *et al.*, 1974). The direct counting (cf. 2.2) and the optical properties are thus in nice agreement, and from this agreement it can also be concluded that the most plausible "mean" value for the refractive index of marine particulates is about 1.05 (cf. 2.1). This result supports the adoption of the Junge (Sheldon) size distribution as an operational hypothesis to analyze the formation of the bulk scattering coefficient.

With this size distribution function (Eq. 5, with $j = 4$), the contribution of the different classes of particles to the total scattering coefficient can be assessed. For the entire population, b is expressed according to the following equation (similar to 20, but for scattering)

$$b=(\pi/4)\int_{d_{\min}}^{d_{\max}} Q_b(d)\,F(d)\,d^2\,d\,d \tag{24}$$

where d can replace ρ in $Q_b(\rho)$, as soon as n and λ have been fixed (1.05 and 421 nm respectively, leading to $\rho = d$); $F(d)$ is the size distribution function supposedly defined between a lower and upper limit (d_{min}, d_{max}). By using the Junge distribution and omitting the constant term, it becomes

$$b=\int_{d_{\min}}^{d_{\max}} S_b(d)\,d^{-j}\,d(d) \tag{24'}$$

where $S_b(d)$ expresses the scattering cross section of all particles having the size d. With the behavior of S_b already described (Fig. 5), it is easy to see that the above integral converges toward a finite value when $d_{min} \to 0$, provided that $j < 7$, and also strictly converges even for an infinite number of particles (when $d_{max} \to \infty$), provided that $j > 3$. The progressive values of the above integral, when j is made variable between 3.2 and 5, are shown in Fig. 9 as a function of d. The different curves indicate the sizes for which a certain percentage of b is attained. For instance, with the exponent $j = 4$, the particles below 0.3 μm contribute less than 5 % of b, half of the scattering is due to particles smaller than about 1.4 μm, and all particles above 14 μm cannot increase the scattering coefficient by more than 5 % (this result is established with d_{min} = 0 and d_{max} = ∞ in Eq. 24'). These values are modified when j varies; lower d values when j increases. When j approaches 3, the upper limit of b is dependent on and logarithmically increasing with d_{max}.

The idea of an "average size" for the population of marine particles (cf. 2.2) is not meaningless from an optical viewpoint, to the extent that most of the scattering (e.g. 80 %, between the curves for 10 % and 90 % of b in Fig. 9) originates from particles in a rather

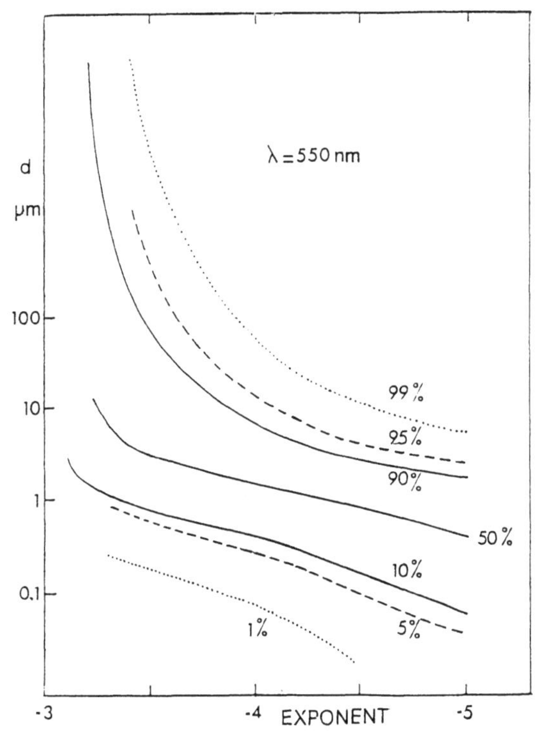

Figure 9. Progressive contribution (%) of particles with a given diameter d (log scale) to the formation of the total scattering coefficient by a population of particles with sizes varying between 0 and ∞ and following a size distribution according to a power function (Junge distribution) of exponent -j, plotted along abcissae (see also Eq. 5 and Eq. 24 where d_{min} = 0 and d_{max} = ∞) (redrawn from Morel, 1973).

narrow size interval (about 0.5 - 5 μm when j = 4), centered on a well-defined value (1.4 μm). Such a conclusion is not greatly altered by the existence of actual physical limits (d_{min}, d_{max}) obviously differing from 0 and ∞, neither by the (frequent) occurences of relative maxima, nor by the present misknowledge of the size distribution in the domain of the very small sizes. The validity of this conclusion could only be questioned if the continuous decrease of the number of particles with increasing size was itself questioned.

Agreement between empirical and analytical approaches

Figure 10 shows as a function of the chlorophyllous pigment concentration, the bulk scattering coefficient (at $\lambda = 550$ nm) resulting from the empirical relationship (Eq. 23a) used with the two extreme values of the coefficient (0.15 and 0.45).

The algal contribution to scattering, b_a, is computed using Eq. 22 with tentative values for b^*, namely 0.4, 0.2 and 0.1 m^2 (mg Chl)$^{-1}$ adopted for oligo-, meso- and eutrophic waters. These decreasing discrete values attempt to acknowledge the fact that big species with high chlorophyll content per cell (leading to low b^* values) seem to dominate in eutrophic

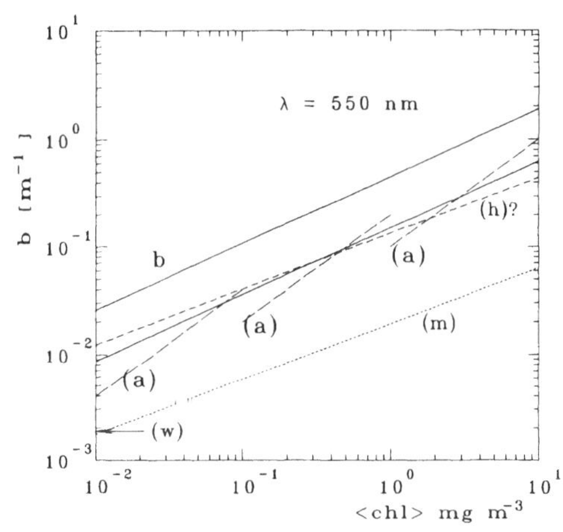

Figure 10. As a function of the chlorophyllous pigment concentration, global scattering coefficient, b, derived from the empirical expression 23a, (the two solid lines). The lines denoted (a) show the contribution of algae (Eq. 22, see text), the dotted line, denoted (m), that of microbes (Eq. 25), and the horizontal line is for pure water. The dashed line denoted (h) is a tentative estimate (see text) of the contribution by all heterotrophic organisms set equal to 7 times that of bacteria.

situations, whereas picoplanktonic cells (with higher b^* values) are the most common in oligotrophic waters. The b_a values plotted in Fig. 10 are not far from accounting for the empirical bulk b-values, at least when they are at their minimum. In spite of their somewhat arbitrary character, the b^* values adopted are believed to be reasonable and lead to the simple conclusion that algae appear to contribute to at least one half of the mean scattering coefficients in oceanic waters.

The bacterial contribution, b_m, can be tentatively estimated by using Cole's expression (Eq. 12) for the "typical" marine bacterium (with a size of 0.55 μm and a relative refractive index of 1.05). With these values, the computed Q_b value is 0.089 and the scattering cross section S_b is 0.021 10^{-12} m^2 (cf. Eq. 11'). After combining with Eq. 19, it results that

$$b_m = 0.019 \langle Chl \rangle^{0.52} \quad \textbf{(25)}$$

also represented on Fig. 10. The contribution of microbes to the formation of b turns out to be rather small. This conclusion is only a qualitative one because of the sensitivity of b_m to the bacterial size. Indeed, the scattering cross section S_g Q_b varies as the fourth power of the size, and b_m should be doubled if the typical size is increased from 0.55 up to 0.65 μm. In addition, the b_m value depends directly on the validity of Eq. 12. It can safely be concluded, however, that microbes are much less efficient than algae in forming the scattering coefficient. This disproportion tends to reduce only in oligotrophic situations.

With the same assumption concerning the size distribution previously used to predict absorption, the contribution of bigger heterotrophs to the formation of the scattering coefficient can be estimated (Eq. 24, where Q_b is computed through Eq. 11 as a function of the size for non-absorbing particles with n = 1.05). The result, very different from that obtained in the case of absorption, shows that the main part of scattering originates from particles with "intermediate" sizes (Fig.9). The hypothesis of a conservative biomass (Sheldon's rule and Junge exponent j = 4) states that the contribution of all heterotrophs with sizes above that of microbes (assumed <0.7 μm) is about 7 times that of microbial organisms. The contribution of particles with size above 10 μm is always negligible. If the living biomass in this range of intermediate sizes is approximately equally shared by both heterotrophs and phototrophs (Azam *et al.*, 1983), their respective contributions would be of the same magnitude. When added, they account for the mean b values observed in oceanic waters (Eq. 23a).

Backscattering coefficient

The backscattering coefficient, b_b, is a crucial parameter to the extent that it governs the reflectance of the ocean (see *e.g.* Morel and Prieur, 1977). The same empirical and analytical approaches as above can be repeated with the aim of understanding the formation of the backscattering coefficient of oceanic waters.

The backscattering ratio $\tilde{b}_b$ (defined as the ratio b_b/b) is of the order of 2 % for oceanic particulate matter (see *e.g.* Petzold, 1972 or Gordon, 1989). It has been actually obvserved that $\tilde{b}_b$ in oceanic waters is not a stable parameter and would decrease from about 2 % in oligotrophic waters down to about 0.2 % in eutrophic waters. For the purpose of modelling the reflectance, an empirical relationship, which accounts for this decrease with increasing C concentrations, has been proposed (Morel, 1988)

$$\tilde{b}_b = 2\ 10^{-3} + 2\ 10^{-2}(1/2 - 1/4 \log C)\,(550/\lambda) \tag{26}$$

This expression combined with Eq. 23a (and letting λ = 550 nm) is used to construct the straight solid line in Fig. 11, representing the variations of b_b along with the pigment concentration.

The $\tilde{b}_b$ ratio for various algae have been experimentally determined (Bricaud *et al.*, 1983; Ahn, 1990). They prove to be within the 10^{-4} - 10^{-3} range, as theoretically expected through Mie theory (see in Morel and Bricaud, 1981b). After a shift by 3 orders of magnitude, the b_a lines of Fig. 10 are redrawn in Fig. 11. The $\tilde{b}_b$ ratio of marine bacteria has also been measured (Morel and Ahn, 1990). It is higher than that of algal cells by at least one order of magnitude. The typical bacterium (d = 0.55 mm and n = 1.05) theoretically has a $\tilde{b}_b$ ratio of 1.1 % at 550 nm. Therefore the b_m values in Fig. 10 are multiplied by 10^{-2} in order to estimate the contribution of microbes to the backscattering coefficient (dashed line in Fig. 11).

The contribution of bigger heterotrophs to b_b can again be computed by using Eq. 24', where Q_b is now replaced by Q_{bb}, the backscattering efficiency factor computed through Mie theory. The comparison of the two partial integrals (from 0 to 0.7 μm, or from 0.7 μm to ∞) shows that the conclusions arrived at when dealing with scattering are completely changed as far as backscattering is concerned. For this coefficient the contribution of the small size fraction (< 0.7 μm) is dramatically dominant, 85 % (if j = 4) and even 97 % (if j = 4.5), with regard to the effect of bigger protists, which can be safely neglected. Therefore only two terms are significant: the contributions of algae and bacteria. Their sum, however, is far from equal to the empirical b_b values and this result implies that an important contribution has not been identified.

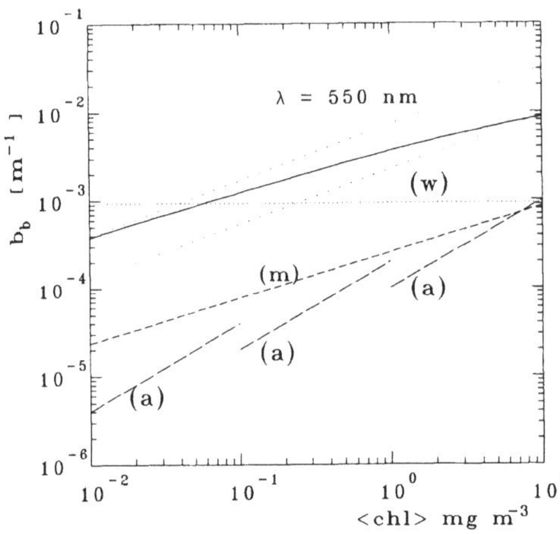

Figure 11. Global backscattering coefficient as a function of the chlorophyllous pigment concentration according to the empirical expression 26 (solid line). The contribution of algae (a) and of microbes (m) to the formation of the backscattering coefficient are also shown (see text).

Theory teaches that only very small particles are efficient back-scatterers. The wide difference between the reconstructed value and the actual b_b value very likely originates from the presence of tiny particles with sufficiently high numerical abundances. Theoretical computations show that it would suffice that the Junge law be valid down to approximately 0.1 μm to account for the major unexplained part of b_b (Morel and Ahn, 1991). Until recently, particles in this size range (0.1 - 0.6 μm) were poorly studied; however, the panorama is changing rapidly. Viruses could be more abundant than bacteria by an order of magnitude (Bϕrsheim *et al.*, 1990; see also Bratback *et al.*, 1990) and thus could produce a sizeable part of the backscattering coefficient. The dry deposition at the surface of the open ocean, giving rise to a population of slowly sinking minerogenic (quartz, illite...) particles (see *e.g.* Blank *et al.*, 1985), is perhaps significant in the backscattering process, even if their optical signature is not detectable when dealing only with absorption or total scattering. Small scales and unidentifiable biodetritic debris are potentially efficient contributors to the backscattering process. It has recently been shown (Koike *et al.*, 1990) that sub-micrometer particles, mainly below 0.6 μm, occur in the upper layers of the ocean in concentration of the order of 10^{13} m^{-3}. Most of them (~ 95 %) are non-living, largely organic material, whereas the remnant fraction is identifiable as bacterioplankton. To the extent that these results prove to be ubiquitous, the open question of the backscattering coefficient formation begins to be answered. This question is a paradoxical establishment of our ignorance, since we are able even from space to detect the light the ocean returns, without having clearly identified the responsible agent (beside the water molecules) for this backscattering phenomenon.

CONCLUSION

The theoretical frame for a comprehensive interpretation of the absorption, scattering, and backscattering properties of various particles is well developeded. Variations in the optical cross sections are predictable. When compared to experimental determinations, predictions are to a large extent, and sometimes perfectly, verified within a wide domain of cell size, from 0.6 to about 25 μm, for both weakly absorbing heterotrophic organisms and variously absorbing phototrophic algae. As a result of such comparisons, the bulk refractive index of particles can be infered, under the proviso that the size or the size distribution is independently determined. While being indirect, this way is perhaps the best to assess the real part of the index of refraction, which is tightly related to the organic carbon content (except in the case of calcite-bearing species).

A weakness of the theoretical approach, actually amenable to improvements, lies in the simplifying assumptions of sphericity and homogeneity made when characterizing the particle. If the spectrophotometric measurements are effected on a collection of particles (on monospecific cultures, for instance), the departure from sphericity does not seem to significantly affect the fit between predicted and experimental scattering properties, because of the random orientation of the cells in suspension (very flat or elongated species admittedly have not been studied). The heterogeneity is not easily detectable and, through the computation, it results in a misestimate of the index, by an unknown factor. This miscalculation is at least suspected when cell-by-cell (microspectrophotometric or flow cytometric) methods are used. With cell orientation not necessarily random, the shape effect could be of importance. This deserves further study, along with other optical properties such as the polarization state of the scattered light, able to provide additional information, which would be valuable in discriminating between the influences of size and index upon the scattering signals.

Optics of marine particles and bulk optical properties of oceanic waters are obviously related, even if the links cannot presently be accurately described. A comparison has been attempted between the cumulated contributions (in absorption, scattering and backscattering) of the various components (those presently identified) and the optical properties of Case 1 waters, as empirically derived from field data. Preliminary conclusions show that absorption and scattering in the upper layers of the ocean are approximately accounted for by the contributions of algae and heterotrophic organisms (of bacteria in particular, in the scattering process), whereas the role of detritus would remain uncertain. When dealing with the backscattering process, the situation is inverted; yhe rather well-identified particles (algal and microbial cells) cannot explain the backscattering coefficient observed at sea. Therefore, the somewhat mysterious, mostly detritic, tiny particles (0.1 - 1 μm) are likely the main contributors to the formation of this coefficient.

ACKNOWLEDGEMENTS

The author would like to thank Y.H. Ahn for his assistance in preparing many figures, and in making available some unpublished results. B. Gentili is duly acknowledged for his efficient help in programing and A. Bricaud for a critical review of a first version of the manuscript.

REFERENCES

Aas E (1981) The refractive index of phytoplankton. Inst Rep Ser, Univ Oslo, 46:61 p

Aas E (1984) Influence of shape and structure on light scattering by marine particles. Inst Resp Ser Univ Oslo 53:112 p

Ackelson SG, Spinrad RW (1988) Size and refractive index of individual marine particulates: a flow cytometric approach Appl Opt 27:1270-1277

Ahn YH, (1990) Propriétés optiques des particules biologiques et minérales présentes dans l'océan. Thèse, Université Pierre et Marie Curie, pp 208

Asano S, Sato M (1980) Light scattering by randomly oriented spheroidal particles. Appl Opt 19:962-974

Azam F, Fenchel T, Field GJ, Gray JS, Meyer-Reil A, Thingstan F 1983) The ecological role of water-column microbes in the sea. Mar Ecol Prog Ser 10:257-267

Bader H (1970) The hyperbolic distribution of particle sizes. J Geophys Res 75:2822-2830

Baker KS, Smith RC (1982) Bio-optical classification and model of natural waters II, Limnol Oceanogr 27:500-509

Beers JR, Reid FMH, Stewart GL (1982) Seasonal abundance of the microplankton population in the North Pacific central gyre. Deep-Sea Res 29:227-245

Bidigare RR, Morrow JH, Kiefer DA (1989) Derivative analysis of spectral absorption by photosynthetic pigments in the western Sargasso Sea. J Mar Res 47:323-341

Blank M, Leinen M, Prospero JM (1985) Major Asian aeolian inputs indicated by the mineralogy of aerosols and sediments in the Western North Pacific. Nature 314:83-86

Brφsheim KY, Bratbak G, Heldall M (1990) Enumeration and biomass estimation of planktonic bacteria and viruses by transmission electron microscopy. Appl Environ Microbiol 56:352-356

Bratbak G, Heldall M, Norland S, Thingstad TF (1990) Viruses as partners in spring bloom microbial trophodynamics. Appl Environ Microbiol 56:1400-1405

Bricaud A (1989) Propriétés optiques du phytoplancton. Etude théorique et expérimentale. Application à l'interprétation de la couleur de la mer. Thèse de Doctorat d'Etat, Univ Pierre et Marie Curie, pp 163

Bricaud A, Morel A (1986) Light attenuation and scattering by phytoplanktonic cells; a theoretical modeling. Appl Opt 25:571-580

Bricaud A, Stramski D (1990) Spectral absorption coefficients of living phytoplankton and non algal biogenous matter: a comparison between the Peru upwelling area and Sargasso Sea. Limnol Oceanogr 35:562-582

Bricaud A, Morel A, Prieur L. (1981) Absorption by dissolved organic matter of the sea (yellow substance) in the U.V. and visible domains. Limnol Oceanogr 26:43-53

Bricaud A, Morel A, Prieur L (1983) Optical efficiency factors of some phytoplankters, Limnol Oceanogr 28:816-832

Bricaud A, Bedhomme AL, Morel A (1988) Optical properties of diverse phytoplanktonic species : Experimental results and theoretical interpretation. J plankton Res 10:851-873

Brown OB, Gordon HR (1973) Two component Mie scattering models of Sargasso Sea particles. Appl Opt 12:2461-2471

Brun-Cottan JC (1971) Etude de la granulométrie des particules marines. Mesures effectuées avec un compteur Coulter. Cah Oceanogr 23:193-205

Butler WL (1962) Absorption of light by turbid materials. J Opt Soc Am 52:292-299

Chisholm SW, Olson RJ, Zettler ER, Goericke R, Waterbury JB, Welschmeyer NA (1988) A novel free-living prochlorophyte occurs at high cell concentration in the oceanic euphotic zone. Nature 334:340-343

Cole JJ, Findlay S, Pace ML (1988) Bacterial production in fresh and saltwater ecosystems; a cross system overview. Mar Ecol Prog Ser 43:1-10

Dubelaar GJB, Visser JWM, Donze M (1987) Anomalous behaviour of forward and perpendicular light scattering of a cyanobacterium owing to intracellular gas vacuoles. Cytometry 8:405-412

Geider RJ (1988) Abundance of autotrophic and heterotrophic nanoplankton and size distribution of microbial biomass in the Southwestern North Sea in October 1986. J Exp Biol Ecol 123:127-145

Gordon Jr DC (1970) A microscopic study of organic particles in the North Atlantic Ocean. Deep-Sea Res 17:233-243

Gordon HR (1987) Bio-optical model describing the distribution of irradiance at the sea surface resulting from a point source embedded in the ocean. Appl Opt 26:4133-4148

Gordon H R (1989) Dependence of the diffuse reflectance of natural waters on the sun angle. Limnol Oceanogr 34:1484-1489

Gordon HR, Morel A (1983) Remote assessment of ocean color for interpretation of satellite visible imagery, A review, in Lecture Notes on Coastal and Estuarine Studies, edited by RT Barber, CNK Mooers, MJ Bowman, and B Zeitzchel, 114 pp., Springer-Verlag, New-York

Gordon HR, Others (1988) A semi-analytical radiance model of ocean color. J Geophys Res 93:10909-10924

Iturriaga R, Mitchell BG, Kiefer DA (1988) Microphotometric analysis of individual particle spectra. Limnol Oceanogr 33:128-135

Junge CE (1963) Air chemistry and Radioactivity, Academic Press, NY, pp 382

Kalle K (1938) Zum Problem der Meereswasser Farbe. Ann Hydrol Mar Mitt 66:1-13

Kerker M (1963) Electromagnetic scattering. Proceedings of the Interdisciplinary Conference, edited by M Kerker. Pergamon Press, pp 592

Kiefer DA, Soohoo JB (1982) Spectral absorption by marine particles of coastal waters of Baya California. Limnol Oceanogr 27:492-499

Kirk JTO (1975a) A theoretical analysis of the contribution of algal cells to the attenuation of light within natural waters. I. General treatment of suspensions of living cells, New Phytol 75:11-20

Kirk JTO (1975b) A theoretical analysis of the contribution of algal cells to the attenuation of light within natural waters. II. Spherical cells. New Phytol 75:21-36

Kirk JTO (1981) Monte-Carlo study of the nature of the underwater light field in, and the relationships between optical properties of, turbid yellow waters. Aust J Mar Freshwater Res 32:533-539

Kishino M, Takahashi M, Okami N, Ichimura S (1985) Estimation of the spectral absorption coefficients of phytoplankton in the sea. Bull Mar Sci 37:634-642

Koike I, Hara S, Terauchi K, Kogure K (1990) Role of sub-micrometer particles in the ocean. Nature 345:242-244

Mitchell BG (1987) Ecological implications of variability in marine particulate absorption and fluorescence excitation spectra. Ph D thesis, Univ Southern California 196 p

Morel A (1973) Diffusion de la lumière par les eaux de mer, résultats expérimentaux et approche théorique, in Optics of the Sea, AGARD Lect. Ser, vol 61, pp 3.1.01-3.1.76, Advisory Group for Aeronautical Research and Development, NATO, Brussels

Morel A (1974) Optical properties of pure water and pure sea water, in Optical Aspects of Oceanography, edited by NG Jerlov and E Steemann Nielsen, pp 1-24, Academic, San Diego, Calif

Morel A (1987) Chlorophyll-specific scattering coefficient of phytoplankton, a simplified theoretical approach. Deep-Sea Res 34:1093-1105

Morel A (1988) Optical modelling of the upper ocean in relation to its biogenous matter content (case 1 waters). J Geophys Res 93:10749-10768

Morel A (1991) Light and marine photosynthesis: A spectral model with geochemical and climatological implications. Prog Oceanogr 26:263-306

Morel A, Ahn YH (1990) Optical efficiency factors of free living marine bacteria : influence of bacterioplankton upon the optical properties and particulate organic carbon in oceanic waters. J Mars Res 48:145-175

Morel A, Ahn YH (1991) Optics of heterotrophic nanoflagellates and ciliates : their optical role in open ocean waters compared to those of bacteria and algal cells. J Mar Res in press

Morel A, Bricaud A (1981a) Theoretical results concerning light absorption in a discrete medium and application to specific absorption of phytoplankton, Deep Sea Res Part A, 28, 1375-131, 1981

Morel A, Bricaud A (1981b) Theoretical results concerning the optics of phytoplankton, with special reference to remote sensing applications. In : Oceanography from Space, J F R Gower edit, Marine Science Series, Vol. 13, Plenum Press, NY, 313-327

Morel A, Bricaud A (1986) Inherent optical properties of algal cells including picoplankton: theoretical and experimental results, in Photosynthetic Picoplankton, T Platt and WKW-Li, eds, Can Bull Fish Aquat Sci 214:521-559

Morel A, Prieur L (1977) Analysis of variations in ocean color. Limnol Oceanogr 22:709-722

Morrow JH, Chamberlin WS, Kiefer DA (1989) A two-component description of spectral absorption by marine particles. Limnol Oceanogr 34:1500-1509

Norland S, Heldal M, Tumyr O (1987) On the relation between dry matter and volume of bacteria. Microb Ecol 13:95-101

Perry MJ, Porter SM (1989) Determination of the cross-section absorption coefficient of individual phytoplankton cells by analytical flow cytometry. Limnol Oceanogr 34:1727-1738

Petzold TJ (1972) Volume scattering functions for selected natural waters. Scripps Inst Oceanogr, Visibility Lab, SIO Ref 72-78

Preisendorfer RW (1961) Application of radiative transfer theory to light measurements in the sea, Monogr. 10, pp. 11-30, Int Union Geod Geophys, Paris

Prieur L, Sathyendranath S (1981) An optical classification of coastal and oceanic waters based on the specific spectral absorption curves of phytoplankton pigments, dissolved organic matter and other particulate materials, Limnol Oceanogr 26:671-689

Rabinowitch E, Govindjee E (1969) Photosynthesis. John Wiley and Sons Inc., pp. 273

Rassoulzadegan F, Sheldon RW (1986) Predator-prey interactions of nanozooplankton and bacteria in an oligotrophic marine environment. Limnol Oceanogr 31:1010-1021

Riley GA (1965) A mathematical model of regional variations in plankton. Limnol Oceanogr 10: R 202-R 215

Schuerman DW (1980) Light scattering by irregular shaped particles. DW Schuerman Edit, Plenum, pp 334

Sheldon RW, Parsons TR (1967) A continous size spectrum for particulate matter in the Sea. J Fish Res Bd Canada 24:909-915

Sheldon RW, Prakash A, Sutcliffe WH Jr (1972) The size distribution of particles in the ocean. Limnol Oceanogr 17:327-340

Sherr ED, Sherr BF, Fallon RD, Newell SY (1986) Small, aloricate ciliates as a major component of heterotrophic nanoplankton. Limnol Oceanogr 31:177-183

Sieburth J McN (1983) Microbiological and organic chemical process in the surface and mixed layers. In : Air-Sea exchange of gazes and particles; PS LISS and WGN SLINN Edit D Reidel Publishing Company, 121-172

Simon M, Azam F (1989) Protein content and protein synthesis rates of planktonic marine bacteria. Mar Ecol Prog Ser 51:201-213

Smith RC, Baker KS (1978) The bio-optical state of ocean waters and remote sensing. Limnol Oceanogr 23:247-259

Smith RC, Baker KS (1981) Optical properties of the clearest natural waters, Appl Opt 20:177-184

Soohoo JB, Kiefer DA (1982) Vertical distribution of phaeopigments. 1. A simple grazing and photooxidative scheme for small particles. 2. Rates of production and kinetics of photoxidation. Deep-Sea Res 29:1539-1551; 1553-1563

Sosik HM (1988) Analysis of chlorophyll fluorescence in marine phytoplankton : Interpretation of flow cytometric signals. MS thesis, Mass Inst Technol, 88 p

Sosik HM, Chisholm SW, Olson RJ (1989) Chlorophyll fluorescence from single cells: Interpretation of flow cytometric signals. Limnol Oceanogr 34:1749-1761

Spinard RW, Brown FJ (1986) Relative real refractive index of marine microorganisms: a technique for flow cytometric estimation. Appl Opt 25:1930-1934

Steele JH, Menzel DW (1962) Conditions for maximum primary production in the mixed layer. Deep-Sea Res 9:39-49

Stramski D, Morel A (1989) Optical properties of photosynthetic picoplankton in different physiological states as affected by growth irradiance. Deep-Sea Res 27:245-266

Stramski D, Kiefer DA (1990) Optical properties of marine bacteria. Proc SPIE, 1302, Ocean Optics 1302:250-268

Stramski D, Morel A, Bricaud A (1988) Modeling the light attenuation and scattering by spherical phytoplankton cells: a retrieval of the bulk refractive index. Appl Opt 27:3954-3956

Tyler JE (Ed.) (1966) Report on the second meeting of the Joint Group of experts on photosynthetic radiant energy. UNESCO Tech Pap Mar Sci 2:1-11

Vaulot D, Partensky F, Neveux J, Mantoura RFC, Llewellyn CA (1990) Winter presence of prochlorophytes in surface waters of the northwestern Mediterranean Sea. Limnol Oceanogr 35:1156-1164

Van De Hulst HC (1957) Light scattering by small particles. Wiley, NY, 470 pp

Yentsch CS (1962) Measurements of visible light absorption by particulate matter in the ocean. Limnol Oceanogr 7:207-217

Yentsch CS (1963) Primary production. Oceanogr Mar Biol Ann Rev 1:157-175

Yentsch CM, Yentsch CS (1984) Emergence of optical instrumentation for measuring biological properties. Oceanogr Mar Biol Ann Rev 22:55-98

Zaneveld JRV, Roach DM, Pak H (1974). The determination of the index of refraction of particles suspended in the ocean. J Geophys Res 77:2677-2680

LIGHT SCATTERING PROPERTIES OF CELLS

G. C. Salzman[1], S. Brito Singham[2],
R. G. Johnston[3], and C. F. Bohren[4]

[1]Life Sciences Division
Los Alamos National Laboratory
Mail Stop M880, P.O. Box 1663
Los Alamos, NM 87545 USA

[2]3681 Townley Road
Cleveland, OH 14122 USA

[3]Chemistry and Laser Science Division
Los Alamos National Laboratory
Mail Stop J565, P.O. Box 1663
Los Alamos, NM 87545 USA

[4]Department of Meteorology
Pennsylvania State University
503 Walker Building
University Park, PA 16802 USA (4)

INTRODUCTION

Light scattering has been associated with flow cytometry from its very beginnings. Its primary use has been in discriminating among cells based on differences in size, shape, refractive index, and the presences of internal structures. The sensitivity of light scattering to all these properties makes it a difficult phenomena to interpret. Various theories are useful for interpreting parts of this powerful tool.

The discussion begins with a presentation of the basic concepts in scattering. The disposition of the elementary units comprising a particle is shown to be as important as the composition of the material from which the particle is made. Polarization concepts are then discussed because they offer a way to improve discrimination among cell types. Some of the standard theories developed to explain parts of the light scattering puzzle are then presented. The coupled dipole approximation is then discussed and examples are given of how this theory can

NATO ASI Series, Vol. G 27
Particle Analysis in Oceanography
Edited by S. Demers

be used to calculate the scattering from small, irregularly shaped cells and other particles. The relatively new idea of phase differential scattering is then presented. The chapter ends with a brief review of the uses of light scattering polarization measurements in flow.

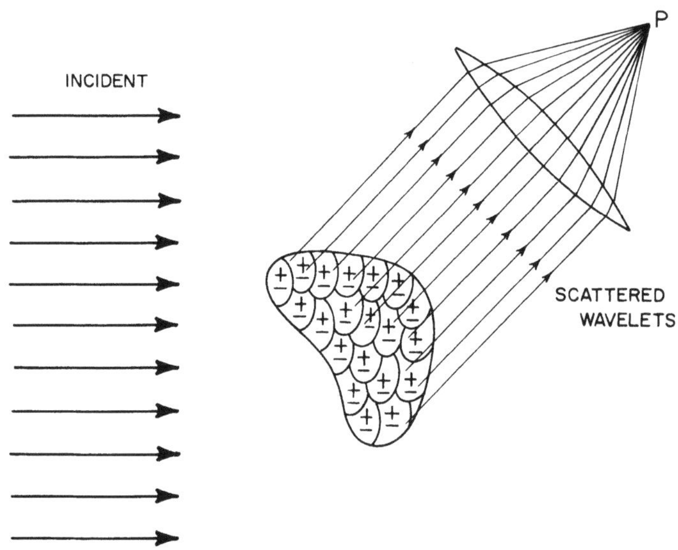

Figure 1. A single particle composed of many, tiny dipoles. The dipoles oscillate under the influence of the electric field of the incident wave. The dipoles reradiate the incident light as scattered wavelets. The light incident on a detector at P is the sum of the scattered wavelets. The phase relations among the scattered wavelets must be taken into account in computing the sum. (Reproduced from Bohren and Huffman, 1983, with permission of the publisher.)

DISPOSITION AND COMPOSITION

Light striking a piece of glass scatters in two directions determined by the laws of reflection and refraction. These laws are two names for the process of scattering. The piece of glass is composed of a tightly packed array of atoms. The incident light wave causes each atom in the array to reradiate the light wave in all directions. These scattered wavelets cancel each other in all directions except those given by the laws of reflection and refraction. If the piece of glass is ground into a pile of tiny fragments, light is scattered in all directions and the pile appears white even though the composition of the piece of glass has not changed. Each

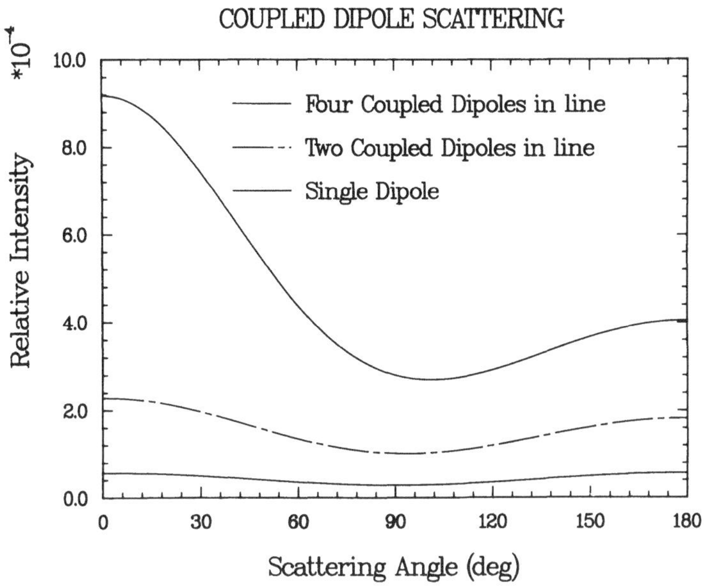

Figure 2. Relative intensity as a function of scattering angle for a single dipole (lower solid curve), two coupled dipoles in line (chain dashed line), and four coupled dipoles in line (upper solid curve). The dipoles were spaced one wavelength apart and the intensity was averaged over all possible orientations of the line with respect to the incident light wave. (Reproduced from Bohren, 1987, p.141, with permission of the publisher.)

fragment has many atoms in an array, but the fragments are no longer in an ordered array, so the scattered waves do not cancel. This example illustrates the point that the disposition of the parts of a particle are as important as the composition of the particle in determining the scattering properties.

Figure 1 shows the scattering of light by a single particle, such as one of the tiny fragments of glass. Imagine that it is composed of a large array of tiny dipolar antennas. A neutral atom can become a dipole when an electric field causes the electrons to become asymmetrically distributed around the nucleus. Equal positive and negative charges separated by a small

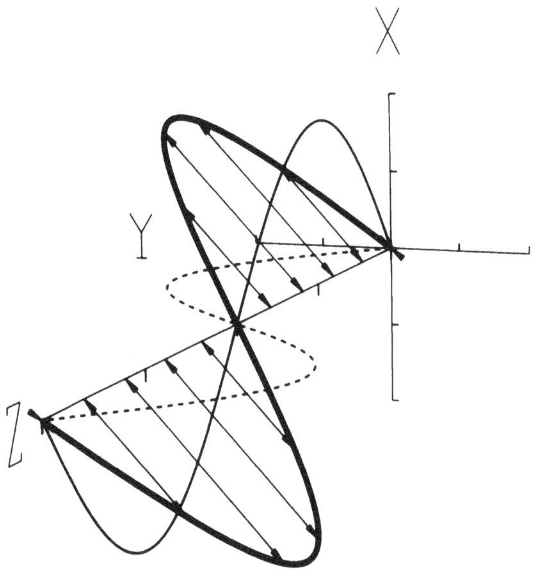

Figure 3. Path of the electric field vector (arrows and heavy solid curve) for linearly polarized light traveling along the z-axis. The plane of the linear polarization is 45 degrees from the x-z plane. The electric field can be resolved into a component parallel to the x-z plane (thin solid curve) and a component parallel to the y-z plane (dashed curve). (Reproduced from Salzman *et al.* (1990) with permission of the publisher.)

distance constitute a dipole. The oscillating electric field of the incident light wave causes these dipoles to oscillate and radiate light at the same frequency as the incident wave. The electric fields of the scattered wavelets add together to determine the intensity of the light at detection point P. The scattered wavelets have phase relations among them that depend on the separation of the dipoles and the angle between the incident direction and the direction toward the detector at P. The intensity at P is the square of the sum of the scattered wavelets and varies as the point P moves in space.

The forward scattering region (the direction of the incident light) is special as illustrated by the scattering from two nearby dipoles. They are excited by the incident wave out of phase because of their finite separation. They also radiate out of phase. In the forward direction, however, the phase difference is the same size but opposite in sign. As a result the scattered waves in the forward direction are exactly in phase. This is as true for many dipoles as it is for two dipoles. The scattered intensity in the forward direction increases rapidly and becomes more sharply peaked as the number of dipoles increases. This is illustrated in Figure 2, which shows the intensity of the scattered light as a function of scattering angle for one, two and four dipoles in a line. In the case of two and four dipoles, the incident wave for each dipole is composed of the external incident wave and the scattered waves from the other dipoles. The dipoles are coupled; they "see" the electric fields of the neighbouring dipoles.

As particle shape changes, the relative locations of the dipoles comprising the particle change and the phase relationships among the scattered wavelets change. In the forward direction, however, the waves are still in phase and the scattered intensity in that direction is least sensitive to particle shape. The same idea applies to particle orientation. To examine particle shape with light scattering, we need to place our detector at scattering angles away from the forward direction. The size of the scattering angle for best sensitivity to particle shape is determined empirically and depends on the size and shape of the particle.

The optical composition of a particle is specified by its refractive index relative to the surrounding medium. The polarizability, which is the magnitude of the induced dipole moment per unit exciting field, is the microscopic counterpart of the macroscopic refractive index. The amplitude of each scattered wavelet is larger for larger dipole polarizabilities. As a result the interaction or coupling among the dipoles is strong for particles composed of dipoles with large polarizabilities. At visible wavelengths, biological particles have low relative refractive indices and are considered to be composed of weakly interacting dipoles.

The wavelength and polarization state are the important characteristics of the incident light. The wavelength determines the scale for the particle size. The polarization state specifies the direction of the electric field of the incident wave as shown in Figure 3. The field is perpendicular to the direction of the wave. This light is linearly polarized because the electric field vectors all lie in a fixed plane. The amount of light scattered by a dipole depends on the orientation of the electric field of the incident wave with respect to the scattering plane. The scattering plane is defined by the direction of the incident wave and a line extending from the particle toward a detector. For light polarized perpendicular to the scattering plane, the induced dipole is also perpendicular to that plane and the intensity of the scattered light is the same for all directions in the scattering plane. For light whose electric field is linearly polarized parallel to the scattering plane, the intensity scattered by the dipole is greatest at 0 degrees and 180 degrees and is zero at 90 degrees with respect to the direction of the incident wave. Scattering by a single dipole or by an array of dipoles depends on the direction of the polarization of the incident light.

Figure 4 shows a circularly polarized light wave moving down the Z axis. The electric field of a circularly polarized wave rotates around the direction of the wave. The heavy line shows the path traced out by the tip of the electric field vector as the wave moves through space.

The Stokes parameters, I, Q, U and V, specify the polarization state of any wave. I is the intensity or irradiance of the wave, Q is the fraction of horizontal linear polarization, U is the fraction of 45 degree linear polarization, V is the fraction of circular polarization. The incident light wave can be represented mathematically as a 4 element Stokes vector with each element having a subscript i; the elements of the scattered light Stokes vector have subscript s. Scattering can be viewed mathematically as the transformation of an incident Stokes vector into a scattered Stokes vector by a 4 X 4 matrix, called the Mueller scattering matrix or just scattering matrix, as

$$\begin{vmatrix} I_s \\ Q_s \\ U_s \\ V_s \end{vmatrix} = \begin{vmatrix} S_{11} & S_{12} & S_{13} & S_{14} \\ S_{21} & S_{22} & S_{23} & S_{24} \\ S_{31} & S_{32} & S_{33} & S_{34} \\ S_{41} & S_{42} & S_{43} & S_{44} \end{vmatrix} \begin{vmatrix} I_i \\ Q_i \\ U_i \\ V_i \end{vmatrix}$$

The 16 real elements of the scattering matrix depend on wavelength and scattering angle. S_{11} is the irradiance of the scattered light; S_{12} measures the extent of scattered horizontal linear polarization. For a single, fixed particle only seven of the scattering matrix elements are independent.

LIGHT SCATTERING MODELS FOR BIOLOGICAL PARTICLES

A biological particle in a flow cytometer laser beam scatters light in all directions. The scattered light is typically detected in a large solid angle in the forward direction and at right angles to the laser beam. This scattering process is described by a number of mathematical theories usually linked to the size of the particle. Each of the theories obscures the underlying physics of the process, which is best represented by the above model of interacting, radiating dipoles.

Mie theory (Bohren and Huffman, 1983) models an uncoated particle as an arbitrary diameter sphere with a homogeneous refractive index. The theory is an exact solution of Maxwell's electromagnetic field equations for a sphere. It has been extended to cylinders and coated spheres (Aden and Kerker, 1951; Kerker *et al.*, 1978). The scattering matrix for a sphere is given by (Bohren and Huffman, 1983)

$$\begin{vmatrix} I_s \\ Q_s \\ U_s \\ V_s \end{vmatrix} = \begin{vmatrix} S_{11} & S_{12} & 0 & 0 \\ S_{21} & S_{22} & 0 & 0 \\ 0 & 0 & S_{33} & S_{34} \\ 0 & 0 & -S_{34} & S_{44} \end{vmatrix} \begin{vmatrix} I_i \\ Q_i \\ U_i \\ V_i \end{vmatrix}$$

The coated sphere model from Mie theory views a cell such as a lymphocyte as homogeneous spherical nucleus surrounded by a concentric spherical coating of cytoplasm. An osmotically

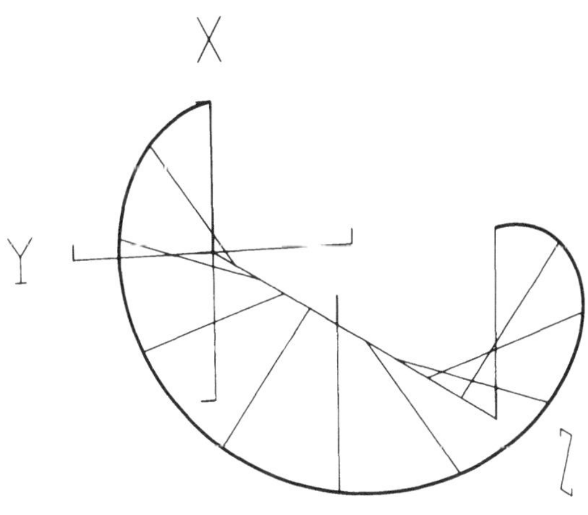

Figure 4. Path of the tip of the electric field vector for a circularly polarized light wave traveling along the positive z-axis. Viewed down the z-axis the path is a circle. The electric field of a circularly polarized light wave can be resolved into linearly polarized components parallel to the x-z and y-z planes and shifted in phase relative to one another by 90 degrees. (Reproduced from Salzman *et al.* (1990), with permission of the publisher.)

swelled human blood cell is viewed as a homogeneous sphere of cytoplasm surrounded by a thin, homogeneous coating of membrane. The coated sphere model was used to calculate S_{11} between 1 degree and 15 degrees for Chinese hamster tissue culture cells (CH0), which are

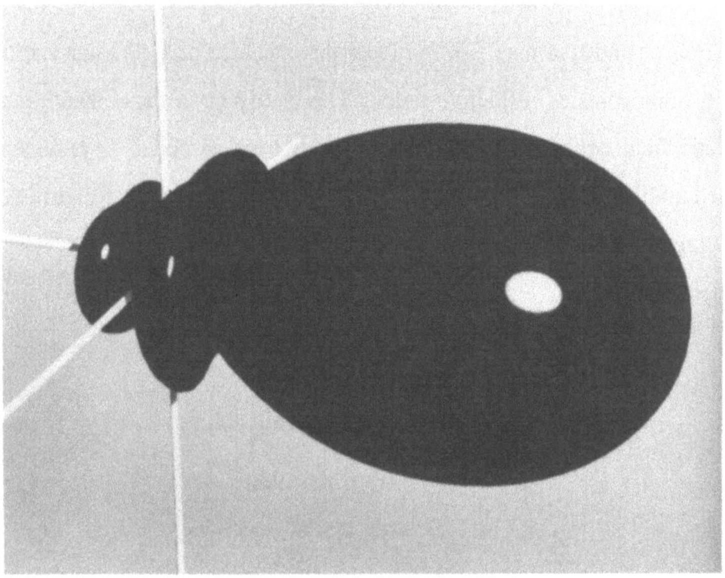

Figure 5. A homogeneous Mie theory calculation of the scattering envelope for a 0.6 micrometer diameter sphere with refractive index 1.37 in water (refractive index 1.33). The wavelength of the illumination is 488 nm in air. The sphere is at the intersection of the white rods and the vertically polarized light beam passes from left to right. The envelope shows the total scattered irradiance for all possible scattering directions. The envelope has been compressed by taking the fourth root of the scattered irradiance. Computer graphic courtesy of Melvin Prueitt, Los Alamos National Laboratory.

spherical and have concentric nuclei (Brunsting and Mullaney, 1972; 1974). The investigators measured the scattered irradiance from suspensions of CHO cells using a film photometer at a wavelength of 633 nm. They also made microscope measurements of the refractive index of CHO nuclei (1.392) and cytoplasm (1.3703). They used a medium of 0.9% sodium chloride with a refractive index of 1.3345. They found that the coated sphere model fit the experimental data better than the homogeneous sphere model.

Figure 5 shows a homogeneous Mie theory calculation of the scattering envelope for a 0.6 micrometer diameter sphere with refractive index 1.37 in water (refractive index 1.33). The wavelength of the illumination is 488 nm in air. The sphere is at the intersection of the white

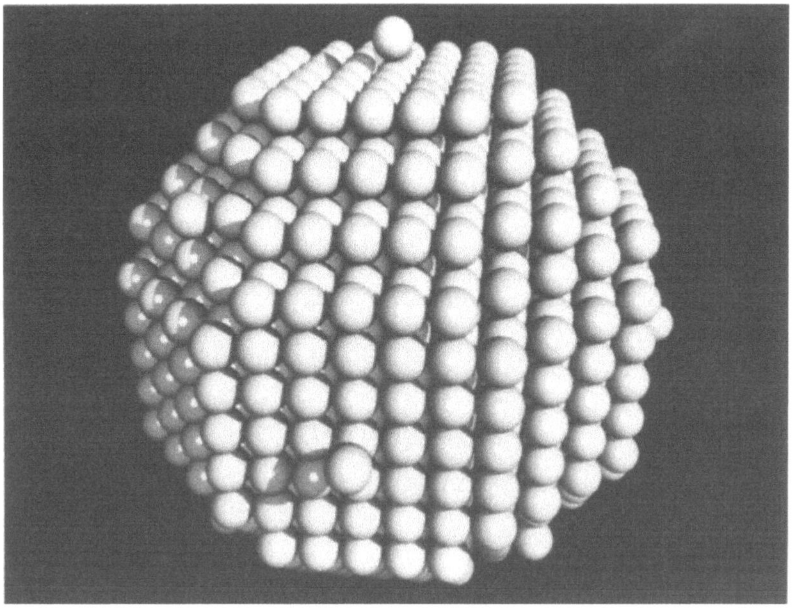

Figure 6. Computer generated model showing a collection of small spherical subunits (dipoles) representing a solid sphere. Computer graphic courtesy of Melvin Prueitt, Los Alamos National Laboratory.

rods and the vertically polarized light beam passes from left to right. The envelope shows the total scattered irradiance for all possible scattering directions. The envelope has been distorted by taking the fourth root of the scattered irradiance. This compresses the envelope in the forward direction and expands it at larger angles so that small details of the envelope can be seen.

Biological cells are certainly not homogeneous spheres. They are irregularly shaped particles filled with optically inhomogeneous materials. There are many approximate models for scattering from particles such as biological cells (Salzman *et al.*, 1990). We choose to describe only one here. The coupled dipole theory described below is an approximate model that may be useful for modelling some small biological particles. The model is tested by comparing its results with those of Mie theory for particles that can bê modeled with Mie theory.

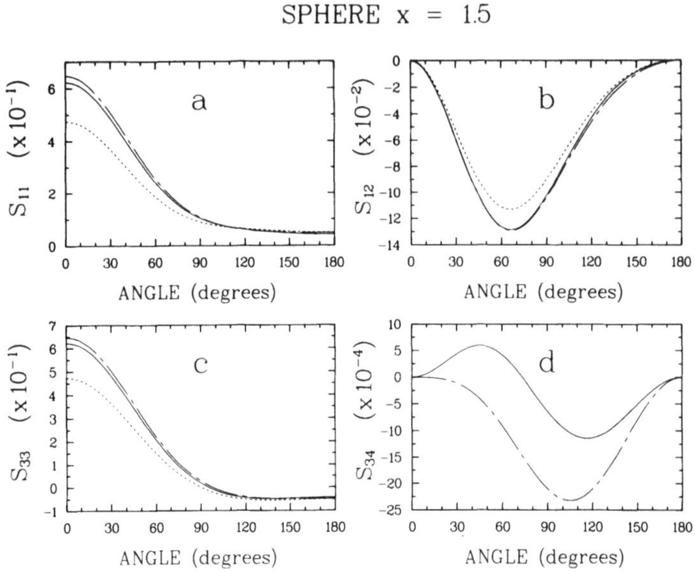

Figure 7. Scattering matrix elements as a function of scattering angle for the homogeneous sphere Mie theory (chain dashed curves) and the coupled dipole approximation (solid curves). The dashed curves are for noninteracting dipoles (no coupling). S_{34} is zero for noninteracting dipoles. Agreement is good except for S_{34}, which is sensitive to the lumpiness of the coupled dipole sphere. (Reproduced from Singham and Salzman, 1986, with permission of the publisher.)

The coupled dipole model was developed to describe light scattering and absorption by irregularly shaped interstellar dust particles (Purcell and Pennypacker, 1973) and was later applied to biological cells (Druger *et al.*, 1979). Fig. 6 shows the coupled dipole model for a sphere as a collection of small spherical subunits (dipoles). Each of the dipoles can be assigned a different polarizability so that particles with complex internal structures can be modeled. The model is very general and enables calculation of all the elements of the scattering matrix.

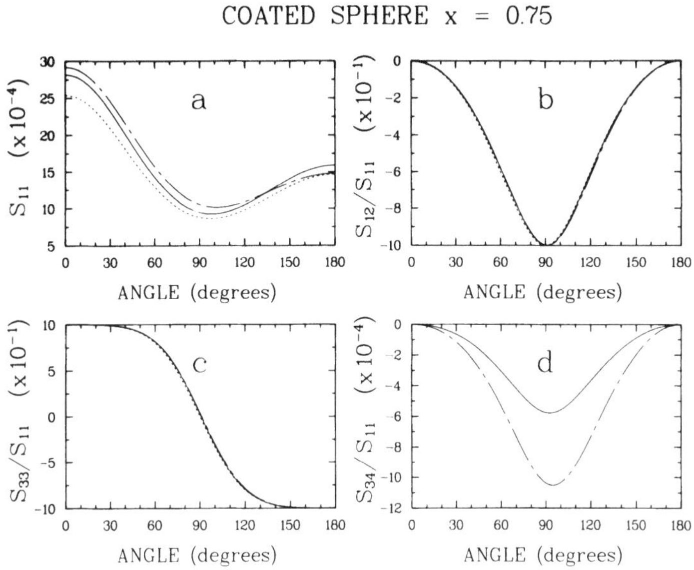

Figure 8. Scattering matrix elements as a function of scattering angle for the coated sphere Mie theory (chain dashed curves) and the coupled dipole approximation (solid curves). The dashed curves are for noninteracting dipoles. The size parameter for the core is 0.56 and that from the shell is 0.75. The agreement is good except for the S_{34} matrix element. (Reproduced from Singham and Salzman, 1986, with permission of the publisher.)

Figure 7 shows a comparison between the coupled dipole approximation (Singham and Salzman, 1986) and Mie theory for some of the scattering matrix elements for a homogeneous sphere. The sphere has a size parameter $x = 1.5$ (sphere circumference divided by wavelength in the suspending medium). The wavelength in the medium was 12.92 micrometers. This wavelength was chosen to obtain the size parameter of 1.5. Each dipole was assigned an isotropic polarizability using the Clausius-Mosotti relation (Jackson, 1975). The 123 dipoles were placed on a cubic lattice. The separation between the outermost dipole and the center of the sphere was 3 micrometers. Figure 8 shows a comparison between the coupled dipole model for a coated sphere and Mie theory. Isotropic, spherical dipoles (149 of them) were placed on a spherical shell 4 micrometers from the origin and 1 micrometer apart. Agreement was good except for the matrix element S_{34}. Note that the off-diagonal matrix elements are normalized to S_{11}.

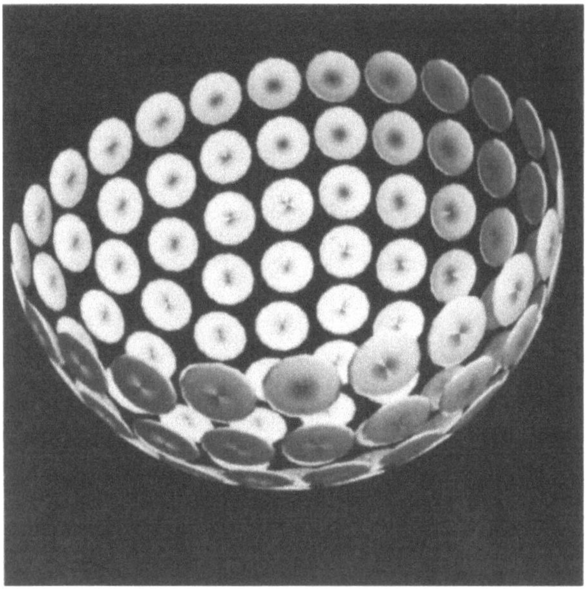

Figure 9. A cutaway view of a spherical shell composed of oblate spheroidal dipoles with anisotropic polarizabilities. Computer graphic courtesy of Melvin Prueitt, Los Alamos National Laboratory.

The use of anisotropic polarizabilities makes it possible to use fewer appropriately shaped dipoles to model a given structure. Figure 9 shows a cutaway view of a spherical shell composed of oblate spheroidal dipoles with anisotropic polarizabilities. Figure 10 shows how well the coupled dipole model with anisotropic polarizabilities works. The calculations are for a spherical shell with size parameter $x = 1.0$. The core is vacuum. The inner radius is 3.443 micrometers and the outer radius is 3.444 micrometers. The wavelength is 21.6 micrometers. Agreement between the coupled dipole approximation and Mie theory for S_{34} is better than that for the solid spheres.

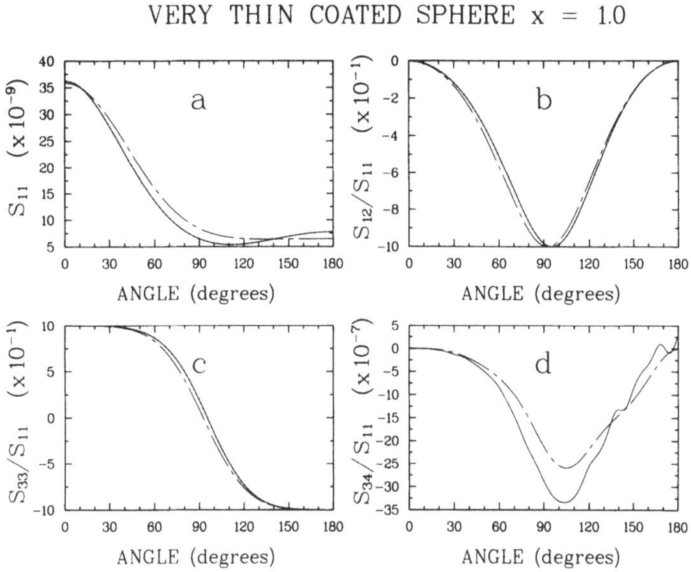

Figure 10. Scattering matrix element values (normalized to S_{11}) as a function of scattering angle for a thin spherical shell modeled with ellipsoidal dipoles with anisotropic polarizabilities. The size parameter for the core is 0.9997 and that for the shell is 1.0. The coupled dipole approximation is shown with the solid curves, the noninteracting dipole approximation is shown with the dashed curves, and Mie theory is shown with chain dashed curves. S_{34} is zero for noninteracting dipoles. Agreement between the coupled dipole model and Mie theory for S_{34} is better here than for the spheres. (Reproduced from Singham and Salzman, 1986 with permission of the publisher.)

Figure 11 shows a model for a thick helix. The subunit dipoles are prolate ellipsoids with anisotropic polarizabilities. Prolate ellipsoids are used here in preference to spheres or oblate ellipsoids because fewer prolate ellipsoids are needed to represent the volume of the helix than would be the case for the other subunit types. Figure 12 shows the coupled dipole approximation for the S_{34} scattering matrix element as a function of scattering angle for values of helix radius (R) and pitch (P) and thickness (T) at a wavelength of 700 nm. This wavelength was chosen to establish the desired size parameter. S_{34} is insensitive to helix thickness but is sensitive to differences in helix pitch and radius.

Figure 11. Model of a thick helix whose subunit dipoles are prolate ellipsoids with anisotropic polarizabilities. Computer graphics courtesy of Melvin Prueitt, Los Alamos National Laboratory.

The coupled dipole approximation requires the solution of 3N linear equations for N dipoles. Computer memory is quickly exhausted if matrix methods are used. A reformulation of the problem in terms of scattering orders has helped to reduce this problem (Singham and Bohren, 1987). The method works well only for small particles with low refractive index relative to the surrounding medium.

POLARIZATION MEASUREMENTS ON BIOLOGICAL PARTICLES

Polarization measurements on biological particles are difficult because most of the interesting scattering matrix elements are much smaller than S_{11}(Salzman *et al.*, 1990). One approach to making suspension measurements of a combination of several of the interesting matrix

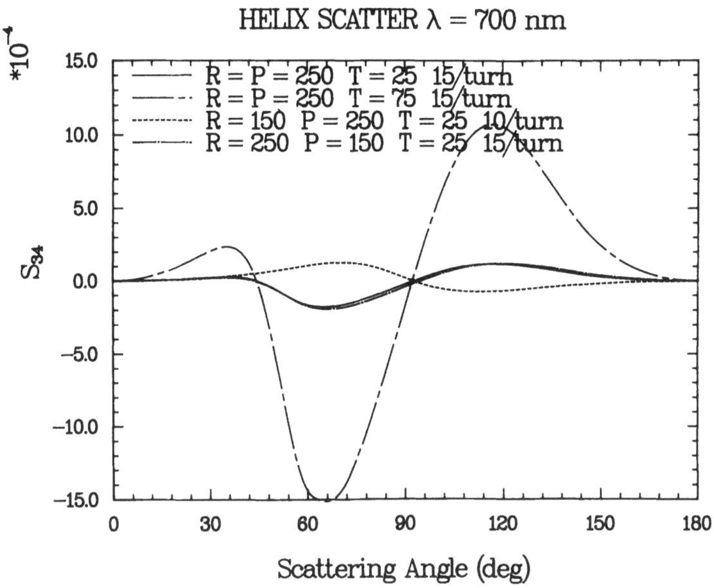

Figure 12. Coupled dipole approximation for the S_{34} scattering matrix element for a helix with different values of radius (R), pitch (P) and thickness (T) at a wavelength of 700 nm. S_{34} is sensitive to helix pitch and radius. (Reproduced from Salzman *et al.* (1990))

elements is phase differential scattering (PDS) (Johnston *et al.*, 1986; Johnston *et al.*, 1987; Johnston *et al.*, 1988). The method uses a laser that produces two, collinear lines at nearly the same frequency (633 nm) with orthogonal linear polarizations. The lines differ in frequency by only 250 kHz. The laser beam is incident on a cuvette suspension of particles.

A detector is placed at a specified scattering angle with a polarizer oriented at 45 degrees from the horizontal scattering plane. A scattering object retards the incident horizontal and vertical polarizations of the laser beam by different amounts. This produces a phase shift in the 250 kHz beat frequency at the detector. In terms of the scattering matrix elements, this phase shift, a, is given by

$$\alpha(45^{o}) = \tan^{-1}\left(\frac{S_{14}+S_{34}}{S_{13}+S_{33}}\right)$$

Figure 13 shows the relative PDS phase as a function of scattering angle for pure suspensions of several different viable bacteria. The discrimination among the bacteria is quite reproducible. It may be possible to make these measurements with a flow cytometer.

A few scattered light polarization measurements have been made with a flow cytometer. The partial depolarization of the light scattered at right angles to the laser beam in a flow cytometer has been used to distinguish human eosinophils from monocytes, lymphocytes, and neutrophils (deGrooth *et al.*, 1987). A 45 degree beam splitter in the right angle scatter detector path reflects part of the scattered light onto a detector. The signal from this

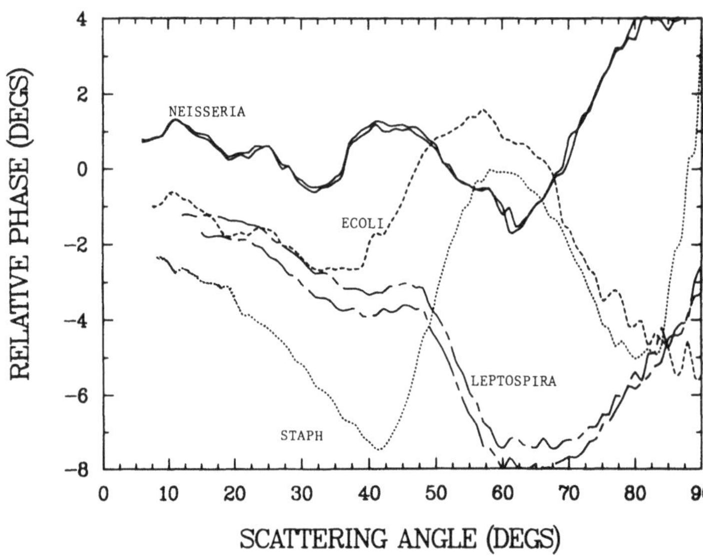

Figure 13. Phase differential scattering from pure aqueous suspensions of different viable bacteria as a function of scattering angle. The bacteria are *Neisseria lactamica*, *Leptospira biflexa*, *Staphylococcus aureus*, and *Escherichia coli B.* (Reproduced from Johnston *et al.* (1988)).

detector is called total orthogonal light scatter and is proportional to the total amount of light scattered into a cone extending from 65 degrees to 115 degrees with respect to the laser beam direction. The light passing through the 45 degree beam splitter passes through a horizontal polarizer and into a detector labeled depolarized light orthogonal light scatter. The discrimination among the cell types is apparent on a bivariate dot plot. This measurement is roughly equivalent to sensing the scattering matrix element S_{12}. For a very small homogeneous particle passing through a flow cytometer with a vertical linearly polarized laser beam, the light scattered at right angles remains vertically polarized. For a larger particle or one with significant internal structure, some of the vertically polarized light is converted into horizontally polarized light.

A method for making scattering matrix element measurements in a flow cytometer has been presented (Sloot *et al.* 1989). The investigators show an optical configuration for making the measurements and develop the theory to take into account the tightly focused laser beam and the use of large area scattered light detectors. Van De Merwe *et al.* (1989) measured scattering angular distributions of S_{34}/S_{11} for cuvette suspensions of *E. coli* bacteria. They showed that the measurements were strongly correlated to bacterial size. They tested the sensitivity of this measurement to changes in internal structure by inducing virus production in the bacteria. S_{34}/S_{11} did not change until the bacteria were lysed by the virus growth process. These discouraging results for cuvette suspensions do not invalidate the use of the technique for single particles in flow. It is clear, however, that a great deal of care must be taken when making such measurements and in interpreting the results in terms of cell structure.

This depolarization phenomenon has been exploited by another group of investigators to distinguish several groups of eukaryotic marine phytoplankton (Olson *et al.* 1989). Separate photodiode detectors were placed above and below the center of the laser beam to intercept scattered light between 10 degrees and 19 degrees. This is apparently far enough from the forward direction for polarization differences to be observable. The upper detector was covered with a vertically oriented polarizer and the lower detector was covered with a horizontally oriented polarizer. Coccolith-covered cells were readily distinguished from naked cells and loose Coccoliths of the coccolithophorid *H. carterae*.

OTHER APPLICATIONS OF LIGHT SCATTERING IN FLOW CYTOMETRY

E. coli bacteria have been analyzed in a microscope-based flow cytometer using light scattering and DNA content measurements with the dyes ethidium bromide and Mithramycin (Boye *et al.* 1983). The light scatter signal was shown to be proportional to bacterial protein content.

Detailed forward scattering measurements in a flow cytometer were used to distinguish among four species of microalgae (Price *et al.* 1978). Scattered light was collected at 32 angles in the forward direction between 0.2 degrees and 20 degrees. Cluster analysis was used to distinguish among the complex scattered light patterns from the different microalgae.

Forward and perpendicular light scatter have been used to provide information on the size and shape of phytoplankton cells (Trask *et al.*, 1982). The investigators also used intrinsic chlorophyll fluorescence and the DNA stains Hoechst 33342 and DAPI. These five measurements on each cell enabled clear distinction among eight algae species.

Forward and perpendicular light scatter in a flow cytometer were used to study intracellular gas vacuoles in blue-green algae (Dubelaar *et al.*, 1987). The number of gas vacuoles in the cells was controlled by external pressure changes. The investigators showed that the presence of gas vacuoles caused a 10-fold increase in perpendicular light scatter and a 5-fold decrease in forward light scatter.

Spinrad and Brown (1986) used a flow cytometer to measure the relative real refractive index of a number of marine microorganisms. They measured the light scattered in the forward direction and near 90 degrees by pure populations of the microorganisms and by calibration beads with known refractive indices. They used the ratios of the light scattered in the two directions to determine the refractive indices of the microorganisms. Their measurements agreed with earlier bulk measurements but showed significant variation among the species of microorganisms. Ackleson and Spinrad (1988) used a Mie theory model of the scattering process to estimate the size and refractive index of several nanoplankton species.

Flow cytometry has been used to study dilution-induced changes in the light scattered by the phytoplankter *D. tertiolecta* (Ackleson *et al.*, 1988). The data show large changes in side scatter at nearly constant cell volume within less than an hour after a large change in the suspension cell concentration. Another group has developed a theoretical model to explain the changes in forward scatter observed when lymphocytes undergo osmotic shock (Sloot *et al.*, 1988).

Flow cytometric forward scattering and orange chlorophyll fluorescence were used to identify a new prochlorophyte that is abundant in the ocean euphotic zone (Chisholm *et al.*, 1988). Button and Robertson (1989) used forward scattering in a flow cytometer to assist in measuring the kinetics of nutrient uptake by microorganisms.

SUMMARY AND CONCLUSION

The physical phenomenon of light scattering is presented in terms of collections of tiny dipoles being excited by the incident light and reradiating the light toward a detector. The scattered wavelets are summed taking into account the phase relationships among them. Linear and circular polarization concepts are introduced. The scattering matrix formalism is presented. The coupled dipole approximation for modelling the scattering from irregularly shaped objects is discussed in some depth and examples given comparing this approximation to Mie theory, a exact model for homogeneous spheres and cylinders. The concept of phase differential scattering is introduced as a method for making measurements of a combination of interesting scattering matrix elements. Finally, several examples are given of the use of scattered light polarization measurements in flow cytometry.

A brief review is then given of some of the other applications of forward and perpendicular light scattering from cells in flow cytometers. These include measurements of refractive index and the effects of osmotic shock, among others.

Flow cytometric light scattering is a useful tool to assist aquatic scientists in understanding the cells they are studying. However, care must be exercised in interpreting the results of light scattering experiments.

ACKNOWLEDGMENT

This work was performed under the auspices of the U.S. Department of Energy and was supported by National Institute of General Medical Sciences grant GM26857, by National Cancer Institute grant CA54518, and by the National Flow Cytometry and Sorting Research Resource Grant RR01315.

REFERENCES

Ackleson SG, Spinrad RW (1988) Size and refractive index of individual marine particulates: a flow cytometric approach. Appl Opt 27:1270-1277

Ackleson SG, Spinrad RW, Yentsch CM, Brown J, Korjeff-Bellows W (1988) Phytoplankton optical properties: flow cytometric examination of dilution-induced effects. Appl Opt 27:1262-1269

Aden AL, Kerker M (1951) Scattering of electromagnetic waves from two concentric spheres. J Appl Phys 22:1242-1246

Bohren CF, Huffman DR. (1983) Absorption and scattering of light by small particles. New York: John Wiley & Sons

Bohren, CF. (1987) Clouds in a glass of beer. New York: John Wiley & Sons

Boye E, Steen HB, and Skarstad K (1983) Flow cytometry of bacteria: a promising tool in experimental clinical microbiology. J Gen Microbiol 129:973-980

Brunsting A, Mullaney PF (1972) Light scattering from coated spheres: models for biological cells. Appl Opt 11:675-680

Brunsting A, Mullaney PF (1974) Differential light scattering from spherical mammalian cells. Biophys J 14:439-453

Button DK, Robertson BR (1989) Kinetics of bacterial processes in natural aquatic systems based on biomass as determined by high-resolution flow cytometry. Cytometry 10:558-563

Chisholm SW, Olson RJ, Zettler ER, Goericke R, Waterbury JB, Welschmeyer A (1988) A novel free-living prochlorophyte abundant in the oceanic euphotic zone. Nature 334:340-343

deGrooth BG, Terstappen LWMM, Puppels GJ, Greve J (1987) Light scattering polarization measurements as a new parameter in flow cytometry. Cytometry 8:539-544

Druger SD, Kerker M, Wang DS, Cooke DD (1979) Light scattering by inhomogeneous particles. Appl Opt 18:3888-3892

Dubelaar GBJ, Visser JWM, Donze M (1987) Anomalous behaviour of forward and perpendicular light scattering of a cyanobacterium owing to intracellular gas vacuoles. Cytometry 8:405-412

Jackson JD (1975) Classical electrodynamics. New York: John Wiley & Sons

Johnston RG, Singham, SB, Salzman GC (1986) Phase differential scattering from microspheres. Appl Opt 25:3566-3572

Johnston RG, Singham, SB, Salzman GC (1987) Zeeman laser scattering: a new light scattering technique. In: Nicolini C, ed. Structure and Dynamics of Biopolymers. Boston: Martinus Nijhoff, 56-65

Johnston RG, Singham, SB, Salzman GC (1988) Polarized light scattering. Comments on molecular and cellular biophysics. 5:171-192

Kerker M, Cooke DD, Chew H, McNulty PJ (1978) Light scattering by structured spheres. J Opt Soc Am 68:592-602

Olson RJ, Zetter ER, Anderson OK (1989) Discrimination of eukaryotic phytoplankton cell types from light scatter and autofluorescence properties measured by flow cytometry. Cytometry 10:636-643

Price BJ, Kollman VK, Salzman GC (1978) Light-scatter analysis of microalgae. Correlation of scatter patterns from pure and mixed asychronous cultures. Biophys J 22:29-36

Purcell EM, Pennypacker CR (1973) Scattering and absorption of light by nonspherical dielectric grains. Astrophys J 186:705-714

Salzman GC, Singham SB, Johnston RG, Bohren CF. (1990) Light scattering and cytometry. In: Melamed MR, Lindmo T, Mendelsohn ML, eds. Flow cytometry and sorting. 2nd ed. New York: Wiley-Liss, 81-107

Singham SB, Bohren CF (1987) Light scattering by an arbitrary particle: a physical reformulation of the coupled dipole method. Opt Lett 12:10-12

Singham SB, Salzman GC (1986) Evaluation of the scattering matrix of an arbitrary particle using the coupled dipole approximation. J Chem Phys 84:2658-2667

Sloot PMA, Hoekstra AG, Figdor CA (1988) Osmotic response of lymphocytes measured by means of forward scattering: theoretical considerations. Cytometry 9:636-641

Sloot PMA, Hoekstra AG, van der Liet H, Figdor CG (1989) Scattering matrix elements of biological particles measured in a flow through system: theory and practice. Appl Opt 28:1752-1762

Spinrad RW, Brown JF (1986) Relative real refractive index of marine microorganisms: a technique for flow cytometric estimation. Appl Opt 25:1930-1934

Trask BJ, van den Engh GJ, Elgershuizen JHBW (1982) Analysis of phytoplankton by flow cytometry. Cytometry 2:258-264

Van De Merwe WP, Huffman DR, Bronk BV (1989) Reproducibility and sensitivity of polarized light scattering for identifying bacterial suspensions. Appl Opt 28:5052-5057

Microbial ecology through individual cell analysis

BACTERIAL INFLUENCE ON THE VARIABILITY IN THE OCEAN'S BIOGEOCHEMICAL STATE: A MECHANISTIC VIEW

F. Azam and D. C. Smith

Marine Biology Research Division
Scripps Institution of Oceanography
University of California, San Diego
La Jolla, California 92093
USA

INTRODUCTION AND STATEMENT OF THE PROBLEM

Traditionally, the only component of the microbial communities studied seriously by biological oceanographers was phytoplankton. Bacteria, protozoa, viruses and fungi were relegated to "specialists" and it was implicitly assumed that, by and large, the ocean's productivity mechanisms and biogeochemical dynamics could be understood and modelled without taking into account the activities of these other components of the pelagic microbiota. This view was challenged by the seminal paper of Pomeroy (1974) and by quantitative studies inspired by it, leading up to the discovery that the pathway DOM --> bacteria --> protozoa --> metazoa (microbial loop; Azam *et al.*, 1983) is a major route for the flows of material and energy in pelagic marine ecosystems. Further, very recently it has been suggested that viruses may also play important ecological roles (Bergh *et al.*, 1989; Proctor and Fuhrman, 1990; Suttle *et al.*, 1990), including a role in the mortality of bacteria and phytoplankton.

Since the heterotrophic bacteria process a large fraction of photosynthetically produced organic matter, it is now widely appreciated that we need to understand the ecosystem-level consequences of bacterial metabolism and growth. We now know that organic matter metabolism by bacteria can have substantial bearing on fundamental issues of foodweb organization, nutrient cycling, primary productivity, and the vertical flux of organic matter. The new challenge is to

NATO ASI Series, Vol. G 27
Particle Analysis in Oceanography
Edited by S. Demers

incorporate these roles of bacteria into mechanistic models designed to understand variability in the ocean's biogeochemical state.

Bacteria had been assigned the roles of remineralizers and particle-formers (via DOM catabolism and assimilation). Recently, however, it has been found that bacteria play variable ecosystem roles depending on the nutritional status of the environment. They can be remineralizers of plant nutrients but also compete for nutrients with phytoplankton (Azam *et al.*, 1983; Wheeler and Kirchman, 1986; Tupas and Koike, 1991). Bacteria can be food for animals, but they also compete with particle-feeding zooplankton for POM by solubilizing POM with hydrolytic exoenzymes (Azam, 1984; Azam and Cho, 1987; Pomeroy and Wiebe, 1988). Obviously, these "role reversals" pose difficulties in conceptualizing and modelling pelagic ecosystems. For example, it would be a major uncertainty indeed if we did not know whether to treat bacteria as a source or a sink for ammonium in attempting to understand biogeochemical dynamics of an ecosystem where primary production was limited by the supply of regenerated nitrogen. Likewise, whether bacteria enrich detritus with respect to nitrogen or whether they enzymatically remove organic nitrogen from the particles will have quite different consequences for biogeochemical dynamics of nitrogen and f-ratios.

How then do we accommodate such qualitative variability in the roles of bacteria into our concepts of the pelagic ecosystems? We think that the problem is symptomatic of a broader issue in the study of pelagic ecosystem dynamics. The traditional approach has been to focus on discipline-specific themes and study them in isolation of other ecosystem components. Consequently, *our models lack the mechanistic complexity of "real world" interactions considered at appropriate space and time scales*. There has been progress in this respect in studies of larger organisms and mesoscale patchiness of animals and their process-rates is now generally accepted. Indeed, there has been a revolution in zooplankton ecology resulting from the discovery that pelagic zooplankton are not just filtering machines, but have sophisticated feeding behaviours including selectivity (Strickler, 1982). However, it remains "business-as-usual" in the study of the ecology of microorganisms; our concepts are based on the assumption that microbes and their nutrient molecules are randomly distributed in space.

Further, the microbes are treated as being devoid of behavioral responses to nutrient and predator-prey fields.

Recent focus on oceanic biogeochemical cycles has made the status quo untenable, by seeking to understand how the biochemical forces influence the *spatial-temporal patterns* of the distribution of N, P, and C atoms. True, biogeochemists are interested in variability at large space and time scale. However, one can justifiably assert (and this is an argument developed in this paper) that the metabolic activities of *individual cells* of phytoplankton, bacteria and protozoa, in large part, form the basis of the ocean basin scale features of biogeochemical dynamics. It is self-evident, for instance, that primary productivity is not a mesoscale process; it is the activity of individual phytoplankton cells, and as such it is regulated at the single cell level by the physicochemical environment in which a particular phytoplankton cell occurs at a given time. Interestingly, our acceptance of the importance of microscale processes is also "patchy". For example, no self-respecting phytoplankton ecologist today would assume that phytoplankton cells are exposed to a constant light field during the light period. On the other hand, paradoxically, our conceptual models still ignore that, in all probability, individual phytoplankton cells experience *remineralized nutrient fields* quite different than those in the bulk-phase. This is in spite of the knowledge that much of the nutrient remineralization is done by bacteria and protozoa and therefore their microscale distribution in relation to the phytoplankton influences the remineralized nutrient field experienced by the phytoplankton cell. Microbes do not interact at mesoscale space; their metabolic interactions occur in spatially intimate microenvironments, and this should result in spatial heterogeneity in the intensity of metabolic processes.

In this paper we examine the proposition that the variability of ecosystem roles of bacteria results from bacteria being in microzones of different nutrient statuses. We suggest that microscale variations in bacterial environments are caused by the presence of discrete loci of organic matter, notably phytoplankton cells and organic particles. We stress the need to focus on the biochemical and behavioral adaptations of bacteria in an ecosystem context, rather than piecemeal study of particular activities of bacteria, if we are to develop a unified concept of the role of bacteria in pelagic marine ecosystems. In other words, the ecosystem role of bacteria should be seen as a consequence of what bacteria, as species with specific biochemical adaptations and evolutionary history, must do to persist in the pelagic ocean environment. The need to focus on the microenvironment is all but obvious, given that bacteria live, grow and die in microenvironments.

ORGANIC MATTER IN THE ENVIRONMENTS OF BACTERIA

In order to develop a unified concept of bacteria-organic matter interactions, and in keeping with our focus on biochemical adaptations, we seek to develop a view of organic matter which is based on what the bacterium "sees" in its environment. Bacteria respond to generalized physicochemical signals and search images and do not make a distinction per se between living and dead materials. Bacteria might metabolically (e.g. with enzymes) attack phytoplankton just as they attack detritus (unless phytoplankton can defend themselves; Pomeroy and Wiebe, 1988). Phytoplankton, detritus, colloids and DOM can be considered as creating an organic matter field within which bacteria operate to optimize survival and persistence. Predators of bacteria (protozoa on free-living and metazoa on attached bacteria) within the organic matter field may further modify the strategies of bacteria to optimize survival and persistence.

In the euphotic zone, a bacterium is within 100s of μm from the nearest phytoplankton (Azam and Ammerman, 1984) and even closer to a detritus particle. Chemotaxis has been demonstrated in marine isolates (Chett and Mitchell, 1976; Paerl and Gallucci, 1985) and natural assemblage (discussed in Azam and Ammerman, 1984; Cho, 1988) although only a small fraction of bacterial assemblages may be chemotactic at any given time (5-35%; Cho, 1988). Swimming speeds in natural assemblages in one observation were on the order of 30 μm sec^{-1} (Ammerman and Azam, 1984). Cells may also become clustered as a result of turbulence and sheer, independent of biological mechanisms (Mitchell *et al.*, 1990). So, bacteria are quite close to the particulate loci of organic matter, and it would be surprising indeed if they did not interact with them. Also, the intervening space contains colloids and polymers which may, in part, be sloughing off from the surfaces of phytoplankton and detritus (below), and with which bacteria may also interact along the way.

It is customary to distinguish between the dissolved and particulate organic matter pools in seawater on the basis of filterability. However, recent reports show that abundant colloidal particles span the size-range between the dissolved and the particulate phases (Koike *et al.*, 1990). For the purpose of understanding the interactions of bacteria with the organic matter, it is useful to think of the organic matter pool as a continuum, containing concentration gradients but no rigid demarcations between phytoplankton, detritus and dissolved polymers. In this view, macromolecules emanating from the cell surface or from dissolving detritus particles merge and mingle with the macromolecules in solution thus creating a polymer field. Phytoplankton cells, especially under nutrient limitation, secrete polysaccharides or mucopolysaccharides (Mykelstad, 1977) which may have the characteristics of a gel. As one moves away from the surface of the phytoplankton cell and into the surrounding water, the polymeric matrix in the seawater would become less concentrated and less viscous. One might even hypothesize that bacteria use viscosity gradients to sense their environment and respond by making "metabolic decisions" (Belas *et al.*, 1986).

This view emphasizes the role of polymers in the spaces between particulate loci. We now know that the concentration (w/v) of polymeric utilizable dissolved organic matter (UDOM) such as protein, polysaccharides, DNA and RNA is roughly an order of magnitude higher than the corresponding monomer pools. Using literature values of the concentration of these polymers in seawater, we calculated, assuming one linear molecule for each pool, the length of DNA, protein, and polysaccharide to be on the order of 2, 310, and 5600 km ml^{-1}, respectively! This is equivalent to a one centimeter cube of seawater being criss-crossed 10^5 - 10^8 times by each of the polymers, creating a polymer meshwork. We are not implying that each polymer pool is composed of a single molecule, and the molecules are certainly folded and branched to some extent, but the calculations give an impression of the intricacy of the matrix with which bacteria interact. Metabolic interactions of bacteria with the polymeric substrates would involve hydrolytic cleavage (discussed later), altering the structure of the matrix.

PHYTOPLANKTON-BACTERIA INTERACTIONS

The "purpose" of phytoplankton exudation

Bacteria-phytoplankton interactions have been a major theme in marine microbiology for decades. There is a large volume of data on the flow of organic matter via exudation by phytoplankton and there is a rich literature on the subject which has been widely debated and reviewed (e.g. Bjϕrnsen, 1988; Williams, 1990). However, despite extensive research there continues to be a lack of consensus on the magnitude and particularly on the ecological and evolutionary significance of DOM exudation by phytoplankton. Bjϕrnsen (1988) has argued that phytoplankton exudation is a "property tax", the price phytoplankton must pay for maintaining high metabolite pools which inexorably tend to leak out into seawater along an extremely high downward concentration gradient. Williams (1990) argued that exudation has evolved to put a substantial part of the photosynthate into the dissolved phase in order to reduce N loss through vertical flux. We have hypothesized that phytoplankton exudation has evolved as a mechanism to establish mutualism with bacteria (Azam and Ammerman, 1984; Azam and Cho, 1987); phytoplankton provide reduced carbon for bacterial metabolism while bacteria produce remineralized nutrients for phytoplankton. Cell surface mucus exudation may also serve to keep bacteria from attacking the phytoplankton cell. The foregoing sampling of ideas illustrates the divergence of views and the lack of a generally accepted theory regarding the "purpose" of exudation by phytoplankton. We will further explore our explanation to see if it may serve as a basis for a theory of phytoplankton-bacteria interaction.

Phytoplankton-bacteria mutualism

Bratback and Thingstad (1985), from observations on P-limited mixed bacteria-phytoplankton chemostats, discovered a paradoxical behaviour of phytoplankton. As P declined, bacteria became strong competitors with phytoplankton, but at that time the phytoplankton secreted carbohydrate which encouraged bacteria to further deplete P, sequestering most of it (Fig 1a). A natural analog of this interaction in N-limited waters is shown in Fig. 2a. In highly oligotrophic waters, bacterial biomass can actually be greater than phytoplankton biomass

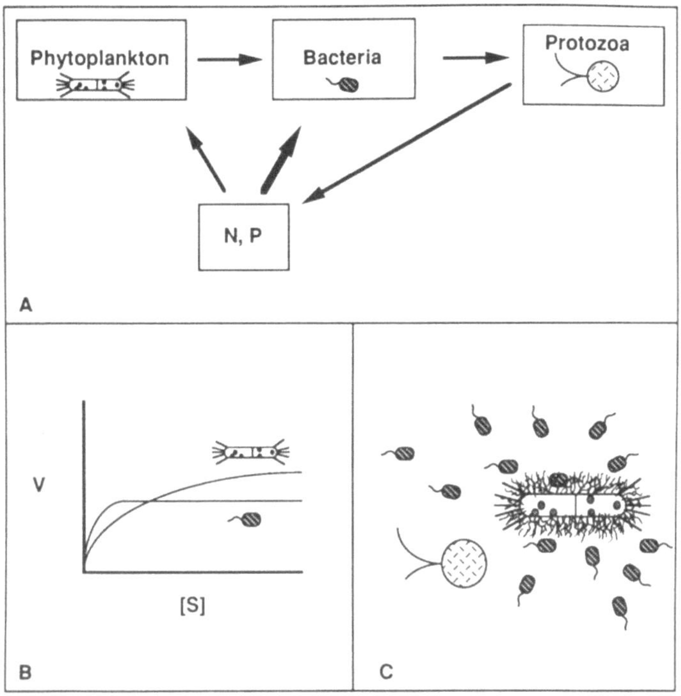

Figure 1. **A**) Diagram depicting competition between bacteria and phytoplankton for nitrogen and phosphorus at low concentrations. Bactivory by flagellates increases the mineralization of N and P. **B**) Uptake *vs* substrate concentation diagram illustrating bacterial competitive advantage at low substate concentrations while phytoplankton outcompete at higher substrate concentrations. **C**) Cartoon showing mineralization via flagellate bactivory in the phytoplankton cell's microenvironment.

(Fuhrman *et al.*, 1989; Cho and Azam, 1990); here most N will occur sequestered in bacteria. This may be because bacteria have low K_m - low V_{max} (low flow) while phytoplankton have high K_m - high V_{max} (high flow) uptake systems for ammonium; consequently, bacteria outcompete phytoplankton for ammonium uptake at low ammonium levels but the opposite holds true at high ammonium levels (Fig 1b). Thus, in low ammonium (micro)environments, using the energy supplied by the phytoplankton, bacteria outcompete phytoplankton for N and thereby dominate the biomass. Parenthetically, we note that high the relative biomass of bacteria in oligotrophic waters (Fig. 2a) may explain the higher than expected diffuse attenuation coefficient on the basis of chl *a* (Phinney and Yentsch, 1985; and Fig 2b); this has implications for remote sensing of phytoplankton biomass.

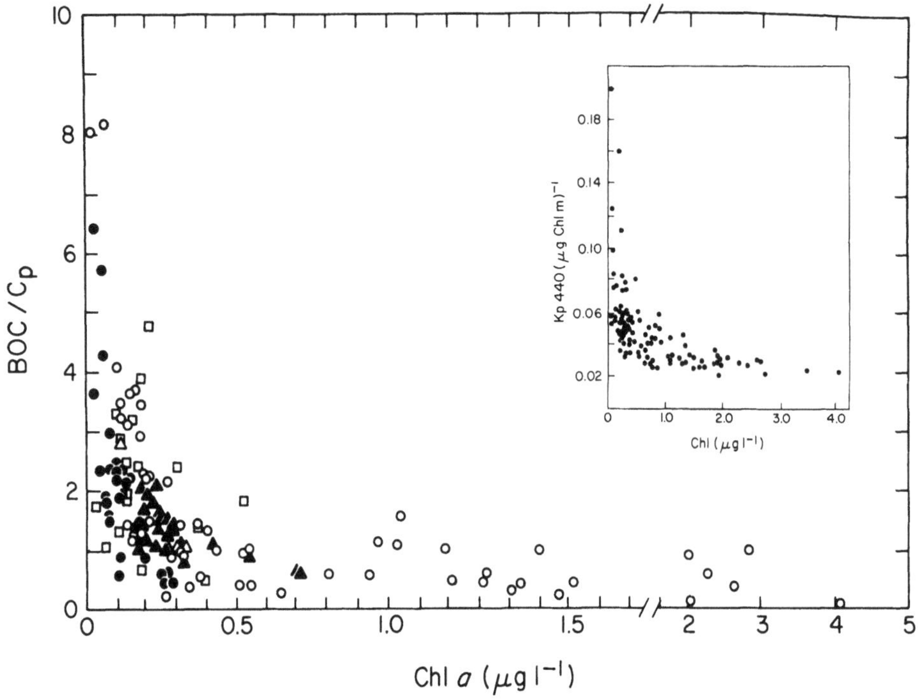

Figure 2. A) Plot of bacterial carbon/phytoplankton carbon (BOC/Cp) *vs* chlorophyll *a* for euphotic zone samples from: central North Pacific gyre (●), Southern California Bight (□: C_p was calculated as POC x 0.158 + POC^2 x 0.0007; ○: C_p was calculated as chl *a* x 50), and along a transect from San Pedro to San Diego (▲: Cp calculated as chl *a* x 50). From Cho and Azam (1990). **B)** Plot of chl *a* vs chl *a*-specific diffuse attenuation coefficient, K_p. Redrawn from Phinney and Yentsch (1986).

Such bacteria-phytoplankton interaction is intriguing only if considered in isolation of the rest of the ecosystem. Bratback and Thingstad (1985) point out that nutrients would be released as a result of predation of protozoa on bacteria (Fig 1c). So the role of bacteria in oligotrophic waters would be to efficiently concentrate the nutrients which become available to phytoplankton upon remineralization. Variation in protozoan predation will influence nutrient cycling; thus it is important to study intact microbial consortia. Interestingly, ammonium remineralization by protozoa may not change the concentration-dependent competitive advantage of bacteria over phytoplankton since the released ammonium will again diffuse to low levels where bacteria have the kinetic edge. So, the paradox of Bratbak and Thingstad persists.

A solution for the phytoplankton cell is to attract bacteria in high abundance in its microenvironment (Azam and Cho, 1987). Intense remineralization (directly by bacteria and due to predation by protozoa on bacteria clustered around a phytoplankton cell) could sustain high ammonium levels in the microenvironment thus switching the kinetic advantage over to phytoplankton. However, Jackson (1989) and Mitchell *et al.* (1985) have argued that exudate diffusion and rapid sinking of phytoplankton prohibit the formation of bacterial clusters. This problem may be solved by phytoplankton exuding surface-bound (rather than diffusible) exudates such as polysaccharides. We suggest that the polysaccharide layer serves two purposes: it attracts bacteria but also keeps them at "arm's length".

Phytoplankton are known to exude cell surface-bound polysaccharides and the exudation increases under nutrient limitation (Mykelstad, 1977). The polysaccharide layer may become enriched with DON and DOP by scavenging protein and other DON and DOP pools from seawater (Azam and Cho, 1987). So, the phytoplankter may, simply by exuding cell-surface polysaccharide, surround itself with a bacterial "culture medium" rich in energy, nitrogen and phosphorus. Bacteria have cell-surface hydrolytic enzymes, such as proteases, polysaccharidases and nucleases, which can hydrolyse the polymers on the phytoplankton surface. Since bacterial exohydrolases are surface-bound, physical contact of bacteria with phytoplankton is required for hydrolysis. However, the phytoplankton remains essentially bacteria-free because polymer hydrolysis cuts bacteria loose from the point of attachment; bacteria attacking the phytoplankter thus find themselves on a mucus treadmill.

Variable roles of bacteria in phytoplankton microenvironments

Our hypothetical scenario can explain how bacteria may be both remineralizers of and competitors for ammonium (Fig. 3a.). Bacteria are net remineralizers when their environment is rich in organic nitrogen and energy sources (Wheeler and Kirchman, 1986; Tupas and Koike,

1990) but compete with phytoplankton at low ammonium. Bacteria feeding in the phytoplankton surface layer would be in a microenvironment conducive to net ammonium release while bacteria in the bulk-phase will be competitors with phytoplankton for ammonium. Further, the high ammonium level in the phytoplankter's microenvironment will kinetically favor phytoplankton because of their high-flow uptake systems; the low-flow systems of bacteria would already be saturated at lower ammonium levels. Enhanced ammonium uptake would alleviate phytoplankton N-limitation and consequently enhance primary productivity.

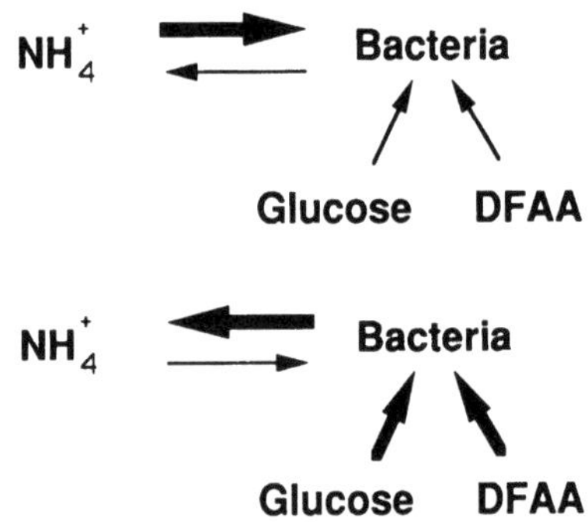

A

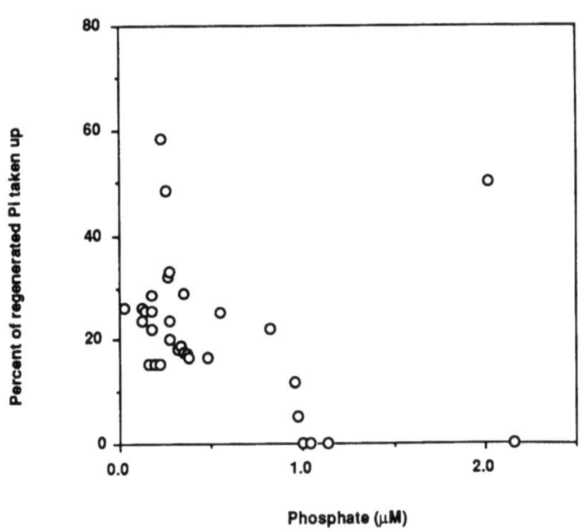

B

Figure 3. **A**) Diagram illustrating the "role reversal" of bacteria in NH_4^+ dynamics: net uptake of NH_4^+ at low glucose and dissolved free amino acid (DFAA) concentrations (top), and net release of NH_4^+ at high glucose and DFAA concentrations (bottom). **B**) Coupled uptake of orthophosphate (Pi) regenerated by 5'-nucleotidase *vs* phosphate concentration. This curve demonstrates the "role reversal" of bacteria in Pi dynamics. Hydrolysis and uptake are tightly coupled at low Pi concentrations but uncoupled at high Pi concentrations. Modified from Ammerman and Azam (in press).

Phosphorus remineralization and its availability to phytoplankton may also involve bacteria-phytoplankton interactions as discussed in relation to the paradox of Bratback and Thingstad (1985). Ammerman and Azam (1985; manuscript) have shown that bacterial 5'-nucleotidase plays a major role in P remineralization. DNA and RNA, which are substantial pools of DOP, are hydrolysed to nucleotides on bacterial cell surface by surface-bound nucleases (Paul *et al.*, 1988). The nucleotides are hydrolysed by the (periplasmic?) 5'-Nase to liberate Pi (so in contrast with N remineralization, the remineralization of P is extracytoplasmic). A variable fraction of the liberated Pi is taken up by the bacterium. Interestingly, the hydrolysis-uptake coupling decreases with increasing Pi concentration in the environment (Fig. 3b) following isotope dilution curves. Essentially, at high environmental levels of Pi, the Pi-permease molecules tend to be occupied by the Pi diffusing into the periplasm from the environment, thus reducing the uptake of periplasmically produced Pi. As the environmental Pi level decreases, the hydrolysis-uptake coupling becomes tighter. This means that bacteria become competitors at low environmental Pi, competing for the environmental Pi but not releasing much Pi produced periplasmically at the expense of DOP. At high environmental Pi, on the other hand, because of loose hydrolysis-uptake coupling, bacteria become remineralizers, releasing a larger fraction of the remineralized Pi into the environment. Once a high net level of Pi is generated in the bacterium's microenvironment (e.g. by predation on bacteria by protozoa; Fig. 1c) then the hydrolysis-uptake coupling becomes loose and the release of remineralized Pi sustains the high Pi level in the microenvironment. Therefore, the system is "autocatalytic" in a kinetic sense. This mechanism explains, at biochemical level, how bacteria can move smoothly between being strong competitors or being efficient net remineralizers of P.

Creation of nutrient rich microzones

It is noteworthy in this connection that phytoplankton grow quite rapidly ($\sim 1\ d^{-1}$) in waters with unmeasurable ($\leq$ few Nm) N (McCarthy and Goldman, 1979; Laws *et al.* 1984). Several explanations to account for this observation have been advanced (e.g. exposure to plumes of zooplankton excretion, McCarthy and Goldman, 1979; and association of phytoplankton with marine aggregates, also known as the "Spinning Wheel Hypothesis", Goldman, 1984). While these hypotheses seek to create nutrient-rich patches via environmental variables, our hypothesis assigns the task to the phytoplankton cell itself: that the phytoplankton cell can by itself by secrete a cell-surface polysaccharide creates and sustains a high ammonium patch in its microenvironment. The advantage to bacteria would be that they are offered a rich "culture medium" on the phytoplankton surface.

We should also point out that enriched microenvironments conducive to ammonium release are, in all probability, not limited to phytoplankton cell surface layers. Particle loci including the pico- and nano-detritus of appropriate nutritional status may also act as loci for ammonium release (discussed later).

Implications for phytoplankton aggregation

The phytoplankton-bacteria interactions described above have implications for phytoplankton aggregation ("marine blizzard" of Smetacek, 1985) as well. Smetacek argues that aggregation and mass sinking represents a survival strategy: upon sinking, phytoplankton make resting forms which may be advected back to the surface later and serve as seed populations. We suggest an alternative hypothesis that, because phytoplankton cells (must) express the surface "mucus" layer, they have a tendency to aggregate continuously. Bacterial exoenzymes minimize aggregation by keeping the polymer layer under check. Further, at low cell densities, collision frequency is low and hence aggregations minimal. Early on in the bloom, aggregation is low despite high cell density because bacteria keep the phytoplankton surface "enzymatically pruned". Also during this period, intense N remineralization in the phytoplankter's microenvironment sustains phytoplankton vitality and growth at the expense of DON. Later, the utilizable DON runs low due to sinking losses of N, and the bacteria, now in the N-poor but energy rich polysaccharide layer on the phytoplankton, become competitors and cease to release ammonium. The N-stressed phytoplankton cells keep secreting polysaccharides, vainly trying to maintain mutualism (Bratback and Thingstad, 1985), and become sticky, aggregate, and sink. Bacteria "move in", colonize the mucus, and with their own glues make the aggregates even more sticky (creating the situation described by Biddanda and Pomeroy, 1988). (Here again one sees a variable role of bacteria: first they prevent aggregation and later in the bloom they enhance it).

In this scenario, and in contrast to the hypothesis of Smetacek (1985) the phytoplankton cells are actually dying under the stress of competition and hydrolytic attack; they may begin to autolyse and lose their organic matter as DOM during descent through the water column. Our scenario is not one of serene descent of phytoplankton escaping the vicissitudes of nutrient limitation in the surface layer, but rather mass mortality on the hands of unforgiving and biochemically

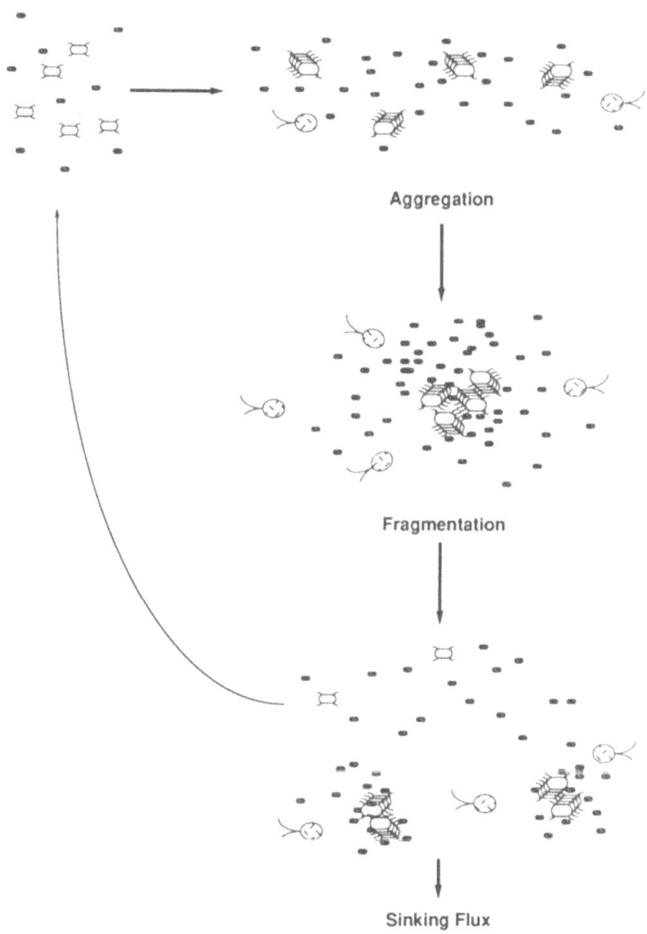

Figure 4. Cartoon depicting phytoplankton cells, producing extracellular mucopolysaccharides under nutrient stress, aggregating and sinking. Bacterial mineralization and protozoan bactivory on these aggregates results in intense mineralization supporting cryptic growth of some algal cells. These cells may avoid sinking out of the euphotic zone as the aggregate fragments due to hydrolytic enzyme activity.

diverse bacteria once the defenses of the phytoplankter are lost. Simon *et al.* (1990) showed that diatom-based marine snow was largely detrital; phytoplankton plus bacteria biomass carbon accounted for only 1-13% of the total aggregate carbon, possibly (but not necessarily) the detritus-dominance reflected mass-mortality of phytoplankton. Yet, the mass-mortality of aggregated phytoplankton may in itself serve as an adaptive strategy via cryptic growth (Fig. 4). Phytoplankton snow aggregates develop high nutrient levels, probably due to intense remineralization of the dying cells. The regenerated nutrients could support robust growth (even luxury uptake of nutrients) of a small fraction of the phytoplankton population in the aggregate while the aggregate is in the euphotic zone. Some of these nutrients replete and the growing fraction of phytoplankton might be released from the aggregate and serve to temporally stretch out the persistence of the phytoplankton species in the environment. (Bacterial progeny may also be released during this period; discussed later). Once below the euphotic zone, the aggregate's metabolism will become dominantly heterotrophic; bacteria will take over and kill most (but perhaps not all) phytoplankton. Some may survive to make resting stages as suggested by Smetacek. Perhaps we should think in terms of multiple strategies of phytoplankton (and bacteria) for survival and persistence in the wake of a bloom.

BACTERIA-DETRITUS INTERACTIONS

Bacteria as competitors with metazoa: biogeochemical implications

Bacterial colonization is often thought of as a way to enrich the detritus with N and P to improve its nutritional quality for detritivorous metazoa. This notion is simplistic when considered from the standpoint of the biochemical adaptations of bacteria. The extensive literature (reviewed by Hoppe, 1984; Pedrós-Alió and Brock, 1983) on the adaptive value for bacteria that attach to particles will not be recounted here. We note, however, that producing progeny on a particle so it might all be eaten by the animals is difficult to rationalize as adaptive. A fundamental departure from this view is the suggestion to consider bacteria not as "helpers" of metazoa but as avid competitors with them for detritus as a common source of nutrition (Azam, 1984; Azam and Cho, 1987; Pomeroy and Wiebe, 1988). In this section we further develop this argument with focus on the probable biochemical adaptations of bacteria in POM utilization in an ecosystem context. In so doing we also attempt to reconcile the variable ecosystem roles of bacteria: while bacteria on particles are potential "metazoan meals", they are also in the process of competing with metazoa for the common resource.

POM may be consumed by animals or by bacteria, or it may sink out of a given water layer. While particle-consuming metazoa utilize POM directly, the osmotrophic bacteria can not. Bacteria must first solubilize the particles and then take up the organic matter in the form of monomers or sometimes oligomers. This absolute requirement for particle solubilization, generally monomerization, imposes a mechanistic constraint on the role heterotrophic bacteria can play as a conduit for the flow of carbon from POM. If bacteria can not exert appropriate and sufficient "biochemical pressure" to solubilize the particles, then the other pathways of carbon flow, namely ingestion by animals and sinking, will be favored. It is of interest to find out whether bacteria manage to act as a significant route for carbon flow in the pathway POM --> DOM --> bacteria, and if so, then what mechanisms couple POM with bacterial uptake systems. Further, do attached (sinking) bacteria use POM in a tightly coupled manner, or does a significant fraction of DOM leak out and is used by free-living (non-sinking) bacteria? Solving these problems is necessary for understanding how, and how much, bacteria influence the pathways of biogeochemical cycling of materials in the ocean's water column including the mesopelagial.

Role of bacterial exohydrolases in particle solubilization and hydrolysis - uptake coupling

Bacteria solubilize particles and polymers with hydrolytic extracellular enzymes, such as proteases and polysaccharidases (Priest, 1985; Gottschalk, 1987). The hydrolases may be secreted (diffusible) or anchored on the cell surface (non-diffusible), and the enzymes may be endohydrolases or exohydrolases (Gottschalk, 1987). Endohydrolases cut the particle or polymer internally (e.g. a-amylase, an endoglucanase and serine protease, an endopeptidase) to generate oligomers or polymers. Exohydrolases hydrolyse monomers or dimers from the end of the polymer or suprapolymer (i.e. POM). Other hydrolases act on specific points in the polymer structure (e.g. debranching enzyme, isoamylase). In some instances, bacteria secrete enzyme complexes (e.g. cellulase with both an endohydrolase and an exohydrolase activity (in the case of cellulase, endo-ß-(1--> 4)-glucanase to internally cut the polymer and exo-ß-(1-->4) glucanase to release dimers from the non-reducing end of the oligomers). The point we wish to make is that there is a varied repertoire of hydrolases with which bacteria can attack most biogenic particles and polymers to hydrolyse them into transportable molecules.

Attack by endohydrolases may cause loose hydrolysis-uptake coupling since polymers are generated which must still be acted upon by exohydrolases to generate transportable monomers and oligomers. Exohydrolases, on the other hand, should cause tighter hydrolysis-uptake coupling (Hollibaugh and Azam, 1983). In a biogeochemical context it is important whether or not attached bacteria tightly couple POM solubilization with uptake of the liberated DOM. If they do, then POM will be solubilized and respired by the attached bacteria which will sink along with the particle. If particle solubilization and DOM uptake are loosely coupled or uncoupled, then DOM will diffuse to the *non-sinking free-living bacteria*. Thus, the nature of coupling between POM and attached bacteria could produce quite different biogeochemical scenarios of carbon flow. We should therefore find out whether or not attached bacteria tightly couple POM hydrolysis with DOM uptake and whether flux into attached bacteria is a significant pathway for carbon flow.

Carbon flow from POM to the free-living bacteria

Several studies have measured carbon demand of particle-attached bacteria, and on that basis calculated whether bacteria were important in particle decomposition (Kirchman, 1983; Hoppe, 1984; Ducklow *et al.*, 1985; Alldredge and Youngbluth, 1985; Jacobsen and Azam, 1984). These studies concluded that attached bacteria grow too slowly to account for significant POC decomposition. Karl *et al.* (1988) concluded that sinking particles were actually unsuitable habitats for bacterial growth. Cho and Azam (1988) determined carbon demand of both attached and free-living bacteria in the mesopelagic zones of 2 Pacific Ocean sites . While attached bacteria were responsible for minor carbon demand (in consonance with the above mentioned studies) *the carbon demand of free-living bacteria was so substantial that it could account for 80% or more of the depth-dissipation of sinking POC.* Most bacteria in the mesopelagial were free-living and over 90% of total bacterial carbon demand was due to these (operationally defined) free-living bacteria. Since bacteria can use only DOM ,our results indicate that *a large fraction of POM somehow becomes DOM (presumably via hydrolases of attached bacteria) but paradoxically a large fraction of the DOM produced is utilized by free-living bacteria.* These results point to a loose hydrolysis-uptake coupling on particles.

This conclusion is supported by a study of marine snow solubilization (Simon, Alldredge, and Azam,unpublished). Hand-collected marine snow from surface waters which was suspended in 0.2 μm filtered seawater from 500 m depth, released an order of magnitude more dissolved combined amino acids (DCAA) into seawater than the carbon demand of attached bacteria (a maximum estimate of DCAA utilization by attached bacteria). These observations also suggest loose hydrolysis-uptake coupling on particles.

In another study of marine snow aggregates (D. C. Smith, A. L. Alldredge, and F. Azam unpubl.), we asked whether colonized particles harbor intense enzyme activities and whether the enzymes were cell-bound there as they are known to be in free-living bacteria. Measurement of enzymatic activity on individual marine snow particles, collected by SCUBA, exhibited activities up to three orders of magnitude higher than in the surrounding bulk water for the enzymes detected (protease, phosphatase, lysozyme, ß-glucosidase, chitosidase, and N-acetyl ß-glucosaminidase). Not all enzyme activities were elevated to the same degree, suggesting that they were not merely the result of higher bacterial densities on the particles but possibly due to the expression of enzymes in response to the particle composition. Interestingly, the particles had very high soluble activity which was released into the dissolved phase; 41.8 $\pm$ 22.7% ($\pm$ S.D.) was found in the supernatant after high speed centrifugation. This soluble activity was 2-3 orders of magnitude higher than the soluble activity in the bulk water. A significant portion of the particle associated activity diffused into the water in approximately one hour. This suggests that these marine snow particles could be a source of enzyme activity into the bulk water. It also implies the existence of within-particle mechanisms for sustained production of soluble enzyme activitics. Marine snow aggregates appeared to be a source of bacteria into the surrounding water, since during the incubation large bacteria morphologically similar to those on the aggregate appeared in the surrounding water. This is consistent with the observations of Jacobsen and Azam (1984) who found that colonized fecal pellets of the copepod *Calanus pacificus* became "baby machines", sloughing off the progeny of the attached bacteria into the surrounding water. Thus the marine snow aggregates can be portrayed as permeant with diffusible hydrolytic enzymes and as a source of DOM as well as bacteria in the environment.

Biochemical strategies of bacteria in POM utilization: the "enzyme pathway" hypothesis

The above observations and arguments have led us to propose the following hypothetical scenario of bacteria-POM interactions: Bacteria attached to pelagic particles secrete diffusible endohydrolases into the particle. The hydrolase secretion dynamically reflects particle composition. Enzyme secretion rapidly solubilizes the particle into polymers, a large fraction of which diffuse out of the particle due to loose hydrolysis-uptake coupling imposed by the diffusible nature of the enzymes. DOM thus becomes accessible to free-living bacteria, particularly the progeny of the attached bacteria which is released into seawater but stays near the particle. Free-living bacteria utilize the polymers emanating from the particle via tight hydrolysis-uptake coupling by using their non-diffusable cell-bound hydrolases.

We express this hypothesis as a material flux pathway (the "enzyme pathway", Fig. 5), structured by the hydrolase activities of attached and free-living bacteria, and by whether the enzyme is a diffusible endohydrolase (E1-type) or a non-diffusible exohydrolase (E2-type): POM-(E1) --> Polymers-(E2) --> Monomers --> attached and free bacteria.

As depicted in Fig. 5, the structure of the enzyme pathway is a necessary consequence of assuming that attached bacteria secrete diffusible hydrolases but the free-living bacteria express cell-associated exohydrolases. Free-living bacteria have repeatedly been shown to have surface-bound hydrolases (Hollibaugh and Azam, 1983; Somville and Billen, 1985; Paul *et al.*, 1988). Attached bacteria may hydrolyse the particle more efficiently by secreting the enzymes. An enzyme molecule affixed to the surface of an attached bacterium has a space of activity merely of molecular dimensions, while the release of the enzyme could enormously increase its space of activity. The ability to *rapidly* solubilize the particle would be critical in the competition of bacteria with metazoa. Solubilization would be even faster if the E1-type (diffusible) enzymes were *endohydrolases*, producing polymers rather than monomers (e.g. serine protease); fewer "cuts" would suffice.

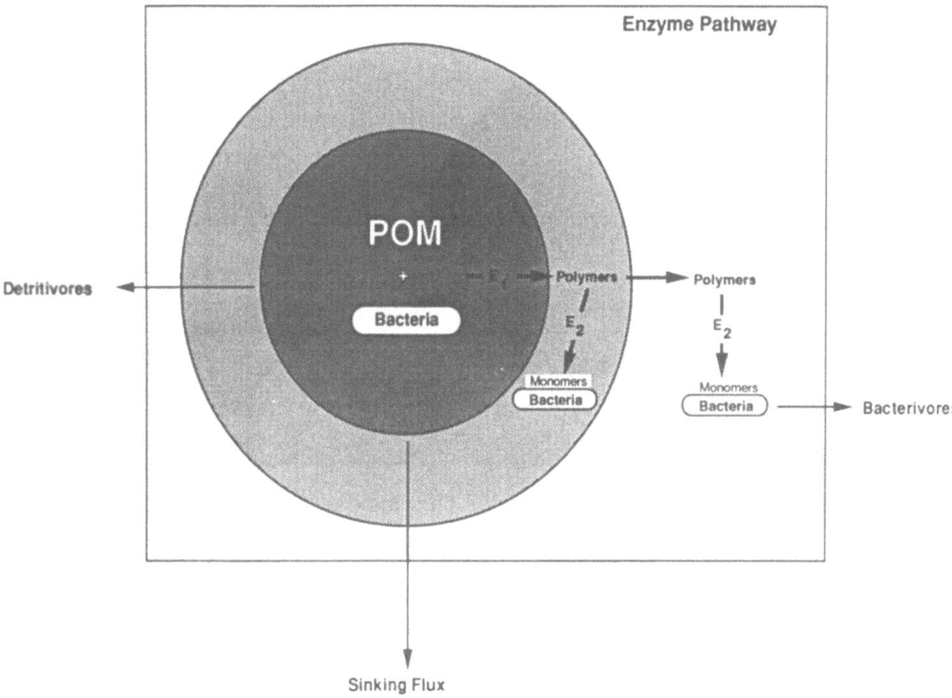

Figure 5. The "Enzyme Pathway" of carbon flow from detritus to free-living and attached bacteria. The figure depicts the hypothesized role of diffusible enzymes (E1), secreted into the particle by the attached bacteria, in particle solubilization. The diffusible nature of E1-type enzymes results in loose hydrolysis-uptake coupling and diffusive losses of DOM from the particle. The enzymes of E2-type are depicted as cell-surface attached and as being able to tightly couple polymer hydrolysis with product uptake. From Smith, Alldredge and Azam (in prep).

Enzyme secretion into the particle matrix should result in loose hydrolysis-uptake coupling. Consequently, some of the DOM produced will diffuse out of the particle and be used by free-but-near bacteria or by bacteria in the bulk-phase. Fig. 5 depicts attached bacteria to also have E2-type enzyme for utilizing polymers which become accessible to attached bacteria. The distinction between the free-but-near bacteria and free-living bacteria in the bulk-phase assumes that colonizer progeny tends not to colonize the particle but stay near it by chemotaxis. This is consistent with the suggestion of Dawson *et al.* (1981) that small starved cells readily attach to surfaces but not to nutrient replete bacteria (those in the particle's microenvironment?). Further, it is likely that antagonistic interactions on the colonized particle tend to reduce bacterial diversity. It would be an adaptive strategy for the attached bacteria to release chemicals to discourage the presence of other species in the vicinity of the particle so that the progeny of the attached bacteria may monopolize the DOM emanating from the particle.

Ingestion of firmly attached bacteria by metazoa: A survival strategy of bacteria?

The scenario of POM utilization by bacteria which we have described does not preclude the possibility, no probability, that at some stage in the life of the colonized particle a metazoan will eat it, bacteria and all. What strategies might bacteria use to "prevent death"? It has been suggested (Robison *et al.*, 1977) that luminous bacteria attached to (nutritionally spent?) particles emit light to announce their presence, so that they may pass into the organically rich fecal material where they may proliferate. A necessary condition the success of this strategy is that at least some bacteria must survive passage through the animal gut. One might hypothesize that an adaptive strategy for particle-attached bacteria would be to make some digestion-resistent variants (a thicker cell wall might suffice; e.g. gram-negative marine bacteria of the genus *Synechococcus* are not digestible by copepods; Johnson *et al.*, 1982; Silver and Bruland, 1982; Silver and Alldredge, 1982; Iturriaga and Mitchell, 1986). The digestion-resistent cells could serve as an inoculum into the metazoan feces then proliferate and propagate the normal (digestible) phenotype. Why has there not been a selection for digestion-resistence, via the production of a thick cell wall, in heterotrophic bacteria? A thick wall may limit DOM uptake, hence its acquisition would be selectively negative in heterotrophic bacteria but not in photosynthetic bacteria.

Biogeochemical implications of POM-bacteria interactions

Large-scale solubilization of POM by bacterial hydrolases could have fundamental implications for the patterns of material flux in the ocean. Carbon flow in the "enzyme pathway" involves major changes in phases and particle size-spectra: POM (particulate) --> polymers (dissolved and colloidal) --> monomers (dissolved) --> attached bacteria (particulate) and free-living bacteria (colloidal and fine particulate). These transitions could greatly increase the surface area of the particulate phases and decrease the sinking rate of particles, with implications for vertical flux of organic matter and surface-reactive materials in the ocean. It is therefore important to elucidate the magnitude and the mechanisms in the flow of carbon from POM to free-living and attached bacteria, particularly in the ocean's interior.

CONCLUDING REMARKS

We have presented the notion that variability in ecosystem roles of bacteria reflects the varied adaptive strategies of bacteria for survival and persistence in the ocean's pelagial. Heterogeneity in the organic matter field imposes variations in metabolism of bacteria (e.g. whether or not metabolic products such as ammonium are secreted) and hence leads to reversals of the ecosystem roles of bacteria. Interactions among bacteria, phytoplankton detritus and protozoa can critically influence nutrient cycling, primary productivity, and sinking flux. In keeping with the theme of this NATO ASI, we emphasize that global ocean-scale biogeochemical dynamics are regulated at the level of individual cell and particle at microscale of space. In order to build mechanistic models of biogeochemical dynamics, then, we need to understand microbial metabolism and interactions at the micrometer scale. This is an exciting challenge and will require the development of methods to study microorganisms as components of intact consortia in their natural setting. Only then will we have a mechanistic basis for incorporating the roles of bacteria into models to understand the variability in the ocean's biogeochemical state.

ACKNOWLEDGEMENTS

The preparation of this manuscript was supported by NSF and ONR grants to F. Azam.

REFERENCES

Alldredge AL, Youngbluth MJ (1985) The significance of macroscopic aggregates (marine snow) as sites for heterotrophic bacterial production in the mesopelagic zone of the subtropical Atlantic. Deep-Sea Res 32:1445-1456

Ammerman JW , Azam F (1985) Bacterial 5'-nucleotidase in aquatic ecosystems: a novel mechanism of phosphorus regeneration. Science 227:1338-1340

Azam, F (1984) The radioecological role of marine bacterioplankton In: Bonnyns-Van Gelder E, Kirchmann R (eds), Role of microorganisms on the behaviour of radionuclides in aquatic and terrestrial systems and their transfer to man. Int. Union Radioecologists, Brussels pp 2-7

Azam F , Ammerman JW (1984) Cycling of organic matter by bacterioplankton in pelagic marine ecosystems: microenvironmental considerations., pp. 345-360 In: Fasham MJR (eds) Flows of energy and materials in marine ecosystems. Plenum Publishing

Azam F, Cho BC (1987) Bacterial utilization of organic matter in the sea., pp. 261-281 In: (eds) Ecology of microbial communities. Cambridge University Press

Azam F, Fenchel T, Field JG, Gray JS, Meyer-Reil L A, Thingstad F (1983) The ecological role of water-column microbes in the sea. Mar Ecol Prog Ser 10:257-263

Belas R, Simon M, Silverman M (1986) Regulation of lateral flagella gene transcription in Vibrio parahaemolyticus. J Bacteriol 167:210-218

Bergh ϕ, Bϕrsheim KY, Bratbak G, Heldal M (1989) High abundance of viruses found in aquatic environments. Nature 340:467-468

Biddanda BA, Pomeroy LR (1988) Microbial aggregation and degradation of phytoplankton-derived detritus in seawater. I. Microbial succession. Mar Ecol Prog Ser 42:79-88

Bjϕrnsen PK (1988) Phytoplankton exudation of organic matter: Why do healthy cells do it? Limnol Oceanogr 33:151-154

Bratbak G, Thingstad TF (1985) Phytoplankton - bacteria interactions: an apparent paradox? Analysis of a model system with both competition and commensalism. Mar Ecol Prog Ser. 25:23-30

Chett I, Mitchell R (1976) Ecological aspects of microbial chemotactic behavior. Ann Rev Microbiol 30:221-239

Cho BC (1988) Significance of bacteria in biogeochemical fluxes in the pelagic ocean. Ph. D. thesis, University of California, San Diego.

Cho BC, Azam F (1990) Biogeochemical significance of bacterial biomass in the oceans euphotic zone. Mar Ecol Prog Ser 63:253-259

Dawson MP, Humphrey B, Marshall KC (1981) Adhesion: a tactic in the survival strategy of a marine vibrio during starvation. Curr Microbiol 6:195-198

Ducklow HW, Hill SW, Gardner WD (1985) Bacterial growth and the decomposition of particulate organic carbon collected in sediment traps. Cont Shelf Res 4: 445-464

Fuhrman JA, Sleeter TD, Carlson C , Proctor LM (1989) Dominance of bacterial biomass in the Sargasso Sea and its ecological implications. Mar Ecol Prog Ser 57:207-217

Goldman JC (1984) Conceptual role for microaggregates in pelagic waters. Bull Mar Sci 35:462-47

Gottschalk G (ed) (1986) Bacterial Metabolism. Springer-Verlag New York

Hollibaugh JT, Azam F (1983) Microbial degradation of dissolved proteins in seawater. Limnol Oceanogr 28:1104-1116

Hoppe H-G (1984) Attachment of bacteria: advantage or disadvantage for survival in the aquatic environment., pp. 283-301 In: Marshall KC (ed) Microbial adhesion and aggregation. Springer

Iturriaga R, Mitchell BG (1986) Chroococcoid cyanobacteria: a significant component in the food web dynamics of the open ocean. Mar Ecol Prog Ser 28:291-297

Jackson GA (1989) Simulation of bacterial attraction and adhesion to falling particles in an aquatic environment. Limnol Oceanogr 34:514-530

Jacobsen JT, Azam F (1984) Role of bacteria in copepod fecal pellet decomposition: colonization, growth rates and mineralization. Bull Mar Sci 35:495-502

Johnson PW, Xu H-S, Sieburth JMcN (1982) The utilization of chroococcoid cyanobacteria by marine protozooplankters but not by calanoid copepods. Ann Inst Oceanogr Paris 58:297-308

Karl DM, Knauer GA, Martin JH (1988) Downward flux of particulate organic matter in the ocean: a particle decomposition paradox. Nature 332:438-441

Kirchman DL (1983) The production of bacteria attached to particles suspended in a freshwater pond. Limnol. Oceanogr.28:858-872

Koike I, Shigemitsu H, Kazuki T, Kogure K (1990) Role of sub-micrometre particles in the ocean. Nature 345:242-244

Laws EA, Redalje DG, Haas LW, Bienfang PK, Eppley RW, Harrison WG, Karl DM , Marra J (1984) High phytoplankton growth rates in oligotrophic Hawaiian coastal waters. Limnol Oceanogr 29:1161-1169

McCarthy JJ, Goldman JC (1979) Nitrogenous nutrition of marine phytoplankton in nutrient-depleted waters. Science 203:670-672

Mitchell JG, Okubo A, Fuhrman JA (1985) Microzones surrounding phytoplankton form the basis for a stratified marine microbial ecosystem. Nature 316:58-59

Mitchell JG, Okubo A, Fuhrman JA (1990) Gyrotaxis as a new mechanism for generating spatial heterogeneity and migration in microplankton. Limnol. Oceanogr. 35:123-130

Mykelstad S (1977) Production of carbohydrates by marine planktonic diatoms. II. Influence of the N/P ratio n the growth medium on the assimilation ratio, growth rate, and production of cellular and extracellular carbohydrates by Chaetoceros affinis var. willei (Gran) Hustedt and Skeletonema costatum (Grev.) Cleve. J. Exp. Mar. Biol. Ecol. 29:161-179

Paerl HW, Gallucci KK (1985) Role of chemotaxis in establishing a specific nitrogen-fixing cyanobacterial-bacterial association. Science 227:647-649

Paul JH, DeFlaun MF, Jeffrey WH (1988) Mechanisms of DNA utilization by estuarine microbial populations. Appl Environ Microbiol 54:1682-1688

Pedrós-Alió C, Brock TD (1983) The importance of attachment to particles for planktonic bacteria. Arch Hydrobiol 98:354-379

Phinney DA, Yentsch CS (1986) The relationship between phytoplankton and light attenuation in ocean waters. SPIE 637 Ocean Optics VIII:321-327

Pomeroy LR (1974) The ocean's food web, a changing paradigm. BioScience 24:499-504

Pomeroy LR, Wiebe WJ (1988) Energetics of microbial food webs. Hydrobiologia 159:7-18

Priest FG (ed) (1984) Extracellular enzymes. Van Nostrand Reinhold Wokingham, U.K.

Proctor LM, Fuhrman JA (1990) Viral mortality of marine bacteria and cyanobacteria. Nature 343:60-62

Silver MW , Alldredge AL (1981) Bathypelagic marine snow: deep-sea algal and detrital community. J. Mar. Res. 39:501-530

Silver MW, Bruland KW (1981) Differential feeding and fecal pellet composition of salps and pteropods, and the possible origin of deep water flora and olive-green "cells". Mar Biol 62:263-273

Simon M, Alldredge AL, Azam F (1990) Bacterial carbon dynamics on marine snow. Mar Ecol Prog Ser 65:205-211

Smetacek V (1985) Role of sinking in diatom life-history cycles: ecological, evolutionary and geological significance. Mar Biol 84:239-251

Somville M , Billen G (1983) A method for determining exoproteolytic activity in natural waters. Limnol Oceanogr 28:190-193

Strickler JR (1982) Calanoid copepods, feeding currents, and the role of gravity. Science 218:158-160

Suttle CA, Chan AM, Cottrell MT (1990) Infection of phytoplankton by viruses and reduction of primary productivity. Nature 347:467-469

Tupas L , Koike I (1990) Amino acid and ammonium utilization by heterotrophic marine bacteria grown in enriched seawater. Limnol Oceanogr 35:1146-1155

Wheeler PA , Kirchman DL (1986) Utilization of inorganic and organic nitrogen by bacteria in marine systems. Limnol Oceanogr 31:998-1009

Williams PJLeB (1990) The importance of losses during microbial growth: commentary on the physiology, measurement and ecology of the release of dissolved organic material. Mar Microbial Food Webs 4:141-256

MOLECULAR SYSTEMATICS, MICROBIAL ECOLOGY AND SINGLE CELL ANALYSIS

E. F. DeLong

Biology Department
Woods Hole Oceanographic Institution
Woods Hole MA, 02543

ABSTRACT

Molecular approaches in taxonomy and systematics are providing a unifying framework for understanding the phylogenetic relationships of diverse biological species. The methods rely on comparison of nucleic acid or amino acid sequences, which can serve as *yardsticks* for measuring evolutionary divergence. Direct sequence analysis can largely circumvent problems inherent in phenotypic comparisons of widely divergent taxa. Macromolecular sequence information is particularly useful in situations where directly comparable phenotypic properties are scarce or difficult to assess. Additionally, macromolecular sequence data bases are valuable resources for determining the phylogenetic affiliations of previously unstudied or uncharacterized organisms. In particular, current understanding of the evolutionary relationships of microbial species has been greatly advanced through molecular phylogenetic comparisons of small subunit ribosomal RNA (rRNA) sequences.

The molecular data employed in systematic and evolutionary studies are also proving useful for ecological studies. By directly retrieving phylogenetically informative gene sequences from mixed microbial populations, it is possible to infer phylogenetic affiliations of individual population constituents. This allows identification of community members without requiring their cultivation, and so avoids some selective biases associated with pure culture methods. In addition, short segments of sequence, such as those found in small subunit rRNA, can be taxa-specific. These sequences may therefore serve as diagnostic markers for particular groups. In conjunction with epifluorescence microscopy, fluor-labeled, rRNA-targeted probes that bind to these diagnostic sequences may be used to determine the phylogenetic identity of individual cells. Thus, macromolecular sequence information can be employed to detect the presence of particular species, and to study their spatial and temporal variability. Recent applications, including molecular phylogenetic analyses of mixed bacterioplankton populations, demonstrate the utility of this approach.

INTRODUCTION

Single cell analyses have proven extremely useful for determining the abundance, activities and variability of microbial species in their natural habitats. In the field of oceanography, recent studies of marine cyanobacteria exemplify the utility of this approach (Waterbury *et al.*, 1986). Their unique size and autofluorescent properties permit the rapid, direct identification and enumeration of marine *Synechococcus* (Waterbury *et al.*, 1979). This in turn has led to detailed understanding of the temporal and spatial distribution and abundance of these

NATO ASI Series, Vol. G 27
Particle Analysis in Oceanography
Edited by S. Demers

phototrophic picoplankters. In the prokaryotic world, however, few other microbes can be so readily identified and quantified. More typically, few distinguishing morphological criteria are available to differentiate diverse microbial species found in environmental samples. Most bacteria are therefore characterized on the basis of biochemical and physiological properties determined on pure cultures. Unfortunately, not all microbes are amenable to cultivation, and so they are difficult to identify, differentiate, and characterize. Molecular biological techniques may offer a means to resolve some of these difficulties and so facilitate the identification and quantitation of individual microbial species in mixed populations. In particular, the use of macromolecular sequence data for systematic, evolutionary and ecological studies has significant potential.

Why study individual bacteria and bacterial species? It is now well accepted that microorganisms play a major role in the biogeochemical cycling of organic and inorganic material in the marine environment. This view stems from the currently recognized high abundance of bacteria, cyanobacteria, and microeukaryotes in seawater (Hobbie *et al.*, 1977; Waterbury *et al.*, 1979; Ducklow, 1983; Fenchel, 1982), and the observation that planktonic bacteria are metabolically quite active (Fuhrman and Azam, 1982; Hagstrom *et al.*, 1979; Karl, 1986). Although the general details of marine bacterial abundance are fairly well characterized, many ecologically important aspects of marine microbial population structure and dynamics are virtually unstudied. There is very little information concerning the complex interactions of different physiological types of marine bacteria, which exist as synergistic microbial consortia in the oceanic environment (Sieburth, 1988). Little is known about the actual species composition of marine bacterial populations, because typically less than a few percent of directly enumerated marine bacteria can be isolated and studied by standard cultivation techniques (Jannasch and Jones, 1959; Ferguson *et al.*, 1984). Variability in the types of marine bacteria which occupy different niches, such as bacteria attached to particulates versus free-living bacterioplankton, also remains largely unstudied. Knowledge of this variability would greatly facilitate correlations of bacterial population structure to the bacterial processes which mediate the flow of energy and matter in the sea.

MOLECULAR SYSTEMATICS

One of the main goals of taxonomy is to provide a hierarchical framework for identifying and classifying diverse biological species. Modern classification systems have an additional goal, beyond imposing order upon biological diversity: phylogenetic systems seek to infer genealogical relationships, which reflect the evolutionary pathways followed during life history. Close phylogenetic relationship implies not only phenotypic or genotypic similarity, but common evolutionary history as well.

Until recently, microbial systematists have relied predominantly on morphological, nutritional and biochemical criteria for taxonomic classification. Phenetic analyses have provided an empirical framework for classifying and identifying physiologically diverse microbial types. Their utility for the identification and characterization of microorganisms is unquestionable. Taxonomic classifications based on phenetic analyses do not necessarily imply evolutionary relationship, however. In addition, most phenetic analyses demand laboratory cultivation, a requirement that may severely limit the range of naturally occurring microbes which can be characterized. Furthermore, identifying common , homologous phenetic traits of evolutionarily diverse taxa can be problematic. Current developments in molecular systematics are now resolving some of these difficulties, and are providing insight into the evolutionary relationships of the most diverse lineages known today (Fox *et al.*, 1980; Woese, 1987; Iwabe *et al.*, 1989; Pace *et al.*, 1986b; Woese *et al.*, 1990).

Molecular approaches to systematics rely on sequence comparisons of informational macromolecules (RNA, DNA, or protein). Comparative sequence analyses can provide a quantitative and relatively unbiased assessment of evolutionary relatedness (Felsenstein, 1982; Felsenstein, 1988; Olsen, 1988; Swofford and Olsen, 1990). In theory, gene sequences provide more evolutionary information than is available from study of the phenotypic properties of the expressed gene (Zuckerkandl and Pauling, 1965). Analyses of highly conserved macromolecular sequences may provide insight about evolutionary relationships over very wide phylogenetic ranges, supplementing phenotypic or morphological studies where few characters are shared between distantly related taxa (Field *et al.*, 1988). Since methods for obtaining and analyzing nucleic acid and protein sequences are continually improving

(Sambrook *et al.*, 1989; Lane *et al.*, 1985; Medlin *et al.*, 1988) sequence data bases are expanding rapidly.

Molecular phylogenetic studies also have application for ecology and oceanography, largely because they provide a means to identify and quantify species in environmental samples. The methods are particularly useful for studying those species which may be difficult or impossible to cultivate in the laboratory. In one application, phylogenetically informative genes are clonally isolated from heterogeneous natural populations, and their sequences determined and compared to homologous genes of known, well-characterized organisms (Pace *et al.*, 1986a). Ideally, these comparative sequence analyses indicate the phylogenetic affiliations of individual population members, and provide a *snapshot* of overall community structure. The same sequence data is also useful for designing taxa-specific nucleic acid hybridization probes, to monitor the spatial and temporal distributions of particular groups (Pace *et al.*, 1986a; Stahl *et al.*, 1988; Giovanonni *et al.*, 1990a). Thus, molecular techniques can provide a means for assessing overall community diversity, and simultaneously provide data useful for studying the temporal and spatial variability of individual species. Nucleic acid probes that bind target molecules with high intracellular concentration (e.g. rRNA) can also be used to determine the phylogenetic affiliations of individual cells (Giovanonni *et al.*, 1988; DeLong *et al.*, 1989a; DeLong and Shah, 1990; Amman *et al.*, 1990a; Tsien *et al.*, 1990).

FROM MOLECULAR SEQUENCE DATA TO ORGANISMAL PHYLOGENY

Molecular phylogenetic analyses use a known quantity, sequence similarity, to estimate an unknown quantity, evolutionary divergence. The inference of organismal phylogenies from molecular sequence comparisons depends partly on a model of the evolutionary process which assumes a random accumulation of mutational change, occurring at a stochastically constant rate (Wilson *et al.*, 1977; Felsenstein, 1988). The random, stochastically constant accumulation of mutations can be viewed as a *molecular clock*. Comparative sequence analyses use these accumulated mutations to estimate the relative divergence of organisms from their most recent common ancestor. These quantitative estimates of evolutionary divergence form the basis of molecular phylogenetic analysis. Molecular clocks *tick* at different rates, so molecular phylogenies based on different macromolecules are not necessarily equivalent. Rapidly evolving macromolecules are mainly useful for measuring short evolutionary distances, while slowly evolving molecules record the relationships between very divergent species. The size and relative degree of conservation (rate of mutational change) in any particular macromolecule studied will largely determine the range and accuracy of phylogenetic inference.

Comparison of strictly homologous features is as crucial to molecular phylogenetic studies as it is to morphological systematics. Structural and functional homology is an absolute requirement for inference of organismal phylogenies from molecular phylogenies. Universal distribution, and the presence of highly conserved, as well as more variable sites, are particularly useful features in informational macromolecules. These properties permit unambiguous comparison of highly divergent sequences, and inference of relationships between both closely and distantly related organisms. The absence of lateral (interspecies) transfer of the genes being studied is another necessary requirement for the inference of phylogenies from molecular sequence data. Genes borne on extrachromosomal elements (e.g. plasmids), which can be transferred within and between species, are inappropriate for inferring organismal phylogenetic relationships.

Historically, both direct and indirect methods have been used to estimate macromolecular sequence similarity (Schleifer and Stackebrandt, 1983). Direct methods involve the comparison of the actual amino acid or nucleic acid sequences. Indirect approaches include immunological techniques, which detect the antigenic similarity of proteins, or nucleic acid hybridization techniques, which measure the similarity of sequences based on the relative stability of nucleic acid heteroduplexes. These approaches yield data in the form of pairwise similarity values. Studies of Baumann *et al.* (1983), focussing on the evolutionary relationships of well-characterized marine eubacteria, demonstrate that good agreement can be obtained between a variety of these indirect techniques. One disadvantage of these indirect approaches, however, is that it is necessary to repeat the entire array of pairwise comparisons whenever a new organism is analyzed.

Analyses of restriction fragment length polymorphisms (RFLPs) is another method for measuring macromolecular similarities. In RFLP analysis, DNA from different organisms is isolated and digested with restriction endonucleases, which cleave the DNA at specific sequences (usually 6 nucleotides in length). This produces DNA fragments of varying lengths, which can be size fractionated by agarose gel electrophoresis. DNA fragments containing sequences homologous to specific genes can be visualized by hybridization with radiolabelled gene probes. The number and size of these DNA fragments should be nearly identical in very closely related organisms, while divergent species will display different DNA restriction fragment patterns (restriction fragment length polymorphisms). Using a variety of restriction endonucleases and hybridization probes, genetic relatedness can theoretically be estimated from these polymorphisms. However, several sources of error inherent in RFLP data render this technique less useful than direct sequence comparisons for phylogeny reconstruction (Swofford and Olsen, 1990).

In contrast to the above approaches, direct acquisition of macromolecular sequence data allows quantitative comparisons of sequences from homologous genes or proteins. Once obtained, archived sequences provide a reference data base which can be directly and immediately compared to newly acquired sequences. The data are first analyzed by aligning homologous nucleotides or amino acid positions of sequences to be compared. Highly conserved regions of sequence can serve as convenient landmarks in the initial alignment process. After the homologous sequence positions are juxtaposed, they are compared and analyzed by any of a variety of available methods (see below). The analyses can be computationally intense because of the large amount of data compared, and the statistical algorithms used to infer phylogenetic tree topology and estimate error. Hence, molecular phylogenetic analyses usually require extensive use of computers. A number of numerical methods for inferring phylogenies from molecular data are currently in use. The methods are rapidly improving, as computer algorithms become more sophisticated and computer facilities more readily available. The major approaches now used for estimating molecular phylogenies are cluster and distance matrix methods, parsimony analysis, and maximum likelihood methods (Swofford and Olsen, 1990). Distance matrix methods calculate a similarity value (S) between each pair of homologous sequences in a given data set. A pairwise *evolutionary distance* value can then be derived from sequence dissimilarity (1-S), which takes into account the possibility of multiple mutations at any given nucleotide position. These pairwise distance values are assumed to be strictly additive in the phylogenetic tree construction. The *best* tree is that which minimizes the difference between the distance values in the branch lengths of the tree, to those distance values of the original data set (Olsen, 1988). Parsimony methods compare sequences on a nucleotide per nucleotide basis. Parsimony analysis seeks a phylogenetic tree topology which minimizes the number of postulated base changes. The *best* tree in parsimony analysis is that which requires the fewest number of nucleotide substitutions. Likelihood methods essentially work backwards, and estimate the likelihood of a given data set occurring, given a specific phylogenetic tree topology. The *best* tree is that which maximizes the likelihood of observing the actual data set, given a specific tree topology. Of the three approaches, likelihood methods are the most computationally intense. A more detailed discussion of the methods of phylogenetic inference is beyond the scope of this chapter. The interested reader is referred to several excellent articles on the methods and assumptions of molecular phylogenetics (Felsenstein, 1982; Felsenstein, 1988; Olsen, 1987; Swofford and Olsen, 1990).

RIBOSOMAL RNA AND *MOLECULAR ECOLOGY*

The application of molecular techniques to ecological and oceanographic studies is still in its infancy. Nonetheless, several studies have recently demonstrated that a molecular approach can help resolve some recurring difficulties in analyses of the diversity and variability of mixed microbial populations. With reference to ecological studies, ribosomal RNA sequence comparisons are particularly useful. This is largely due to their universal distribution, utility in nucleic acid hybridization assays, and the growing data base of reference rRNA sequences. The following section will focus on the development of rRNA cloning, sequencing and hybridization techniques for ecological applications.

Ribosomal RNAs contain a wealth of evolutionary information, which can be extracted by the process of comparative sequence analysis. This process relies only on our ability to isolate and analyze individual rRNA sequences. Pace *et al.* (1986a) first suggested that this approach can remove a major stumbling block to gaining information about naturally occurring microbes, many of which are recalcitrant to cultivation. These workers realized that by recovering nucleic acids from complex microbial assemblages, and clonally retrieving individual rRNA genes from mixed population DNA, it is possible to extract information about the phylogenetic relationships of resident microbes, even those which resist cultivation. This information may be used to infer the likely physiological or biochemical properties of population members, and design experiments to test these hypotheses. In addition, the sequence information can be used to construct organism-specific nucleic acid hybridization probes, for monitoring the abundance and distribution of individual microbial species.

Isolation of small subunit rRNA genes from complex populations provides a way to identify members of the original consortia, without requiring isolation of individual organisms in pure culture. The approach hinges on the construction of recombinant DNA libraries, from which rRNA genes can be identified, sorted and sequenced. Several strategies are now available for preparation of recombinant DNA libraries from mixed population nucleic acids (Pace *et al.*, 1986a; Weller and Ward, 1989; DeLong *et al.*, 1989b; Giovanonni *et al.*, 1990a). The first to be proposed (Pace *et al.*, 1986a) involves the initial isolation of fairly large quantities (>100 μg) of high molecular weight, total population DNA. This mixed population DNA is partially cleaved with a restriction endonuclease. DNA fragments in the 10-20 kilobase size class are then purified, and inserted into a well characterized cloning vector (eg. bacteriophage lambda). These cloning vectors allow the stable propagation and in vivo amplification of the isolated DNA fragment. The end product is a recombinant DNA *library*, consisting of hundreds of thousands of bacteriophage, each of which contain a DNA fragment originating from the mixed population DNA. Individual clones containing rRNA genes can be identified using radioactively labeled, broad-spectrum rRNA probes (Pace *et al.*, 1986a). These probes, by virtue of Watson-Crick base pairing, bind specifically to recombinant DNA clones which contain rRNA genes. Individual clones which contain rRNA gene inserts are next isolated and purified. The rRNA sequences of cloned rRNA genes are then determined, and compared to a data base of sequences from well-characterized organisms. In this fashion, the phylogenetic identity of individual population members can be ascertained (Pace *et al.*, 1986a; DeLong *et al.*, 1989b; Schmidt *et al.*, submitted).

Alternative methods for retrieving small subunit rRNA genes from natural populations have recently been developed. Weller and Ward (1989) proposed the initial isolation of mixed population rRNA, rather than DNA. Using the mixed population rRNA as a template, ribosomal DNA (rDNA) can be generated enzymatically using the enzyme reverse transcriptase. Individual rDNA genes can then be clonally isolated and sequenced as described above. Successful application of this method was used to study the species diversity of hot spring microbial mats (Ward *et al.*, 1990). An advantage of this method is that in theory all clones obtained contain rRNA genes. Very vigorous lytic protocols can be used, which may significantly enhance the extraction of nucleic acids. There is, however, a drawback to the method. Production of full-length cDNAs is problematic, because reverse transcriptase tends to *stop* at modified bases or complex secondary structural elements. Thus full length rRNA clones may be difficult to obtain (Ward *et al.*, 1990).

Another promising technique involves the specific amplification of small subunit rRNA genes from mixed population DNA. This is achieved via the polymerase chain reaction (PCR), which exploits the thermostable properties of Thermus aquaticus DNA polymerase (Taq polymerase, Sakai *et al.*, 1988). Oligonucleotide primers, which hybridize to the distal ends of a specific, targeted gene, initiate the synthesis of new DNA strands. Taq polymerase catalyzes the synthesis of nascent strands of the targeted gene, which is flanked by the oligonucleotide primers. Repeated cycles of DNA denaturation (@ 94°C), reannealing of oligonucleotide primers (@ 40°C), and primer extension (@ 72°C), are possible, since thermostable Taq polymerase can withstand repeated heating cycles at 94°C. The denaturation/reannealing/primer-extension cycle can be repeated up to 40 times, resulting in an exponential increase of the DNA from the targeted gene. Very small amounts of starting total DNA (typically nanograms) can be used to amplify microgram quantities of the specific, targeted gene(s) of interest.

For subsequent phylogenetic analysis, amplified small subunit rRNA genes can be cloned, and the rRNA genes of individual clones sequenced (Medlin *et al.*, 1988; Giovanonni *et al.*, 1990a). This technique has been exploited to amplify small subunit RNA genes from both pure cultures and mixed populations. The PCR approach has the advantage of being extremely rapid, since all the clones obtained will contain rRNA genes. One disadvantage of the approach is that, depending on the reaction conditions, Taq DNA polymerase can have a relatively high error rate (Ennis *et al.*, 1990; Eckert and Kunkel, 1990), which may be detected when amplification products are cloned, and the individual clones are sequenced.

Also, the amplification of heterologous DNAs can produce hybrid DNAs (chimeras) which contain elements of two different templates (Meyerhaus *et al.*, 1990).

At the present time, all of the above-described alternatives for isolating rRNA genes from mixed populations have advantages and disadvantages (Table 1). The *shotgun* cloning approach has been applied successfully to natural samples, and may avoid some complications introduced by PCR. However, this approach is very time consuming and labor intensive, even for the analysis of just a single sample. Direct isolation of rRNA, and production and cloning of ribosomal DNA is a tractable alternative, but full length rRNA clones appear to be rarely recovered (Ward *et al.*, 1990). Enzymatic amplification of rRNA genes using PCR has also been successfully applied, and is much more sensitive and rapid. PCR may, however, introduce artifacts due to the problems with Taq polymerase error described above. All the techniques can potentially provide information about the phylogenetic diversity of mixed populations, without requiring cultivation of the resident microbes. To demonstrate the utility of these techniques, a brief review of some recent investigations, which shed light on the phylogenetic identity of oceanic bacterioplankton, follows.

RECENT APPLICATIONS IN OCEAN SCIENCE

Giovanonni *et al.*, (1990a) recently reported on the genetic diversity of Sargasso Sea bacterioplankton. In these studies, large volumes of surface sea water were concentrated by tangential flow filtration to obtain hundreds of milligrams wet weight of bacterioplankton cells (Giovanonni *et al.*, 1990b). DNA was extracted from this mixed cell population, and small subunit rRNA genes were subsequently amplified from the mixture using the polymerase chain reaction. The amplified rRNA genes were cloned, and phylogenetic analysis of small subunit rRNA sequences performed. The sequences retrieved from the oligotrophic bacterioplankton consortia fell into two main clusters: a cyanobacterial cluster, and a cluster most closely affiliated with the alpha subdivision of the Proteobacteria (Stackebrandt *et al.*, 1988). The cyanobacterial sequences, not surprisingly, were very similar to those obtained from well-characterized, motile, phycoerythrin containing Synechococcus cultivated from similar habitats (Waterbury *et al.*, 1985; Waterbury *et al.*, 1986; D. Distel and J. Waterbury, unpublished data). Although none of the cyanobacterial sequences retrieved from the Sargasso

Sea PCR clone library exactly matched those from cultured species, all were specifically related to cultured marine cyanobacteria, some at the near-species level. Most likely, no exact matches were found because the data base of marine cyanobacterial reference sequences is still

Table I. Comparison of different approaches used for cloning rRNA genes from mixed microbial populations.

ADVANTAGES	DISADVANTAGES
RIBOSOMAL cDNA LIBRARY	
RIGOROUS LYSIS PROTOCOL POSSIBLE	RNA PRONE to CHEMCIAL and ENZYMATIC DEGRADATION
"ALL" RECOMBINANTS rDNA CLONES	PROBLEMS In RECOVERY OF FULL LENGTH cDNA
MAY SELECT "ACTIVE" COMMUNITY MEMBERS	POTENTIAL BIAS DUE to VARIABLE rRNA/CELL, and PRIMER ANNEALING and CLONING STEPS
PCR rDNA LIBRARY	
RIGOROUS LYSIS PROTOCOL POSSIBLE	POTENTIAL FOR TAQ DNA POLYMERASE ERROR
"ALL" RECOMBINANTS rDNA CLONES	POSSIBILITY of "SHUFFLE GENES" or CHIMAERIC PCR PRODUCTS
EXTREMELY SENSITIVE-NEED VERY LOW AMOUNTS of DNA	EXTREMELY SENSITIVE-BACKGROUND CONTAMINATION A POTENTIAL PROBLEM
CAN SELECTIVELY AMPLIFY SPECIFIC GROUPS	POTENTIAL SELECTIVE BIAS DURING PRIMER ANNEALING and AMPLIFICATION
"SHOTGUN" DNA LIBRARY	
UNBIASED BY PRIMER EXTENSION/AMPLIFICATION STEP	NEED GENTLE LYSIS PROTOCOL
	NEED LARGE AMOUNTS of DNA
LIBRARY EASILY STORED	POTENTIAL SELECTIVE BIAS IN ENDONUCLEASE DIGESTION, CLONING AND CLONE IDENTIFICATION STEPS
POTENTIAL for CHROMOSOME 'WALKING' and ISOLATION of OTHER GENES	LABOR INTENSIVE

rather small. The other cluster of sequences observed by Giovanonni and colleagues (dubbed *Sar 11 cluster*) grouped most closely with the alpha subdivision of Proteobacteria, a subdivision which contains both phototrophic and oligotrophic genera. However, no known, cultivated organisms are very closely affiliated with this cluster. To assess the relative abundance of the *Sar 11* group, Giovanonni *et al.*, used the newly acquired sequence data to design an oligonucleotide hybridization probe which specifically recognizes the *Sar 11* group. Hybridization of this group-specific probe to mixed population nucleic acids indicated that Sar 11 cluster rRNA comprised ca. 15% of the total eubacterial rRNA in the original sample. Based on these data, the authors suggest that this novel group of organisms likely plays an important ecological role in the Sargasso Sea. This hypothesis has been recently extended by data obtained in the Central North Pacific Ocean, as described below.

A slightly different approach was used to analyze small subunit rRNA genes from oligotrophic bacterioplankton populations of the Central North Pacific (DeLong *et al.*, 1989b; Schmidt *et al.*, submitted). Bacterioplankton samples were collected at a Hawaii Global Ocean Flux Study site (Station Aloha, Drs. David Karl and Chris Winn, project coordinators), located approximately 100 km north of Oahu. High molecular weight DNA was extracted from the concentrated bacterioplankton, and used to construct a clone library in bacteriophage lambda. A total of ca. 1 x 10^7 recombinant bacteriophage clones were recovered. About 3 x 104 of these were screened for rRNA genes, resulting in the isolation of 40 rRNA gene-containing recombinant clones. Partial small subunit rRNA sequencing revealed the phylogenetic affiliations of specific population members. The diversity of small subunit rRNA genes recovered spanned a fairly broad phylogenetic range: clones representative of the cyanobacteria, the alpha, and the gamma subdivision of the Proteobacteria, as well as eukaryotic dinoflagellates, were all present in the library.

Preliminary comparisons of partial small subunit rRNA sequence data (ca. 190 nucleotides) from Central North Pacific bacterioplankton (Schmidt *et al.*, submitted), with those obtained by Giovanonni *et al.*, in the Sargasso Sea (1990a), reveals striking similarities (Figure 1). Cloned cyanobacterial sequences obtained from two different ocean gyres displayed similarity values ranging from 0.955 to 0.985. The cyanobacterial sequences form distinct clusters, represented in both the Sargasso Sea and the Central North Pacific (Figure. 1a). These groups are quite closely affiliated with cultured and well-characterized open-ocean cyanobacteria (Giovanonni *et al.*, 1990a; Waterbury and Distel, unpublished data). Notably, several Proteobacterial clones from Central North Pacific bacterioplankton are nearly identical to the novel, deep-branching *Sar 11 cluster* from the Sargasso Sea (Giovanonni *et al.*, 1990a; Schmidt *et al.*, 1990). Partial sequence similarity values between representatives of this cluster, originating from two different ocean gyres, are remarkably high (0.969 to 0.998, see Figure 1b).

The preliminary data indicates that at least some of the bacterioplankton from these two central oceanic gyres are quite closely related, at the near-species level, and that they are likely globally distributed (Figure 1). These data suggest that phylogenetic distance between related bacterioplankton within a given ocean gyre is at least as great as that found between closely related species of different oceanic provinces. These data also support Giovanonni's hypothesis that the *Sar 11* group is an important component of the bacterioplankton. This novel group may well be cosmopolitan in oligotrophic gyres of the world's oceans. These results also demonstrate that both the PCR and the *shotgun library* approaches are viable options for assessing the phylogenetic diversity of mixed microbial populations. The two different cloning strategies, employed at different times of year on opposite sides of the globe, have yielded remarkably similar results. The utility of accumulating sequence data bases is also demonstrated in the comparison of these two data sets. Much more data needs to be collected to characterize the diversity, variability, activities and abundance of bacterioplankton species. The use of rRNA specific hybridization probes is one potential tool for detecting and identifying particular taxa, and estimating their abundance and distribution.

HYBRIDIZATION PROBES AND SINGLE CELL ANALYSIS

The nucleotide sequence of small subunit rRNAs consists of both very conserved, as well as more variable, nucleotide positions. Some short regions of rRNA sequence are invariant between distantly related taxa (Lane *et al.*, 1985; Giovanonni *et al.*, 1988). Other more variable regions are unique to individual species (Stahl *et al.*, 1988; DeLong *et al.*, 1989a; Amman *et al.*, 1990a; Tsien *et al.*, 1990). These short, contiguous segments ($\approx$20 nucleotides in length) of rRNA sequence can serve as targets for rRNA-specific, nucleic acid hybridization probes. The more conserved sequence regions are useful for discriminating between broad phylogenetic groups, such as eukaryotes and eubacteria (Giovanonni *et al.*, 1988). The more variable regions of rRNA are useful targets for genus,species and subspecies

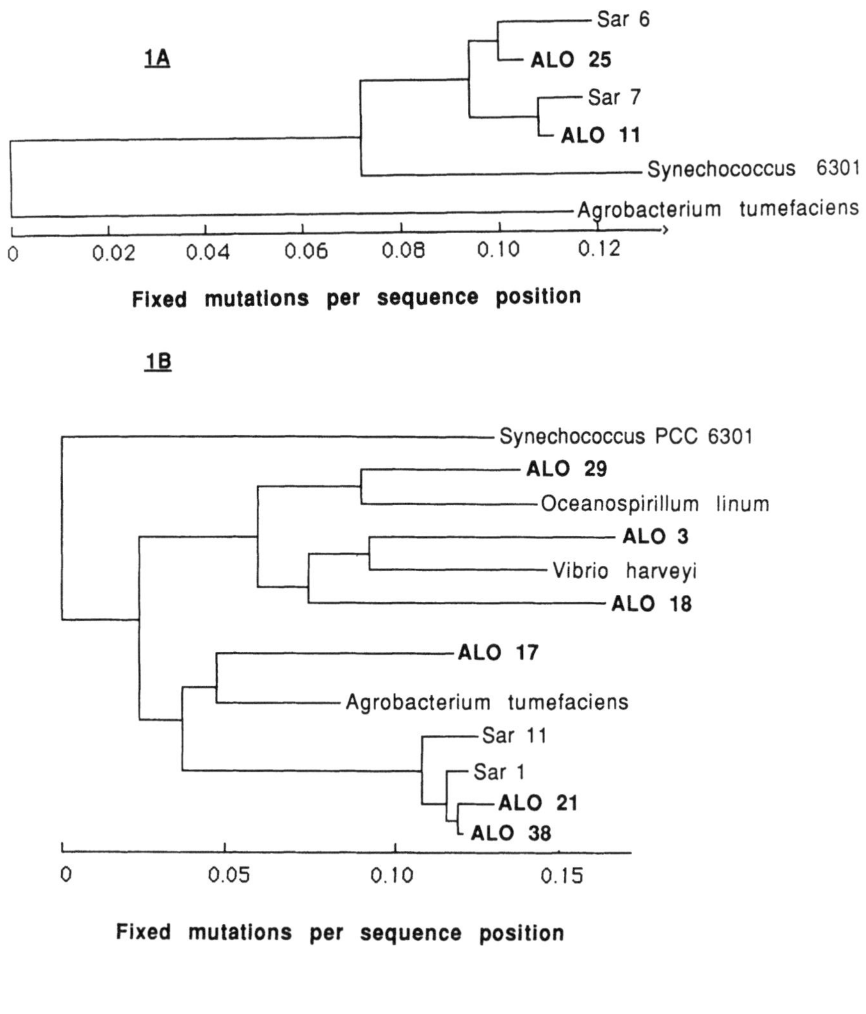

Figure 1. Evolutionary relationships of bacterioplankton population constituents of the Sargasso Sea and the Central North Pacific. 1A, Phylogenetic relationships of naturally occurring cyanobacteria from the two oceanic gyres. 1B, Phylogenetic relationships of naturally occurring Proteobacteria (*purple bacteria*) from the two oceanic gyres. Pairwise evolutionary distances were estimated by comparing partial small subunit rRNA sequences (ca. 190 nucleotides, bases 305-495, *E. coli* numbering system) cloned from mixed population DNA. Phylogenetic tree topology was inferred using a least squares distance matrix method (Olsen, 1988). Data sources: *Synechococcus* PCC 6301, Tomioka and Sugiura, 1983; *Agrobacterium tumefaciens*, Yang *et al.*, 1985; *Oceanospirillum linum*, Dr. Carl R. Woese, unpublished, GenBank Accesion # M22365; *Vibrio harveyi*, Drs. David Lane and Reinhardt Rosson, unpublished; Sargasso Sea (Sar) cloned sequences, Giovanonni *et al.*, 1990a; Aloha Station (ALO) cloned sequences, Schmidt *et al.*, submitted)

specific hybridization probes. Since short DNA hybridization probes form stable duplexes with their rRNA target only if there is near-perfect Watson-Crick base pair complementarity, probes can be tailored for high specificity.

Computer assisted analyses of aligned rRNA sequences can be used to identify regions of nucleotide sequence that are unique to a given species (or other taxon). Initially, an approximately 20 nucleotide sequence (or longer) is identified that is unique to a particular group or species. Oligonucleotide probes complementary to these diagnostic sequences are then prepared, using an automated DNA synthesizer. Ideally these oligonucleotides should form perfect double-stranded nucleic acid duplexes with the target sequence, but have several mismatches with the homologous rRNA sequence of *nontarget* organisms. The specificity and fidelity of probe binding is confirmed by hybridization of the probe with a collection of rRNAs from a variety of different species. Typically, the probe is radiolabelled and hybridized to rRNA immobilized on a nylon filter. Probe binding is detected by autoradiography. Hybridization and wash temperatures are determined empirically for each new probe. A rough estimate for the dissociation temperature (Td) of short oligonucleotide probes is: Td $\approx$ 4°(number of G+C residues) + 2°(number of A+T residues) (Wallace and Miyada, 1987). In general, hybridization and wash temperatures about 10°C below the Td of a perfectly matched duplex yield strong hybridization signals with good specificity. A well designed probe should give a strong positive hybridization signal with rRNA from the targeted organism, but demonstrate no background binding to rRNA of heterologous species.

Radiolabelled rRNA-specific probes have been used in a number of environmental studies. Quantitative estimates of the representation of specific groups are obtained by hybridizing radiolabelled probes to nucleic acids extracted from natural samples. Relative binding of the specific probe is normalized to that of a *universal* probe, which binds to all rRNA species, and therefore provides an estimate of total rRNA. This approach has been taken to estimate the abundance of particular microbial species present in bovine rumen, and their response to environmental perturbation (antibiotic challenge, Stahl *et al.*, 1988). Giovanonni *et al.* (1990a) have estimated the abundance of specific marine bacterioplankton groups with this method.

A particularly useful feature of ribosomal RNAs is their cellular abundance. Because of their central involvement in protein synthesis, the rRNAs account for 70% or more of total cellular RNA in both prokaryotes and eukaryotes. Due to this high concentration of intracellular target sequence, in situ hybridizations with rRNA-specific probes are quite sensitive. Phylogenetic identification of individual microbial cells using rRNA specific probes was first demonstrated by Giovanonni and coworkers, using radioactively labeled oligonucleotides (Giovanonni *et al.*, 1988). However, this approach is time consuming and technically demanding, and not well suited for routine clinical or environmental studies. An extension of this technique uses fluorescently labeled oligonucleotide probes, for phylogenetic identification of individual microbial cells by epifluorescence microscopy (DeLong *et al.*, 1989a; Amann *et al.*, 1990a; Tsien *et al.*, 1990). The method is technically straightforward and, with fluorescent probes in hand, may be performed in a few hours.

Internal positive and negative controls may be included in parallel samples, to ensure that cells contain intact, accessible rRNA, and that background noise is minimal. A *universal* probe, which binds to the rRNA of all organisms studied so far, serves as a positive control. Simultaneous hybridization is performed with this internal control probe, labeled with a fluorochrome which excites and emits at a different wavelengths than the specific probe (DeLong *et al.*, 1989a). Bright, uniform, cellular fluorescence obtained with the positive control demonstrates that the cells are intact, and that ribosomal RNA binding sites are accessible. A negative control, in which no probe is added to the hybridization buffer, serves as a check for excessive autofluorescence. Another negative control involves the use of a fluor-labeled probe which does not bind to rRNA sequences. This control detects any nonspecific binding of nucleic acid probes.

Methods that have been used to detect and quantitate the fluorescence of cells tagged with fluorescently labeled rRNA probes include confocal microscopy (DeLong *et al.*, 1989a) and flow cytometry (Amann *et al.*, 1990b). Confocal microscopy allows *optical sectioning* of the sample, so that fluorescently labeled cells can be localized three dimensionally in tissue sections. Video images obtained with the confocal microscope are stored on disk for documentation, and subsequent image analysis can provide quantitative estimates of fluorescence (DeLong *et al.*, 1989a). Image analysis of cells labelled with fluorescent, rRNA-targeted probes is proving useful for identifying and localizing bacterial endosymbionts in their host's tissue (Dan Distel, John Waterbury and Ed DeLong, unpublished data).

A recent application of fluorescent rRNA probes is the analysis of labeled cells by flow cytometry (Amann *et al.*, 1990b). Hybridizations are performed with fixed cells in solution, rather then with smears or sections. After hybridization with the fluorescent probe, the cells are diluted in buffer and quantified by flow cytometric analysis. Initial studies show that probe binding is highly specific, and that the technique readily identifies probe-binding cells, even when present as a small fraction (<1%) of the total cell population. The coupling of flow cytometry with the use of rRNA-targeted probes has great potential. Given adequate fluorescent signal, specific cell types could be identified and quantified very rapidly (thousands of cells per second).

The fluorescent signal resulting from in situ hybridization with fluor-labeled rRNA probes is largely dependent on the rRNA content of the targeted cell. For any given cell type, fluorescence intensity can vary, and depends in part on the physiological status of the cell at the time of fixation (Figure 2). Slowly growing cells, which contain fewer ribosomes, yield proportionally lower fluorescence signal intensities (DeLong *et al.*, 1989a; Figure 2). The low signal obtained from cells which contain few ribosomes, coupled with the inherent autofluorescence found in many natural samples, are factors which currently limit the use of the technique in ecological studies. A number of potential methods for amplifying the signal obtained with fluorescently labeled rRNA hybridization probes exist. One approach is to use several different probes of identical specificity, targeted to different regions of the same rRNA molecule (Amann *et al.*, 1990 a,b). A two or three-fold increase in fluorescent signal intensity should be readily achievable using this strategy. Another approach involves indirect labelling with a fluor-labeled *reporter* groups (E.F. DeLong, unpublished data). An rRNA probe, which is covalently linked to a polythymidine *tail*, is first hybridized with the sample. After hybridization and washing, the fluorescently labeled reporter group (eg., fluorescently labeled polyadenosine) is added. Since several fluorescent *reporters* can bind to each *tail*, the fluorescent signal may be significantly enhanced. Preliminary experiments with cultured bacteria and protozoa suggest that a several fold increase in signal is achievable using this method.

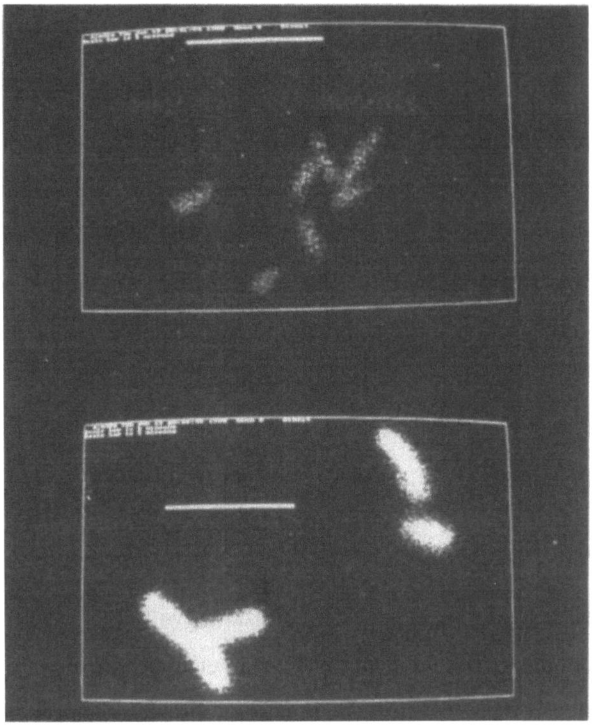

Figure 2. Effect of growth rate on cellular fluorescence derived from the binding of fluor-labeled rRNA targeted, *universal* probe. Video images of labelled *E. coli* cells which were growing slowly (top panel, doubling time ≈ 120 min.) or rapidly (bottom panel, doubling time ≈ 20 min.). The fluorescent signal reflects the cell size and cellular abundance of rRNA. Methods are described in DeLong *et al.* (1989a).

CONCLUSION

A long range goal of the above described studies is to provide a means to study diverse microbial species as they occur in mixed assemblages. Marine Synechococcus species are perhaps the best understood of the marine bacterioplankton, largely because of their inherent properties which allow rapid identification and enumeration. Similar but more general techniques are greatly needed to understand the ecology of less easily identified microbial species. Sensitive and specific methods, which exploit macromolecular sequence differences, may offer a means to monitor the abundance and variability individual taxa. Application of such techniques to microbial ecology and oceanography will lead to a better understanding of predator-prey dynamics, the spatial and temporal variability of species, and syntrophic microbial interactions, as they occur in mixed populations.

ACKNOWLEDGEMENTS

I would like to thank to Dr. Norman Pace for advice and support, Dr. Thomas Schmidt for collaborative effort in the analysis of Central North Pacific bacterioplankton, and Dr. David Karl for hospitality and ship time at the Aloha Station. This work was funded in part by Office of Naval Research Young Investigator Award N00014-90-J-1917 to EFD. This is Woods Hole contribution number 7564.

REFERENCES

Amann RI, Krumholz L, Stahl DA (1990a) Fluorescent oligonucleotide probing of whole cells for determinative, phylogenetic, and environmental studies in microbiology. J Bacteriol 172: 762-770

Amann RI, Binder BB, Olson RJ, Chisholm SW, Devereux, Stahl DA (1990b) Combination of 16S rRNA-targeted oligonucleotide probes with flow cytometry for analyzing mixed microbial populations. Appl Environ Microbiol 56:1919-1925

Baumann P, Baumann L, Woolkalis MJ, Bang SS (1983) Evolutionary relationships in Vibrio and Photobacterium: a basis for a natural classification. Ann Rev Microbiol 37:369-398

DeLong EF, Wickham GS, Pace NR (1989a) Phylogenetic stains: Ribosomal RNA-based probes for the identification of single cells. Science 243:1360-1363.

DeLong EF, Schmidt TM, Pace NR (1989b) Analysis of single cells and oligotrophic picoplankton populations using 16S rRNA sequences. In: Hattori T, Ishida Y, Maruyama Y, Morita RY and A.Uchida, (eds) Recent Advances in Microbial Ecology. Jap Sci Soc Press, Tokyo, p 697

DeLong EF, Shah J (1990) Fluorescent ribosomal RNA probes for clinical application: a research review. Diagnos Clinc Test 28:41-44.

Ducklow HW (1983) Production and fate of bacteria in the oceans. Bioscience 33:494-501

Eckert KA, Kunkel TA (1990) High fidelity synthesis by the Thermus aquaticus DNA polymerase. Nuc Acid Res 18:3739-3744

Ennis PD, Zemmour J, Salter RD, Parham P (1990) Rapid cloning of HLA-A,B cDNA by using the polymerase chain reaction: frequency of errors produced in amplification. Proc Natl Acad Sci 87:2833-2837

Felsenstein J (1988) Phylogenies from molecular sequences: inference and reliability. Annu Rev Genet 22:521-565

Felsenstein J (1982) Numerical methods for inferring evolutionary trees. Quart. Rev. Biol. 57:379-404

Fenchel T (1982) Ecology of heterotrophic microflagellates: quantitative occurrence and importance as consumers of bacteria. Mar Ecol Prog Ser 9:35-42

Ferguson RL, Buckley EN, Palumbo AL (1984) Response of marine bacterioplankton to differential filtration and confinement. Appl Environ Microbiol 47:49-55

Field KG, Olsen GJ, Lane DJ, Giovanonni SJ, Gheslin MT, Raff EC, Pace NR, Raff RA (1988) Molecular phylogeny of the animal kingdom. Science 239:748-550

Fox GE, Stackebrandt E, Hespell RB, Gibson J, Maniloff J, Dyer TA, Wolfe RS, Balch WE, Tanner R, Magrum L, Zablen LB, Blakemore R, Gupta R, Bonen L, Lewis BJ, Stahl DA, Luehrsen KR, Chen KN, Woese CR (1980) The phylogeny of prokaryotes. Science 209:457-463

Fuhrman JA, Azam F (1982) Thymidine incorporation as a measure of heterotrophic bacterioplankton production in surface waters: evaluation and field results. Mar Biol 62:1-12

Giovannoni S J, DeLong E F, Olsen G J, Pace N R. (1988) Phylogenetic group-specific oligodeoxynucleotide probes for identification of single microbial cells. J Bacteriol 170:720-726.

Giovanonni SJ, Britschgi TB, Moyer CL, Field KG (1990a) Genetic diversity in Sargasso Sea bacterioplankton. Nature 345:60-63

Giovanonni SJ, DeLong EF, Schmidt TM, Pace, NR (1990b) Tangential flow filtration and preliminary phylogenetic analysis of marine picoplankton. Appl Environ Microbiol 56:2572-2575

Hagstrom A, Larsson U, Horstedt P, Normark S (1979) Frequency of dividing cells, a new approach to the determination of bacterial growth in aquatic environments. Appl Environ Microbiol 37:805-812

Hobbie JE, Daley RJ, Jasper S (1977) Use of nuclepore filters for counting bacteria by fluorescence microscopy. Appl Environ Microbiol 33:1225-1228

Iwabe N, Kuma K, Hasegawa M, Osawa S, Miyata T (1989) Evolutionary relationship of archaebacteria, eubacteria, and eukaryotes inferred from phylogenetic trees of duplicated genes. Proc Natl Acad Sci USA 86:9355-9359

Jannasch HW, Jones GE (1959) Bacterial populations in seawater as determined by different methods of enumeration. Limnol Oceanog 4:128-139

Karl DM (1986) Determination of *in situ* biomass viability metabolism and growth. In Bacteria in Nature, J. S. Poindexter and E. R. Leadbetter, eds., pp 85-176, Plenum Press

Lane, DJ, Pace B, Olsen GJ, Stahl DA, Sogin ML, and Pace NR (1985) Rapid determination of 16S ribosomal RNA sequences for phylogenetic analyses. Proc Natl Acad Sci USA 82:6955-6959.

Meyerhaus A, Vartanian J, Wain-Hobson S (1990) DNA recombination during PCR. Nuc Acid Res 18:1687-1691

Medlin L, Elwood HJ, Stickel S, Sogin ML (1988) The characterization of enzymatically amplified eukaryotic 16S-like rRNA-coding regions. Gene 491-499

Olsen, GJ (1988) Phylogenetic analysis using ribosomal RNA sequences. Meth Enzymol 164: 793-812

Olsen, GJ (1987) Earliest phylogenetic branchings: comparing rRNA-based evolutionary trees inferred with various techniques. Cold Springs Harbor Symp Quant Biol 52:825-837

Pace, NR, Stahl DA, Lane DJ, and Olsen GJ (1986a) The analysis of natural microbial populations by ribosomal RNA sequences, In K. C. Marshall (ed.), Advances in microbial ecology, vol 9, Plenum Press, New York, p 1

Pace NR, Olsen GJ, Woese CR (1986b) Ribosomal RNA phylogeny and the primary lines of descent. Cell 45:325-326

Sakai RK, Gelfand DH, Stoffel S, Scharf S, Higuchi R, Horn GT, Mullis KB, Erlich HA (1988) Primer-directed enzymatic amplification of DNA with a thermostable DNA polymerase. Science 239:487-493

Schleifer KH and Stackebrandt E (1983) Molecular systematics of prokaryotes. Ann Rev Microbiol 37: 143-187

Schmidt TM, DeLong EF and Pace NR (1990) Comparison of marine picoplankton communities: phylogenetic analysis of cloned rDNAs from a natural picoplankton population, submitted

Sieburth JM (1988) The trophic roles of bacteria in marine ecosystems are complicated by synergistic consortia and mixotrophic cometabolism. Prog Oceanog 21:117-128

Stackebrandt E, Murray RGE, Truper HG (1988) Proteobacteria classis nov., a name for the phylogenetic taxon that includes the *Purple bacteria and their relatives* Int J Sys Bacteriol 38:321-325

Stahl DA, Flesher B, Mansfield HR, and Montgomery L (1988) Use of phylogenetically based hybridization probes for studies of ruminal microbial ecology. Appl Environ Microbiol 54:1079-1084

Swofford DL and Olsen, GJ (1990) Phylogeny reconstruction In D. M. Hillis, and C. Moritz (eds.) Molecular systematics, Sinauer Assoc. Inc. Sunderland, MA.,pp 411-501

Tsien HC, Bratina BJ, Tsuji K, and Hanson RS (1990). Use of oligodeoxynucleotide signature probes for identification of physiological groups of methylotrophic bacteria. Appl Environ Microbiol 56:2858-2865

Tomioka, N, Sugiura, M (1983) The complete nucleotide sequence of a 16S ribosomal RNA gene from a blue-green alga, *Anacystis nidulans*. Mol Gen Genet 191:46-50

Wallace, R B and Miyada CG (1987) Oligonucleotide probes for the screening of recombinant DNA libraries. Meth Enzymol 152:432-433

Ward DM, Weller, R, Bateson MM (1990) 16S rRNA sequences reveal numerous uncultured microorganisms in a natural community. Nature 345:63-65

Waterbury JB, Watson SW, Guillard RRL, Brand LE (1979) Wide-spread occurrence of a unicellular, marine, planktonic, cyanobacterium. Nature 277:293-294

Waterbury JB, Wiley JM, Franks DG, Valois FW, Watson SW (1985) A cyanobacterium capable of swimming motility. Science 230:74- 76

Waterbury JB, Watson SW, Valois FW, Franks DG (1986) Biological and ecological characterization of the marine unicellular cyanobacterium *Synechococcus*. In:Platt T, Li WKW, (eds.) Photosynthetic picoplankton. Can Bull Fish Aquat Sci vol 214, p71-120

Weller R, Ward DM (1989) Selective recovery of 16S rRNA sequences from natural microbial communities in the form of cDNA. Appl Environ Microbiol 55:1818-1822.

Wilson AC, Carlson SC, White TJ (1977) Biochemical evolution. Ann Rev Biochem 46:573-639

Woese CR (1987) Bacterial Evolution. Microbiol Rev 51:221-271.

Woese CR, Kandler O, and Wheelis ML (1990) Towards a natural system of organisms: proposal for the domains Archaea, Bacteria, and Eukarya. Proc Natl Acad Sci USA 87:4576-4579

Yang DC, Oyaizu Y, Oyaizu H, Olsen GJ, Woese CR (1985) Mitochondrial origins. Proc Natl Acad Sci USA 82:4443-4447

Zuckerkandl E, Pauling L (1965) Molecules as documents of evolutionary history. J Theo Biol 8:357-366

From cell to Oceans

FROM INDIVIDUAL PLANKTON CELLS TO PELAGIC MARINE ECOSYSTEMS AND TO GLOBAL BIOGEOCHEMICAL CYCLES [1]

L. Legendre[2] and J. Le Fèvre[3]

[2]Département de biologie,
Université Laval,
Québec, Québec, Canada, G1K 7P4

[3]Laboratoire d'Océanographie Biologique,
Université de Bretagne Occidentale,
29287 Brest Cedex, France

ABSTRACT

It is often assumed that the biological CO_2 pump in the oceans is essentially driven by the upward flux of NO_3^-, so that the potential export of biogenic carbon from the upper ocean is stoichiometrically equivalent to N-derived phytoplankton new production. The steady-state/stoichiometry model does not consider that ecosystem dynamics have a significant influence on the export or sequestration of biogenic carbon. Contrary to this assumption, there are several cases where ecosystems do have a significant effect on the export and sequestration of biogenic carbon. These include the fixation of nitrogen gas in the upper layer of the ocean (phytoplankton and coral reefs), the export of carbon by organisms with high carbon content (thecate dinoflagellates, coccolithophores, foraminifers, pteropods), the production of long-lived dissolved organic matter, and the fact that the pathways of export may differ in length and complexity. This is considered within the context of a general typology of pelagic marine ecosystems, based on the various possible combinations linking standing stock to production of phytoplankton: (1) production and standing stock dominated by large cells (e.g. upwelling, ice-edge and episodic blooms); (2) production by small and large cells, standing stock dominated by large cells (e.g. exceptional blooms); (3) production and standing stock of small and large cells (e.g. spring bloom in the North Atlantic); (4) production by small and large cells, standing stock dominated by small cells (e.g. Alaskan Gyre in the North Pacific); (5) production and standing stock dominated by small cells (e.g. oligotrophic ocean). These five types of ecosystems correspond to different modes of phytoplankton production (as controlled by hydrodynamics) and different structures (as reflected in the standing stocks), which influence the pathways of carbon export and sequestration as well as the renewable marine resources.

INTRODUCTION

One of the most pressing issues presently confronting the scientific community, worldwide, is the possible rapid global warming of the Earth. The recent anthropogenic increase in atmospheric carbon dioxide may be responsible for about half this global warming

---

[1]Contribution to the programme of GIROQ (Groupe interuniversitaire de recherches océanographiques du Québec) and Unité de recherche D13630, Centre National de la Recherche Scientifique (France)

NATO ASI Series, Vol. G 27
Particle Analysis in Oceanography
Edited by S. Demers

(Ramanathan *et al.*, 1985). This is not to say that the Earth climate was at any time in a steady state, but rather that the present rate of change might be much more rapid than normally resulting from geophysical events (e.g. glaciations, plate tectonics, etc.) Terrestrial ecosystems (e.g. forests) may be unable to cope with changes occurring over decades instead of centuries or millennia. In addition, rising sea level combined with subsiding continental masses and increasing frequency of storms might endanger many coastal areas and cities.

It has been estimated that up to about half the CO_2 released since the beginning of the industrial era has probably been absorbed by the oceans (Sundquist, 1985). Carbon dioxide transferred into the deep oceanic waters is effectively removed from the atmosphere for centuries, thus reducing the magnitude of global warming. Carbon dioxide enters the upper ocean by gas exchange across the air-sea interface or as dissolved compounds in river waters; it can be transferred to deep waters by two general pathways: (1) ocean-scale transport of ***dissolved*** inorganic and organic carbon by deep convection, and (2) local sinking of biogenic ***particles*** from the surface into the deep sea. The second pathway is orders of magnitude faster than the first one (days/weeks vs. decades/centuries; see Legendre and Gosselin, 1989). Close to the surface (euphotic layer), solar light fuels the photosynthetic incorporation of inorganic carbon into organic molecules by microalgae, a fraction of which sinks to deep waters (as intact cells, faecal pellets, marine snow) or is actively transported by vertically migrating organisms. This export pathway is known as the "biological CO_2 pump" (Volk and Hoffert, 1985). Understanding and quantifying this biological pump, as well as the other CO_2 export pathways, is central to the Joint Global Ocean Flux Study (JGOFS; e.g. Brewer *et al.*, 1986; Platt *et al.*, 1989a), which is an integral part of the International Geosphere-Biosphere Programme.

Another area of growing concern for the scientific community is the present state of exploited marine renewable resources. Many authors have pointed out that the models used for stock management are inadequate, and that the general relationship between primary production and fish yield, which is probably mediated by the success of larval recruitment to the adult population, still eludes understanding (see Lasker, 1988, and Rothschild, 1988), potentially

Table 1. Glossary of some oceanographic terms, as understood in the present paper.

Biological CO_2 pump: Photosynthetic incorporation of inorganic carbon into organic molecules by microalgae, followed by export to depth (see phytoplankton production export); the biologically driven downwards flux of carbon causes a proportional flux of carbon dioxide from the atmosphere into the ocean.

Bloom of phytoplankton: A rapid increase in phytoplankton biomass, caused by locally enhanced primary production, that result in an abnormally high cell concentration (Legendre, 1990).

Hydrodynamic singularity: Abrupt change in hydrodynamic characteristics, e.g. pycnocline, eddy, front, upwelling area, temporal transitions in vertical stability of the water column (Legendre and Le Fèvre, 1989).

Hydrodynamic trap: Any hydrodynamic structure (e.g. convergent circulation associated with fronts, eddies, Langmuir convection cells) that can accumulate particulate material, originating from a wider area, into a restricted location. The particles may be detrital material or living organisms and the accumulation may be purely passive, e.g. through a combination of buoyancy and downwelling, or partly active, e.g. upwards swimming and downwelling (Le Fèvre and Frontier, 1988).

Nitrogenous nutrients (see Fig. 1)

allochthonous: imported into the surface waters from depth (mainly as nitrate) and also from the atmosphere (including nitrate, ammonium and N_2 gas fixed by cyanobacteria);

autochthonous: excreted in the surface waters by the heterotrophic food web (including ammonia, urea and amino acids).

Picoplankton: Fraction of the plankton ranging in size between 0.2 and 2 μm (Sieburth *et al.*, 1978).

Phytoplankton production (see Fig. 1)

export: flux of biogenic material from surface waters to depth, that include long-lived dissolved organic matter as well as sinking and swimming particles (up to fish and whales);

new: production derived from allochthonous nutrients, imported into the surface waters from depth and also from the atmosphere; often estimated from the uptake of ^{15}N-labelled nitrate;

recycled: production which does not leave the surface waters;

regenerated: production fuelled by autochthonous nutrients, regenerated in the surface waters by the heterotrophic food web; often estimated from the uptake of ^{15}N-labelled ammonia (and sometimes urea);

total: new + regenerated, or export + recycled; often estimated from the fixation of $H^{14}CO_3^-$.

Production: Flow of primary energy into any trophic compartment (Legendre *et al.*, 1986); in the oceans, the dominant form of primary energy is the photosynthetically available solar radiation and the main primary producers are phytoplankton; secondary production corresponds to the flow, into another trophic compartment, of energy biochemically stored by primary producers; and so forth along the food webs.

Sequestration of CO_2: Removal of dissolved inorganic CO_2 from the atmosphere and the surface waters for a period of interest to global warming (i.e. at least a few hundred years). This includes the sinking or downwards transport of biogenic dissolved and particulate carbon followed by burial in sediments or dissolution in deep waters, the long-term fixation of carbon into calcareous material, and the binding of carbon into long-lived dissolved organic matter.

Ultraplankton: Fraction of the plankton with size smaller than 5-10 μm (Sverdrup *et al.*, 1942).

leading to overexploitation. Fishing practices such as the use of drift nets in the North Pacific Ocean may also have devastating effects on marine stocks and ecosystems. In addition, the increasing frequency of exceptional phytoplankton blooms (e.g. red tides, brown tides, green tides, etc.) may imperil coastal fisheries and aquaculture operations (e.g. White, 1984; Gosselin *et al.*, 1989). Recent reviews (e.g. Parsons and Lalli, 1988; Cushing, 1989; Legendre, 1990) have suggested that one of the key factors in the control of exploited marine populations may be phytoplankton production and its export pathways.

Flow cytometry and image analysis, which provide estimates of particle size together with simultaneous cell-by-cell measurements of various characteristics, are playing an increasingly important role in biological oceanography (see the review by Legendre and Yentsch, 1989). So far, a large share of the effort has been devoted to the development of techniques suitable for analyzing aquatic particles. As suggested by Legendre and Yentsch (1989), the next phase, which is already under way in several laboratories, will be to use these techniques for developing and testing theories in biological oceanography. Within this context, the present paper will show that a significant component of the biologically driven flux of carbon dioxide into the oceans is determined from the level of individual cells up, and mediated by ecosystem properties. This is significant for the sequestration of carbon in the oceans and also for exploited renewable marine resources. It will be concluded that relevant characteristics of biogenic particle dynamics and of marine ecosystems must be incorporated into the models developed for the management of exploited stocks and for the prediction of the global flux of carbon into the oceans[4].

ROLE OF PELAGIC MARINE ECOSYSTEMS IN THE SEQUESTRATION OF CARBON

It is often assumed, at least implicitly, that the biological CO_2 pump is essentially driven by the upward flux of nitrate, so that the potential export of biogenic carbon from the upper ocean would be stoichiometrically equivalent to phytoplankton new production (e.g. Eppley and Peterson, 1979), as estimated from the uptake of $^{15}NO_3^-$ (Dugdale and Goering, 1967) or the upward flux of NO_3^- (e.g. Lewis *et al.*, 1986). This approach assumes that ecosystems are

---

[4] Table I gives glossary of some oceanographic terms, as understood in the present paper.

in a steady-state condition over the significant spatio-temporal scales; it also implies that variations in the proportion of carbon and nitrogen assimilated (and released) by marine plankton (see Lancelot and Billen, 1985, and references therein) only have short term relevance and that the stoichiometric balance is conserved at the steady-state scale. However: "*...the underlying hypothesis of a dynamic steady state for the biogeochemistry of the pelagic region of the ocean is to be considered as no more than an idealisation of a system that, in reality, is perturbed intermittently such that the condition of instantaneous equilibrium is rarely, if ever, attained*" (Platt *et al.*, 1989b). One of the best-known marine environments in the world, on a long-term basis, is the English Channel and the North Sea and, in addition to the usual interannual variations, major changes over decades are widely documented there. These include variations in the total winter stock of inorganic nutrients, yearly primary production, zooplankton maximum standing stock (by at least one order of magnitude) and taxonomic composition, and commercial fish stocks (see Russell *et al.*, 1971; Southward, 1980, and references therein). The observed changes were at first considered cyclical (the "Russell Cycle"; Cushing and Dickson, 1976), but the cyclical pattern has recently weakened (Southward, 1983) and non-periodic changes have been described in association with pluri-annual climatic phenomena (Cushing, 1988). With the help of historical fisheries records, such changes can be traced back in the past for several centuries (e.g. Southward, 1974). Longer-term changes are of course associated with the glaciations, so that the proper time-scale for a steady-state balance in production processes can be elusive indeed.

The usual steady-state/stoichiometry model does not consider that ecosystem dynamics may have a significant influence on the export or sequestration of biogenic carbon. Given a certain upward flux of nitrate, it is assumed that there will be a proportional downward flux of biogenic carbon, if the fluxes are integrated over proper spatio-temporal scales (Platt *et al.*, 1989b). Contrary to this usual assumption, we shall first review several cases where ecosystems do have a significant effect on the export and sequestration of biogenic carbon (the next four sections), and then consider these various cases within the context of a general typology of pelagic marine ecosystems.

NITROGEN FIXATION (AND DENITRIFICATION)

Direct fixation of N_2 gas by phytoplankton is one mechanism by which marine ecosystems may influence the export of carbon to depth. Until recently, it had been assumed that direct fixation of N_2 by cyanobacteria was negligible in surface waters, because this process does not normally take place in the presence of oxygen. However, recent evidence (reviewed by Legendre and Gosselin, 1989) suggests that the upper layer of the ocean may offer a variety of reducing environments, potentially favourable for N_2 fixation, which include oxygen-depleted microzones associated with surfaces of organic and inorganic aggregates and internal microzones within aggregates (bundles) of the filamentous N-fixing cyanobacterium *Oscillatoria* spp. In addition, Ohki and Fujita (1988) have shown that N_2 fixation could sometimes occur in the presence of relatively high oxygen concentrations. Under steady state conditions, phytoplankton production derived from N_2 fixation should lead to carbon export. Indeed, Lewis *et al.* (1986) pointed out that estimates of nitrogen supply by vertical turbulent transport, in the oligotrophic ocean, are too low to meet the demand implied by production estimates derived from indirect methods such as annual changes in oxygen concentration, and that part of the discrepancy could be accounted for if the amount of N_2 fixation were higher than previously thought. It must actually be stressed that, because N_2 is the most abundant gas dissolved in seawater, the export of biogenic carbon fuelled by nitrogen fixation is independent of any hydrodynamically or atmospherically driven influx of NO_3^- or NH_3/NH_4^+. It follows that conditions enhancing N_2 fixation may be of major significance for the export of biogenic carbon from the surface waters, especially in oligotrophic oceanic waters where nitrate-driven new production is usually thought to be low (Eppley and Peterson, 1979), although the most recent estimates tend to be somewhat higher than the previous ones (see Eppley, 1989, and Platt *et al.*, 1989b).

Eukaryotic cells provide an ideal reducing environment for N_2 fixation by cyanobacteria. In the N-depleted oligotrophic waters of the central North Pacific Ocean, for example, the endophytic cyanobacterium *Richelia intracellularis*, which has been shown to be N_2-fixing (Mague *et al.*, 1974), is frequently associated with several species of the diatom genus *Rhizosolenia*. Venrick (1974) found it in more than 90% of the cells in three *Rhizosolenia* species that were responsible for localized blooms in the North Pacific Central Gyre, and she hypothesized that nitrogen fixation was instrumental in the development of the blooms. More recently, Heinbokel (1986), using epi-fluorescence microscopy[5], found that *Richelia* were much more frequent than previously thought (i.e. present in about 80% of the hosts' cells) in two species of the diatom genus *Hemiaulus* from the Pacific Ocean, and concluded that the significance of this kind of symbiosis in the nitrogen dynamics of the ocean may have been largely underestimated in previous studies. Indeed, Carpenter (1983), in evaluating the global contribution of N_2 fixation effected by marine pelagic cyanobacteria (which he put at about $5\ 10^{12}$ g a^{-1}), pointed out the unfortunate lack of information on the contribution of fixed nitrogen by *Richelia*; he emphasized that taking into account this contribution could result in a major upward revision of the estimates.

Coral reefs are also thought to account for a significant share of global N_2 fixation (Capone, 1983) and to play an important role in carbon sequestration. They develop amid an environment that is particularly poor in nitrogen compounds, and have nevertheless be found to be net exporters of N-nutrients, at a rate that could exhaust the reserves in the local biomass pool within months (Webb *et al.*, 1975). Interactions between the members of coral reef communities are extremely complex (see, for instance, Mann, 1982); one of the major links is the association between corals and zooxanthellae (endosymbiotic dinoflagellates), which allows the corals to globally behave as functional photoautotrophs. In the same way as phytoplankton, however, the algal symbionts require a conventional source of N-nutrients, which cannot be supplied by the surrounding oligotrophic waters. Nitrogen fixation in coral reefs is effected by the cyanobacterium *Calothrix crustacea*, that occurs in the form of a thin film or a feltlike mat over different portions of the reefs (Wiebe *et al.*, 1975). The process has been estimated to take place at as high a rate as has been found in any other system, marine or terrestrial (see Capone, 1983), and to be the most important nitrogen source for the coral community, through various steps, which are still incompletely elucidated and may include grazing by fish, excretion, nitrification by bacteria, etc. Alternatively, it has been hypothesized (Rougerie and Wauthy, 1986) that the major source is an upward flow of nutrient-rich deep water driven, within the porous structure of the reefs, by the heat flux from the subsiding volcanoes on which atolls generally grow. Whatever their nitrogen source,

[5]Large cells containing chloroplasts and endophytic cyanobacteria would be ideal tagets for flow-cytometric studies.

corals precipitate large amounts of dissolved carbonate, a process in which the symbiosis with zooxanthellae is instrumental, and are therefore quite effective in sequestering carbon, directly in the euphotic layer, despite the widely held view that their "net community production" (i.e. their capacity to export organic carbon) is low (see Mann, 1982). According to Bé *et al.* (1982), the symbiotic relationship with zooxanthellae allows corals and foraminifers (see below) to be the foremost producers of biogenic calcite in the circumglobal warm-water belt, the former over the shallow platforms and the latter over the deep sea bed. The possible effects of calcite precipitation by coral reefs and other organisms on atmospheric CO_2 is discussed at the end of the next section. In addition, coral reefs are thought to be net exporters of nitrate and could be indirectly fuelling the planktonic carbon pump in surrounding oligotrophic waters.

Denitrification is the antagonistic process to nitrogen fixation, by which oxidized forms of combined nitrogen (nitrate, nitrite and nitrous oxide) are reduced to free nitrogen gas. As in the case of nitrogen fixation, it is generally considered that denitrification only occurs under anaerobic or low-oxygen conditions (see Codispoti, 1989). In the open ocean, sites favourable for denitrification would be oxygen-deficient portions of the water column, below the euphotic layer (e.g. Ward and Zafiriou, 1988). More generally, denitrification could take place in reducing environments (e.g. marine snow) similar to those potentially favouring N_2 fixation (see above). According to Codispoti (1989), denitrification and nitrogen fixation may be approximately balanced on a time scale of about 107 years, but imbalances in which denitrification may exceed nitrogen fixation could reduce export production, and thus potential sequestration of carbon, by 20 to 30% for periods of several thousands years.

PLANKTON ORGANISMS WITH HIGH CARBON CONTENT

Estimates of carbon export from measurements of nitrogen-derived new production are generally based upon the so-called "Redfield ratio" (e.g. Redfield, 1958; Redfield *et al.*, 1963), i.e. in terms of atoms C/N/P = 106/16/1 (hence C/N = 6.6), which corresponds to the average cellular composition of marine plankton[6]. However, many plankton organisms export an extra share of carbon, either as organic thecae or in the form of calcareous skeletal

[6]All elemental ratios will henceforth be quoted in atoms.

structures, consisting mainly of $CaCO_3$ with small and variable proportions of other substances such as $MgCO_3$. Carbon export from ecosystems with such organisms is thus in stoichiometric imbalance relative to nitrogen, i.e. each atom of nitrogen derived from allochthonous nutrients effectively causes, under steady state conditions, the export of more atoms of carbon than predicted from the Redfield ratio.

Many dinoflagellates are enclosed in a theca, usually consisting mainly of cellulose (Loeblich, 1970), but sometimes of other carbohydrates (e.g. a non-cellulosic polyglucan in the case of *Peridinium cinctum*; Nevo and Sharon, 1969). A thick cell wall contributes to increasing the carbon content of thecate dinoflagellates (Gonyaulacales and most Peridiniales are armoured with several thick plates). Wynne *et al.* (1982), for instance, found a C/N ratio of 13.9 for *Peridinium cinctum* under bloom conditions in Lake Kinneret, Israel. Similarly, Haug *et al.* (1973) reported that, during the annual phytoplankton annual cycle in Trondheimsfjord, Norway, insoluble carbohydrate accounted for 32-52% of the organic matter in those samples rich in dinoflagellates whereas the figure was usually $< 10\%$ when dinoflagellates were scarce, and they ascribed the difference to dinoflagellate thecal carbon.

Plankton with calcareous tests include photosynthetic cells (mainly coccolithophores), protozoans (foraminifers) and metazoans (mainly thecosomatous pteropods). Unfortunately, no maps of global carbonate productivity have been established (Bishop, 1989). However, according to Lisitzin (1972, as quoted by Bishop, 1989), coccolith and foraminiferal productivity is maximal between 50°N and 50°S, in general coincidence with waters with a temperature above 10°C; the highest concentrations of suspended particulate carbonate are found in surface waters in two mid-latitude belts, i.e. near the Antarctic Convergence (40-60°S) in the southern hemisphere and near the polar front in the northern hemisphere, and along the continental margins. In the North and tropical Atlantic Ocean, Honjo (1980) and Takahashi and Bé (1984) report average sinking velocities between 65 and 2500 m d^{-1} for foraminiferal and pteropod tests, with a value > 5000 m d^{-1} for the larger pteropod shells (ca. 10 mm); coccoliths, in that case, were mainly carried in faecal pellets, and thus had a sedimentological behaviour similar to large particles. Calcareous tests may therefore be very efficient in exporting biogenic carbon to depth.

High concentrations of suspended particulate carbonate along continental margins (Lisitzin, 1972; see above) might well correspond to coccolithophore blooms, as observed by Holligan *et al.* (1983) along the continental shelf edge in the Celtic Sea. According to Holligan (1987), blooms of coccolithophores occur under conditions of intermediate turbulence, maintained by physical instabilities such as tidal mixing and upwelling. An example of the significance of coccolithophore blooms for the vertical transport of carbonate is provided by Honjo (1982) who observed, during a bloom of the coccolithophore *Umbellicosphaera sibogae*, a mass flux of particles of 876 mg m^{-2} d^{-1} with carbonate accounting for 62% of the total flux at 3560 m. The production and export of coccolithophores in oceans, as well as their significance for the global biogeochemical cycles, are discussed in detail by Holligan and Balch (this volume).

An important aspect of foraminiferal production is that several planktonic species (belonging to the genera *Globigerinoides* and *Orbulina*) consistently possess symbiotic dinoflagellates (zooxanthellae), and a species of the genus *Hastigerina* harbours commensal dinoflagellates (refs. in Bé *et al.*, 1982). Experimental study of photosynthesis in the planktonic foraminiferan *Globigerinoides sacculifer* (Jϕrgensen *et al.*, 1985) has shown that the uptake of dissolved nitrogen and phosphorus by the organism is diffusion-limited, so that the capture of prey is necessary to supply nutrients to the dinoflagellate symbiont; however, symbiotic photosynthesis can cover the energy requirements of the foraminifers and allow efficient recycling of the nutrients. In addition, Bé *et al.* (1982) have shown that the symbionts enhance calcification of their host's skeleton, as in the case of corals (see above). In the tropical oligotrophic ocean, however, foraminifers are found in very low numbers, so that their contribution to the total carbon production is small (e.g. 0.1-0.2%; Jϕrgensen *et al.*, 1985); given their calcareous tests and their high sinking velocities, the contribution of foraminifers to the export and sequestration of carbon may nevertheless be proportionally higher. The significance of the phenomenon in other environments remains to be quantified.

While apparently no data are available on the partition of carbon between organic matter and carbonate in coccolithophores or foraminiferans, some information on the matter can be found in the literature for thecosomatous pteropods. The values given by Gorsky *et al.* (1988) convert to a C/N ratio of 6.5 for the organic matter of *Cavolinia inflexa*, i.e. almost exactly the Redfield ratio, while Curl (1962) ascribes 2.74% of the dryweight in *Limacina* sp. to carbonate carbon and 25.6% to organic carbon. If these figures are combined to yield typical pteropod values, the resulting C/N ratio for the whole animal is 7.2, i.e. an increase by about 11% with respect to the value in the organic matter. Omori (1969) tabulated weight measurements and elemental analyses for a number of zooplankton species in the North Pacific, including four pteropods; the highest C/N value, for *Limacina inflata*, is 12.9, a 95% increase over the Redfield ratio. Whether or not the pteropod specimens were decalcified prior to analysis is not explicitly stated, but indirect clues suggest that they were not, and that it is reasonable to assume that the reported discrepancy with respect to the Redfield ratio is largely due to carbonate carbon. The same is true for Beers (1966) data from the Sargasso Sea, where the C/N ratio for pteropods ranges from 6.4 to 9.7, with a mean value of 8.2, i.e. a 24% increase over the Redfield ratio.

At least part of the thecosomatous pteropods occupy a specialized niche in the pelagic trophic webs. Together with gelatinous organisms (e.g. salps and appendicularians), they are among a number of forms that use ciliary-mucous mechanisms to collect their food; these mechanisms are part of a filtration system in the gelatinous plankters and take the form of an external net in the pteropods. As a result, these organisms are most often able to feed on very small particles, down to bacterial size in the case of appendicularians (King *et al.*, 1980). The ability to feed on protozoan-sized plankton is well-documented in the pteropod genus *Limacina* (e.g. Longhurst and Williams, 1979), although the larger species, such as *L. helicina* also use their net to capture metazoans such as small copepods (Gilmer and Harbison, 1990). Large populations of small pteropods tend to develop by feeding on protozoans preying upon bacteria, in situations where the latter are recycling accumulated particulate organic matter (see Le Fèvre and Frontier, 1988). In northern temperate seas, pteropods can accordingly be expected to dominate in autumn, when significant recycling of the biomass built up in the productive season takes place. Indeed, in the Norwegian Current region, Peinert *et al.* (1989) report that *Limacina* are quite abundant at the time when the copepod populations significantly decrease and that *Limacina* shells are a major component of the material recovered in sediment traps (at a depth of 1000 m) from September to November (and at that time of the year only). In European waters, high abundance of *Limacina* has also been reported earlier in the annual cycle, in situations where large phytoplankton stocks were apparently being decomposed by bacteria rather than being grazed upon by herbivores (Le Fèvre, 1986). More generally, the pteropod link in food webs, and the resulting sequestration of carbon in the form of carbonate (in excess of the Redfield ratio by at least 10% and possibly up to 100%) should be favoured by conditions under which preexisting organic matter is being recycled by microheterotrophs. Overall, it has been estimated (Berner and Honjo, 1981) that the pelagic aragonite flux (contributed to mainly by pteropods), largely from shallow waters but with significant lateral transport, accounts for at least 12% of the worldwide carbonate flux to the deep ocean.

An interesting aspect of carbonate export from the surface layer into deep waters is the dissolution of calcareous tests with depth, so that calcite (produced, among pelagic organisms, by coccolithophores and foraminiferans) is not found in sediments deeper than 4000-5000 m and aragonite (produced by pteropods) in those deeper than 1000-2500 m. In deep-water areas, carbonate-carbon is thus progressively released into the water column, so that its fate (i.e. rapid release to the atmosphere vs. sequestration for tens to hundreds of years in the deep waters) depends on sinking velocity and on the deep circulation. In shallow waters, on the other hand, carbonate sediments may sequester carbon for millions of years (e.g. sediment accumulation up to several kilometres deep near the continents and on the flank of ridges, limestone deposits from epicontinental seas accumulated on the continents). Areas shallower than 3000 m and 4000 m represent more than 20% and 40%, respectively, of the total surface of the oceans. These, and especially the continental margins, should therefore be carefully investigated when assessing the global flux and sequestration of carbon in the oceans. Another reason for taking the continental shelves into account is the possible significant export of both dissolved and particulate organic matter from the ocean margins into the deep sea (e.g. Walsh, 1989).

Concerning the biogeochemical cycle of carbon, the production of plankton organisms with calcareous tests leads to rather complex results. Limestone deposits were derived, through biological processes, from atmospheric CO_2 but it has been postulated (e.g. Volk, 1989) that, at some geological time scales (e.g. glacial/interglacial oscillations), the precipitation of carbonate by marine organisms results in higher atmospheric CO_2 through changes in the carbonate chemistry of surface waters (i.e. lower alkalinity and concentration of total dissolved CO_2). This effect has been discussed for coccolithophores (e.g. Dymond and Lyle,

1985; see also Holligan and Balch, this volume) as well as corals (Berger, 1982). On the other hand, plankton organisms with calcareous tests generally sink faster than other organic particles produced in surface waters (see above), thus reducing the amount of carbon respired during their downward transit, and the adsorption of organic matter onto calcite particles (Suess, 1973) may contribute to increase the burial of organic carbon in ocean sediments (see Holligan and Balch, this volume). The production and export of biogenic carbon by calcareous organisms may therefore lead to high sequestration of inorganic and organic carbon, but the overall effect of calcite precipitation by organisms could, at least on some time scales, be an increase in atmospheric CO_2. The phenomenon is therefore of major significance for the biogochemical flux of carbon, on both the short and long terms. As far as thecate dinoflagellates are concerned, their production and export, although poorly documented in terms of actual C/N ratios at sea, should favour high sequestration of carbon in the oceans.

LONG-LIVED DISSOLVED ORGANIC MATTER

It has been recently proposed that the concentrations of dissolved organic carbon (DOC) and nitrogen (DON) in oceanic waters are much higher than previously thought (Sugimura and Suzuki, 1988). Toggweiler (1989) has discussed the implication of these new values, and reported results from a modelling exercise suggesting that about half the new production could be exported to depth as long-lived DOM (with a characteristic lifetime of 200 years). Legendre and Gosselin (1989) have shown that the partitioning of new production into POM and DOM (Fig. 1), under the present conditions of global change, may effectively uncouple the export of biogenic material to depth from the usually measured "f-ratio" (ratio of new to total production). However, it must be realized that the long-lived DOC could chemically retain (i.e. sequester) carbon in the surface layer, for periods of interest to global change (i.e. 100s years), before this carbon is transferred and released as CO_2 (respired) into the deep waters. Toggweiler (1989) hypothesized that the long-lived organic compounds reported by Sugimura and Suzuki (1988) may be formed by condensation reactions between carbon-rich phytoplankton exudates and nitrogen-rich bacterial enzymes. Even if conditions leading to the production and breakdown of the long-lived DOM are still poorly known, they obviously are of major interest for the study of global carbon fluxes in the oceans.

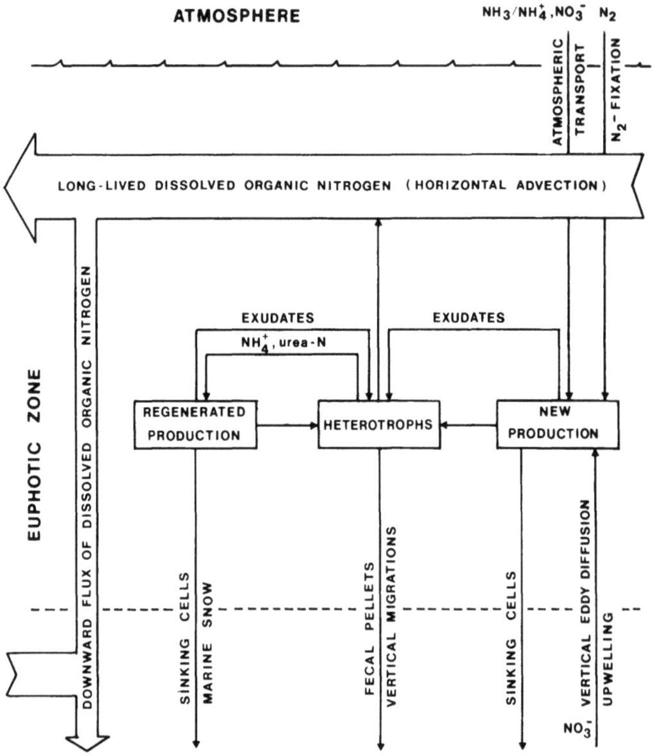

Figure 1. Export of dissolved and particulate organic matter from the euphotic layer and sources of allochthonous inorganic nitrogen for phytoplankton production; in the euphotic layer, relationships between new and regenerated production, pathways of nitrogen recycling, and advection of long-lived dissolved organic nitrogen. From Legendre and Gosselin (1989).

EXPORT PATHWAYS

Finally, variations in the pathways through which production is exported may also contribute to uncouple the actual sequestration of biogenic carbon from the influx of allochthonous nitrogenous nutrients. The word "export" is used here in a broader sense than generally understood by geologists and geochemists, who usually consider only the particulate material sinking from the euphotic layer, in other words "*the amount of fallout* [that the ocean] *generates from the photic zone: that is (...) the amount of organic matter which will accumulate in a trap set below this zone*" (Berger *et al.*, 1989). Material passed on to the food web also is, in the ecological sense, export from the primary production system. It may end up being exported from the production area either horizontally, through passive transport associated with circulation patterns or active migration of large animals, or vertically, again passively, through sedimentation of living or detrital particles, or actively, through vertical migrations. The production of long-lived dissolved organic matter should also be considered (see above). Not all export pathways result in the sequestration of carbon; commercial harvesting of fish, for instance, will eventually lead to carbon recycling through respiration, as will some other types of export to the food webs. The fraction of the original carbon production that will eventually be sequestered is largely controlled by ecosystem processes, which may result in a variable extent of uncoupling, in time and in space, between production and export in the geochemical sense. Legendre and Le Fèvre (1989) have reviewed the various export pathways from the primary producers, their hydrodynamic control, and their significance in terms of particulate organic matter (POM) fallout from the surface layer (Fig. 2). The possibilities include the sinking of large intact cells, the grazing of large cells by herbivores and production of faecal pellets as well as active transport by vertically migrating organisms, the grazing of detrital biogenic material (accumulated in hydrodynamic traps) by microphagous feeders, the incorporation of small cells into sinking marine snow, and the direct grazing of ultraplankton by metazoans. As defined in Legendre and Le Fèvre (1989), large phytoplankton cells ($> 5\ \mu m$, including chains of small cells) generally belong to diatoms, dinoflagellates and coccolithophores, while small cells ($< 5\ \mu m$) comprise both eukaryotic and prokaryotic photoautotrophs.

Massive sedimentation of large intact cells, at rates that may exceed 100 m d^{-1} (refs. in Goldman, 1988), mainly occurs under bloom conditions. Legendre's (1990) review shows that microalgal blooms largely depend on the balance between phytoplankton production (resulting from hydrodynamic forcing) and grazing by herbivorous zooplankton (see also Parsons and Lalli, 1988). Under low grazing pressure, most of the large cells may sediment. Many cases are reported in the literature of large fallout of often intact diatom cells following a bloom, either in upwelling regions, or at certain phases of the annual plankton cycle. Smith *et al.*

(1983), for instance, report direct fallout of diatoms to the sea floor (at a depth of 145 m) on the continental shelf in the Peru upwelling area. They found that the diatom cells on the

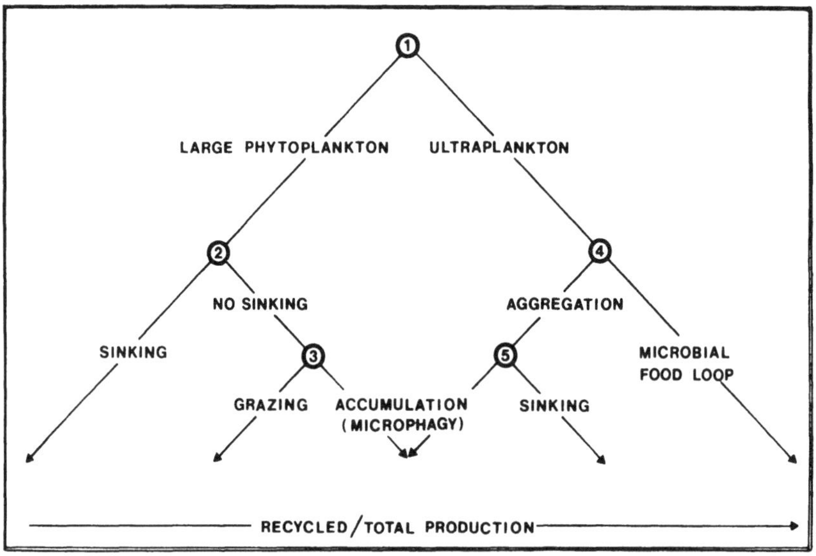

Figure 2. Model of export production (downwards arrows) in oceans. At each bifurcation (numbers), part of the production may be channelled into export pathways, which does not preclude coexistence with recycling pathways. The length and complexity of the food webs involved in the export of biogenic carbon increase towards the right. According to Legendre and Le Fèvre (1989), hydrodynamic conditions control the five bifurcations. Reprinted with permission from the Dahlem Workshop Report LS44, p. 51, John Wiley and Sons, Chichester.

bottom still contained significant amounts of chlorophyll pigments and that the interfacial sediment contained labile fatty acids, known to occur in diatoms and not usually thought to survive water column transit. The process gives rise to underlying deposits of a diatomaceous ooze, rich in organic matter and highly anoxic. Similar sediments are known elsewhere, and the same authors mention for comparison samples from Walvis Bay (Benguela upwelling) still rich in organic matter and mineralogically similar to Peru sediments, although they were older (70 years) and much poorer in labile compounds. It should be noted that reducing environments, such as diatomaceous oozes, should favour the long-term sequestration of organic carbon. Smith and Nelson (1985) and Wilson *et al.* (1986) described a large diatom bloom at the receding ice edge in the Ross Sea (Antarctic Ocean), which probably remained in the water column only for a short time; the authors emphasized that such production events (which may have gone unnoticed in past studies) can be instrumental in the deposition of the large amounts of diatomaceous ooze found in the Southern Ocean. Episodes of phytoplankton sedimentation following blooms are also well known on temperate continental shelves. They can be associated with blooms occurring regularly in the annual cycle, as exemplified by the Kiel Bight of the Baltic Sea, which is particularly well documented (e.g. Graf *et al.*, 1982, 1984; Smetacek *et al.*, 1984), or with episodic blooms resulting from summer mixing events, as reported a number of times, for instance, from coastal areas in the northeastern Pacific Ocean (e.g. Iverson *et al.*, 1974; Takahashi *et al.*, 1977). In the shallower areas, such as the Kiel Bight, the fallout from the blooms is likely to be utilized by the benthos rather than sequestered in the sediments.

Active grazing by herbivores, on the other hand, leads to sedimentation of faecal pellets. Copepod faecal pellets are wrapped in a protective membrane that prevents degradation of the content during at least part of the downwards transit and they package fine material (including coccoliths) into larger, and therefore faster-sinking, particles. The sinking velocity of copepod faecal pellets is of the order of 100 m d^{-1} and shows variations with water temperature; they will sink faster at higher temperature which, on the other hand, will tend to accelerate the microbial degradation of their contents (e.g. Honjo and Roman, 1978), so that there are limits to the effectiveness of copepod faecal pellet contribution to the export and eventual sequestration of biogenic carbon. Microheterotrophs (bacteria and their protozoan predators) will develop, for instance, by taking advantage of ageing biomass and detrital material not being directly consumed by metazoans, especially if the material is being accumulated in hydrodynamic traps, and, more generally, in situations that do not favour the buildup of a full-fledged herbivore food chain (see Legendre and Le Fèvre, 1989). As pointed out above, the specialized consumers of microzooplankton are gelatinous organisms (e.g. salps, doliolids, appendicularians) and pteropods. All of these produce faecal pellets that sink faster than those of copepods or euphausiids (Alldredge, 1984). In addition, gelatinous organisms are instrumental in the production of marine snow, where ultraplankton cells (which would normally not sink at all) are aggregated together with other material and thus made available for export to depth (e.g. Silver *et al.*, 1986). As regards pteropods, a major aspect of their contribution to carbon sequestration is sedimentation of their calcareous shells (see above).

In general, the length and complexity of the food webs involved in the export of biogenic carbon increase towards the right of Fig. 2. Rapid sedimentation of intact cells or other biogenic particles (e.g. some faecal pellets) into deep waters would favour the sequestration of carbon, since the amount of organic material respired during the downwards transit (and potentially rapidly returned to the atmosphere) is largely determined by sinking velocity. The export pathways thus often strongly influence the amount of carbon sequestered by the ocean.

TYPOLOGY OF PELAGIC MARINE ECOSYSTEMS

Several authors (e.g. Miller and SUPER Group, 1988; Parsons and Lalli, 1988; Peinert *et al.*, 1989; Legendre, 1990) have recently pointed out that two contrasting types of ecosystems may be found in temperate and subpolar waters, despite relatively similar annual and/or maximal levels of carbon fixation by the phytoplankton in both cases, i.e. some ecosystems present a well-established spring phytoplankton bloom while others show little increase in the average seasonal chlorophyll *a*. For the subarctic Atlantic and Pacific Oceans, as an example, Parsons and Lalli (1988) explain the observed difference in standing stocks by the fact that, in the North Pacific, both microzooplankton and large-sized copepods react almost immediately to any increase in primary production while, in the North Atlantic, there is a lag between phytoplankton production and grazing by zooplankton. As a consequence, ***standing stocks*** of phytoplankton in the North Pacific are dominated by small cells, while both large and small cells are found in the North Atlantic during the summer, even though the ***production*** of large and small cells might be similar in the two systems. Such contrasting ecosystems have also been described in the Bering Sea (see Legendre, 1990), while Peinert *et al.* (1989) report the existence, in the region of the Norwegian Current, of an ecosystem rather similar to the one described by Parsons and Lalli (1988) in the Pacific.

There are pelagic marine ecosystems that exhibit phytoplankton production and standing stock characteristics quite different from those described in the previous paragraph. For example, in upwelling and other systems with intense blooms, large phytoplankton cells may dominate both the primary production and the biomass (e.g. upwelling and ice-edge blooms). At the other end of the spectrum, in oligotrophic tropical waters, picoplankton ($< 2\ \mu m$) dominate both the production (50-60%) and biomass ($> 50\%$ and sometimes $> 90\%$) of primary producers. Finally, some systems may exhibit a very large biomass of cells > 5 μm, often pertaining to species avoided by grazers (e.g. *Phaeocystis, Gyrodinium aureolum*; Holligan, 1987). This would be the case of "exceptional" blooms (sensu Tett, 1987), where phytoplankton biomass in surface waters may reach > 100 mg m^{-3} in terms of chlorophyll *a* concentration.

Using the various possible combinations of phytoplankton production and standing stock, as dominated by large or small cells or resulting from a combination of various cell sizes, it is possible to design a typology of pelagic marine ecosystems, as shown in Fig. 3. This figure schematizes the five possible patterns linking standing stock to production of phytoplankton. In any given environment, some types of pelagic ecosystems may exist for only part of the year and, in some cases, two or perhaps several types may coexist in the same general area. It will be shown that this typology has an operational character, since the production of small vs. large cells is under hydrodynamic control (e.g. Legendre and Le Fèvre, 1989), while the size-distribution of the standing stock is influenced by ecosystem structure, as well as by phytoplankton production. The relationship between the size-distributions of phytoplankton production and standing stock appears to be a fundamental characteristic of the structure of the overall pelagic ecosystem, with consequences for fisheries and for the export and sequestration of biogenic carbon.

TYPE 1. PRODUCTION AND STANDING STOCK DOMINATED BY LARGE CELLS

Large and small cells develop together in most planktonic ecosystems, as in the classical case of temperate waters where plankton production undergoes a well-marked annual cycle (see Type 3 below). Large cells, generally consisting of diatoms, are favoured at the initial stages of species succession, when the nutrient supply is abundant and vertical mixing (one form of "auxiliary energy" *sensu* Margalef, 1978) is strong enough to maintain non-motile cells in the euphotic layer, while other types of large cells, together with small ones, take over when the environment becomes hydrodynamically less energetic. Primary production dominated by (and thus biomass consisting mainly of) large cells generally corresponds to a diatom bloom, especially in a transient situation that does not allow the further stages of the succession to take place. Such a bloom will occur when a major change in hydrographic conditions (a strong hydrodynamic singularity) suddenly frees the algae from the previously prevailing limitation, which may either be irradiance, as in Sverdrup's (1953) critical depth model of the triggering

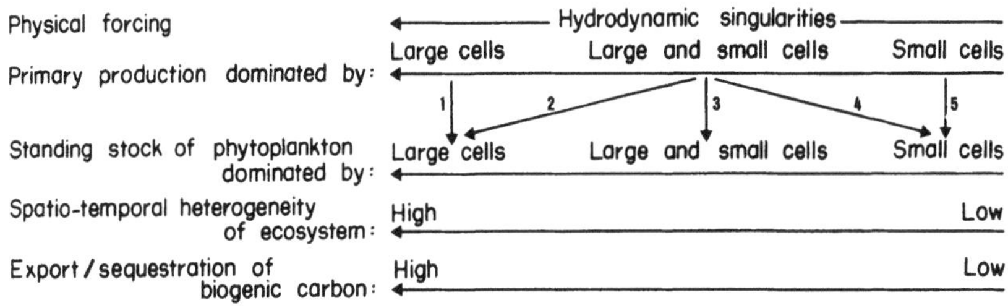

Figure 3. Five types of pelagic ecosystems, as defined by the size distributions of phytoplankton involved in primary production vs. standing stock. Schematic physical forcing of primary production, and major characteristics of ecosystem structures with consequences for the export and sequestration of biogenic carbon. Examples of the five types of ecosystems, as shown below: (1) ice-edge bloom (Fig. 4); (2) exceptional bloom (Fig. 5); (3) spring bloom in the North Atlantic (Fig. 6); (4) Alaskan Gyre in the North Pacific (Fig. 7); (5) oligotrophic ocean (Fig. 8).

of the spring bloom, or the nutrient supply, as can happen in the case of episodic upwelling or wind-mixing events.

Phytoplankton biomass and production dominated by large cells have been reported in upwelling blooms; Bishop *et al.*, (1980), for instance, found that the diatom *Skeletonema costatum* accounted for most of the biomass and production at the level of the nutricline when active upwelling was taking place at the shelf break off Georgia (Atlantic coast of the USA). An example is also provided by the ice-edge bloom described in the Antarctic by Smith and Nelson (1985) and Wilson *et al.* (1986), where a single large (mean length 28 μm) diatom species in the genus *Nitzschia* accounted for up to 85% of cell counts and 90% of the production, with two other species in the same genus being responsible for the rest (Fig. 4). Another possible case might be the episodic blooms of large cells (e.g. *Ethmodiscus rex*, up to 2000 μm diam.), hypothesized by Goldman (1989) to occur at the bottom of the euphotic layer in oligotrophic oceanic waters, where short-lived and localized mixing events would allow rapid bursts of growth. These ephemeral blooms would remain mostly undetected given their transient and local nature, and would be found side-by-side with the microbial food loop where primary production and biomass is dominated by very small cells. According to Goldman, episodic blooms of large cells might be responsible for a large fraction of the (new) primary production in deep oceanic waters (see Type 5 below).

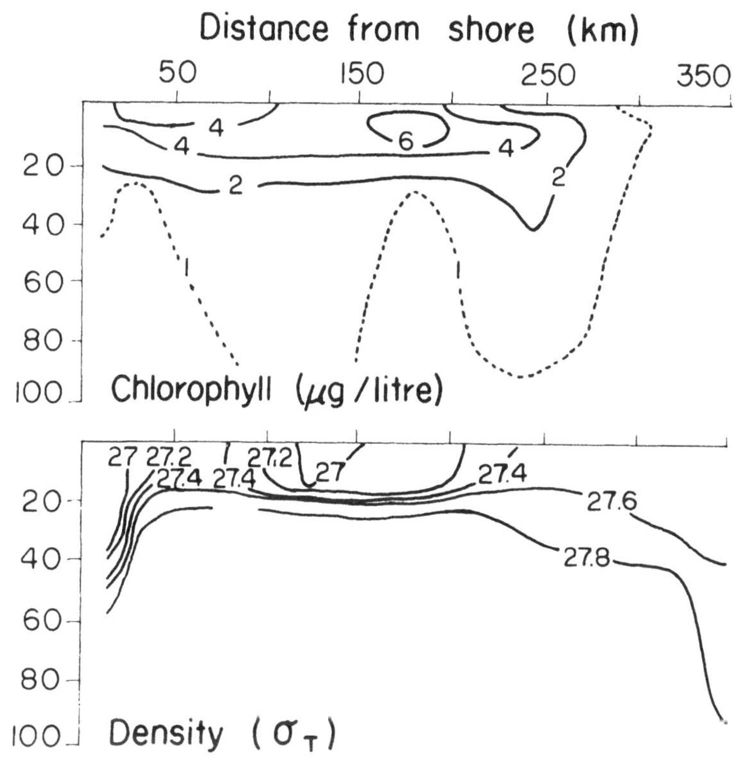

Figure 4. Type 1 ecosystem: bloom of diatoms (mainly in the genus Nitschia) extending 250 km from the receding ice edge, Ross Sea, Antarctic Ocean. The bloom is restricted to waters where the melting ice had resulted in reduced density. Adapted from Smith and Nelson (1985).

The very development of such diatom blooms implies that no or little grazing is impeding the biomass buildup (see Legendre, 1990), a condition that is best met with when no preexisting herbivore stock is ready to take advantage of it, i.e. most often when little primary production was taking place beforehand. Under such circumstances, the most likely fate of the diatoms is their rapid sinking to depth, as documented for receding sea-ice edges (e.g Smith & Nelson,

1985; Nelson and Smith, 1986; Olson, 1980) and upwelling areas (e.g. Cronin and Morris, 1982; Poutanen and Morris, 1983; Smith *et al.*, 1983; Barlow, 1984), and hypothesized for episodic blooms of large cells in open oceanic waters (Goldman, 1989). Rapidly settling cells, as explained above, are efficient in transporting carbon to depths where it may be effectively sequestered.

TYPE 2. PRODUCTION BY SMALL AND LARGE CELLS, STANDING STOCK DOMINATED BY LARGE CELLS

Examples of this type of ecosystem may be the so-called "exceptional" blooms (Fig. 5), including red tides and similar phenomena, where chlorophyll concentrations in surface waters may reach > 100 mg m^{-3} (Tett, 1987). Some exceptional blooms involve carbon-rich thecate dinoflagellates. According to Holligan (1987), the physical conditions leading to exceptional blooms include: for diatom outbursts, strong vertical stratification (e.g. freshwater runoff), low turbidity conditions in vertically mixed waters, wind-induced upwelling, nutrient pulses and eutrophication; for blooms of coccolithophores, intermediate turbulence maintained by

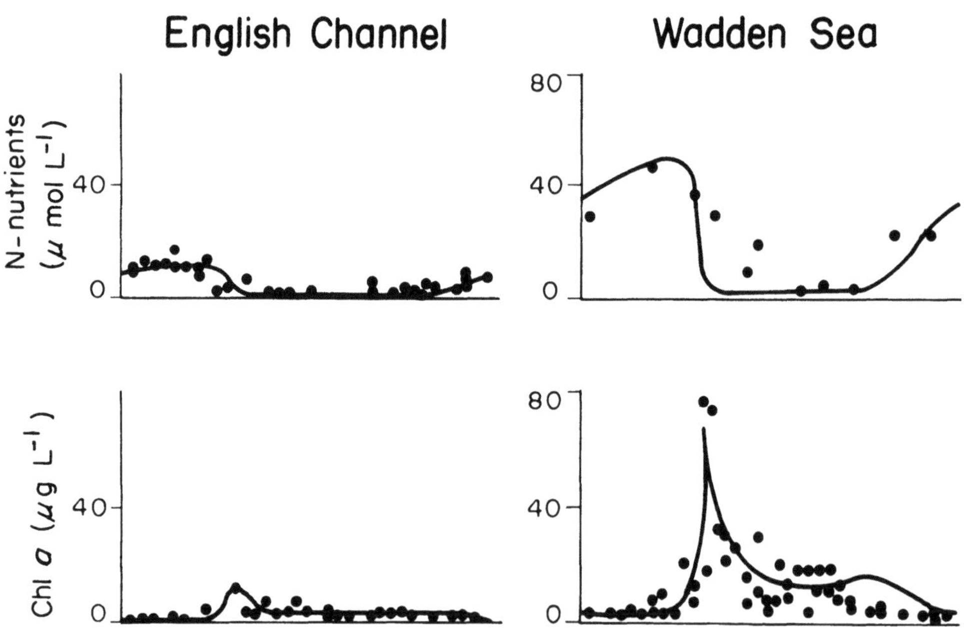

Figure 5. Type 2 ecosystem: seasonal variations in nitrate concentration and phytoplankton biomass, in the English Channel and in the Wadden Sea. The points refer to observations, and the curves are the results of a simulation. Nutrient enrichment in the Wadden Sea mostly results in a short-lived, explosive bloom of phytoplankton. Adapted from Lancelot *et al.* (1987).

physical instabilities such as tidal mixing and upwelling; for large-scale dinoflagellate blooms, relatively shallow pycnocline/nutricline, controlled by tidal mixing, upwelling or geostrophic effects. He also points out that exceptional blooms often involve phytoplankton species avoided by grazers (e.g. *Phaeocystis, Gyrodinium aureolum*) or develop in areas where grazing pressure is low. As a result, the biomass of phytoplankton becomes dominated by ungrazed large cells, even in the face of a significant contribution to primary production by small cells. In a review of nuisance phytoplankton blooms, Paerl (1988) came to similar conclusions as to the factors most likely leading to bloom-sensitive waters. In general, the combination of high nutrients with strong vertical stability is only found in transient situations or in areas subjected to continental inputs (Margalef, 1978). Legendre (1990) has briefly reviewed the reasons for the increasing occurrence of exceptional blooms, which often consist of noxious species. The reasons invoked are still conjectural, but it has been suggested that the general eutrophication of the coastal zone by human activity (inorganic and organic effluents, etc.) is a contributing factor; fish farming may also be involved in some blooms, but the question is still under debate.

Boalch (1984) has reviewed the effects of exceptional blooms on commercially exploited species in the English Channel and elsewhere. These effects include, in the case of toxic species, poisoning and kills of vertebrates (fish, birds and humans) and, more generally, such consequences as mortality resulting from oxygen depletion and mechanical damage to animals (e.g. clogging of fish gills). He concluded that no case is recorded of an exceptional bloom that was of advantage to fishing. Concerning the export of carbon to depth, exceptional blooms can result in significant sedimentation (e.g. Lancelot *et al.*, 1987). According to Peinert *et al.* (1989), rapid sedimentation has been reported for blooms of most of the large cells, which include diatoms, coccolithophores and such flagellates as *Phaeocystis*. The ageing standing stock at the final phase of a bloom due to species unpalatable to herbivores is likely to be recycled by decomposers. These will be grazed upon by microheterotrophs which, in turn, may fall prey to microphagous metazoans (including carbonate-rich pteropods) under certain conditions (see above). Actual sequestration of carbon would largely depend on the depth of the waters into which the particles will eventually sink.

TYPE 3. PRODUCTION AND STANDING STOCK OF SMALL AND LARGE CELLS

This type may be among the best known in classical oceanography, since it is exemplified, among other cases, by the annual cycle in the temperate waters of the North Atlantic (Fig. 6). One of the major characteristics there is the development of the spring bloom of phytoplankton, which is accounted for by Sverdrup's (1953) critical depth model (see also Smetacek and Passow, 1990). The hydrodynamic environment in the North Atlantic is quite energetic and strong wind-induced vertical mixing prevails in winter, both in the open ocean

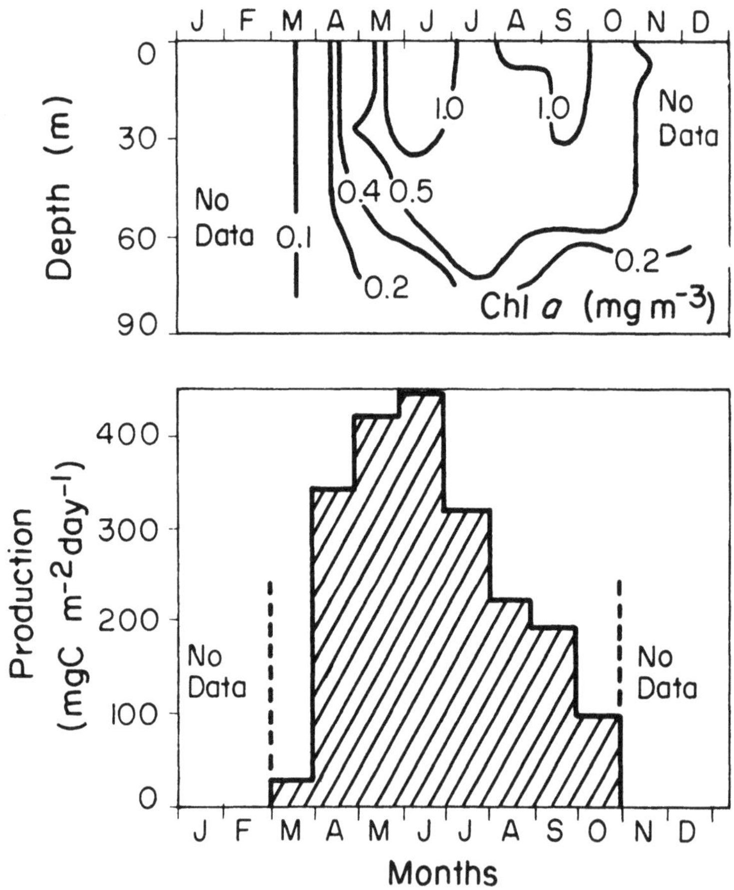

Figure 6. Type 3 ecosystem: phytoplankton chlorophyll a biomass and production (0-50 m) at Ocean Weather Station I, in the North Atlantic Current. Adapted from Parsons and Lalli (1988).

and on the continental shelves, where, in addition, "*tidal streams (...) equivalent to hurricane-force winds in the atmosphere blowing regularly twice per day*" (Simpson, 1981) occur in some places. Because phytoplankton are mixed down below the critical depth, irradiance limitation prevails throughout the winter so that no bloom can develop. Stabilization of the water column in the spring favours a diatom bloom, which takes place sooner or later in different locations, depending on the time when mixing no longer exceeds the compensation depth (e.g. Pingree *et al.*, 1976; Morin *et al.*, 1985). More moderate vertical mixing favours a coccolithophore bloom, often as the second stage after the diatom bloom in the annual cycle. This bloom is apparently most intense over the European continental slope (Holligan *et al.*, 1983), but it is also a major characteristic of the production cycle over large areas of the open ocean (e.g. Holligan *et al.*, 1988a) and occurs on the continental shelves as well (e.g. Holligan *et al.*, 1988b). In summer, when full stratification has developed, zooplankton biomass has increased and the nutrients are exhausted in the surface mixed layer, primary production is mainly effected by dinoflagellates and smaller flagellates (e.g. Holligan and Harbour, 1977). In shelf areas where the depth is shallow enough and/or the tidal streams strong enough for the water column to remain well-mixed year round, the tendency is for a single phytoplankton maximum, occurring in early summer (e.g. Grall, 1972a, 1972b; Boalch *et al.*, 1978; Sournia *et al.* 1987). Diatoms usually dominate this summer bloom, with a significant contribution by small flagellates and other nano/picoplankton (Atkins, 1945; Grall, 1972b; Klein and Sournia, 1987). In places where a seasonal thermocline develops, a second bloom may occur in autumn, involving diatoms whose growth responds to the increase in vertical mixing. One specific case where enhanced phytoplankton growth takes place in late summer and autumn is the continental slope, where, as exemplified by the European shelf-break, nutrient supply is brought about by the interaction between wind-induced mixing and internal tides (e.g. Pingree and Mardell, 1981; Mazé *et al.*; 1986; Le Fèvre and Frontier, 1988; Serpette and Mazé, 1989). The same apparently exists on the North American side of the ocean, where similar physical forcing is found (e.g. Sandström and Elliott, 1984).

Secondary production also follows a distinctive annual cycle. The situation in the open ocean has been reviewed by Parsons and Lalli (1988), mainly from data collected at Ocean Weather Station I, in the North Atlantic Current. The herbivore biomass there is dominated by the copepod *Calanus finmarchicus*, which overwinters below the euphotic zone at the last copepodite stage and reproduces in spring. For reproducing, however, the animals require a supply of energy which is not available before primary production has begun to take place. A time lag ensues before they can respond to the onset of phytoplankton outgrowth, which enables the spring phytoplankton bloom to occur. The food supply is accordingly large when the offspring of the overwintering animals eventually develop, so that they can mature fast and reproduce in summer. This second generation feeds mainly on regenerated production (effected mainly by flagellates) before migrating to deep waters for overwintering. The lack of diatoms in their diet is a likely reason why they do not accumulate sufficient reserves, especially in the form of lipids, to reproduce early and suppress the next spring bloom. According to Parsons and Lalli (1989), the relatively high winter temperature at OWS I would also cause significant metabolic losses in the overwintering copepod population and contribute to their being in poor condition at the beginning of the productive season. A time lag between phytoplankton outgrowth and the increase in herbivore biomass seems to be general in Atlantic temperate waters, even in shelf areas that remain well-mixed throughout the year and where a single phytoplankton biomass peak is found in summer (e.g. Wafar *et al.*, 1984).

An annual cycle of this type leaves ample room for the export and sequestration of biogenic carbon. As long as the herbivore stock is too low to graze them, the diatoms from the spring bloom will largely sink, as will part of the material from the subsequent coccolithophore bloom. In autumn, recycling of the summer biomass pool by microheterotrophs may result in the buildup of large pteropod populations which will export carbon in their calcareous shells. Similar processes can also take place at other times of the year in hydrodynamic traps, where ageing organisms or detrital material accumulate (e.g. summer tidal fronts on the European continental shelf). Annual cycles of this type are largely driven by variations in hydrodynamic conditions, which initiate a response at the level of primary producers that may take time to be reflected in the secondary producers. The succession of blooms and biomass peaks in animal populations thus corresponds to a dynamic imbalance and, given interannual variations, the system probably never reaches steady state conditions.

Such systems with high amplitude production cycles (i.e. blooms of large cells, followed by peak production of herbivores) favour a food web leading to large animals and fish (see

Ryther, 1969) in specific locations where primary production dynamics are tuned to the needs of herbivores (see Le Fèvre and Frontier, 1988; Legendre and Le Fèvre, 1989). Thus, a

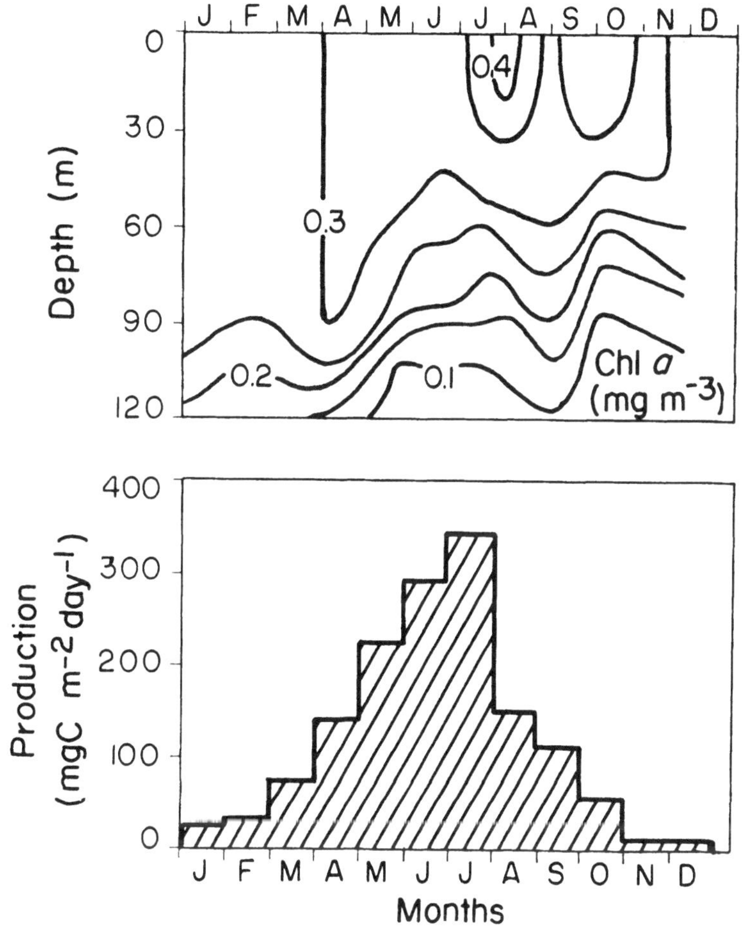

Figure 7. Type 4 ecosystem: phytoplankton chlorophyll a biomass and production (0-50 m) at Ocean Weather Station P, in the Alaskan Gyre of the North Pacific Ocean. Adapted from Parsons and Lalli (1988).

hydrodynamically structured environment is a major factor in aggregating biomass, in both space and time, which facilitates commercial exploitation. In the case of the North Atlantic, major fisheries are actually found on the continental shelves and slopes on both sides, where the degree of hydrodynamic heterogeneity is largest. More generally, these factors would explain, according to Cushing (1989), why the great fisheries of the world are based on diatoms and their > 5 μm successors in the spring bloom (and, in upwelling areas, the larger flagellates).

TYPE 4. PRODUCTION BY SMALL AND LARGE CELLS, STANDING STOCK DOMINATED BY SMALL CELLS

A good example of this type of ecosystem is found in the oceanic waters of the Alaskan Gyre in the North Pacific Ocean (Ocean Weather Station P), as recently reviewed by Parsons and Lalli (1988) and Miller and SUPER Group (1988). One of the major features in the annual cycle there (Fig. 7) is that little increase in chlorophyll *a* standing stock takes place with respect to the winter background level, although the integrated annual production is similar to that measured in the North Atlantic at OWS I, where a well-defined spring bloom occurs (Fig. 6). At any time, the phytoplankton standing stock near OWS P mainly consists of small flagellates; larger diatoms, however, are also present in small numbers, and eventually dominate samples incubated in the absence of zooplankton (McAllister *et al.*, 1960). Biomass increase by large diatoms is therefore suppressed by grazing, but the large cells do effectively contribute to primary production. The annual increase in primary production takes place rather progressively; not inconsistently with Sverdrup's (1953) critical depth model, it is associated with a change in the depth of the mixed layer but, because there is a permanent halocline at about 100 m, this change is much less sharp than in areas with a spring bloom. The relatively weak temporal hydrodynamic heterogeneity of the environment would contribute to explaining why significant primary production is found year-round.

Production by small cells is normally transferred to microheterotrophs. In the Alaskan Gyre, large protozoan standing stocks are present in winter but, according to Parsons and Lalli (1988), they are eaten away by large copepods belonging to the genus *Neocalanus* at the onset of the productive season. As in Type 3 above, these copepods overwinter at the last copepodite stage and reproduce in the spring. What is peculiar to the Alaskan Gyre, according to Parsons and Lalli (1988), is that these copepods have a single generation per year, they are in particularly good condition at the end of the overwintering period, and their reproduction is in phase with the annual increase in primary production (diatoms). One reason for their good condition may be the low water temperature, which minimizes metabolic losses. The protozoan biomass also provides an initial food supply that allows the animals to reproduce at the very beginning of the production increase; the continued production will be exploited later by their offspring. This strategy for survival and reproduction might also be favoured by an accumulation of lipid reserves, derived from diatoms. According to Miller and SUPER Group (1988), the actual picture could be somewhat more complex than described by Parsons and Lalli (1988). The euphotic layer appears to be subdivided into two strata, with taxonomic differences in both primary producers and large herbivores. One of the factors contributing to the absence of blooms would be inefficient utilization of the nitrate pool by phytoplankton (in both strata). Control by grazing is not disputed by Miller and SUPER Group (1988), but their view is that microzooplankton play a more important role than previously acknowledged, since protozoans consume at least as large a share of the annual primary production as do major copepod grazers.

This type of ecosystem is apparently not restricted to the subarctic Pacific Ocean. *Calanus finmarchicus*, the dominant copepod grazer in subarctic and cold-temperate Atlantic waters, may sometimes follow the same life-cycle strategy as the *Neocalanus* species in the Alaskan Gyre (e.g Norwegian Current, Peinert *et al.*, 1989; Labrador Sea, Parson and Lalli, 1988). Other species act in such a way that they are ready to take advantage of phytoplankton outgrowths as soon as they occur, but may not always be efficient enough to suppress blooms (e.g. *Calanoides carinatus* in the Somali upwelling, Indian Ocean; Smith, 1982).

In such an ecosystem, where the activities of producers and consumers, large and small, are largely coupled in time, the export of new primary production will tend to be channelled towards grazing and subsequent steps in the food chains rather than towards sedimentation. Miller and SUPER Group (1988) report that phytoplankton losses to cell sinking seldom exceed 10%, the major part being partitioned between microzooplankton and macrozooplankton grazing, the latter (accounting for 10-51% of phytoplankton losses) producing some additional fallout of material to deep waters. In addition, most of the carbon fixed by phytoplankton may be respired during the long downwards transit of the biogenic material, and returned to the atmosphere. This might accordingly result in lower sequestration of carbon than expected from the influx of allochthonous nitrogenous nutrients. Because of the predominance of grazing over sedimentation, the North Pacific should produce more pelagic fishes than the North Atlantic (Parsons and Lalli, 1988). The background biomass pool also constitutes a dynamic sink for carbon, which is possibly larger in the Alaskan Gyre (at least as regards plankton) than in environments with stronger temporal variations; this living carbon stock, however, can easily be broken down if the conditions undergo long term changes and does not provide an efficient buffer in the global biogeochemical cycles. In addition, because the environment is rather weakly structured in time and space, the biomass is likely to be dispersed, which can hinder commercial exploitation of the stock (see Cushing, 1989). Fishing techniques in large areas of the Pacific Ocean are actually designed to cope with a dispersed resource (at the price of their being highly destructive in some cases, e.g. mammoth drift nets). One of the exceptions is the catch of salmon as they migrate back into their native rivers; fishing in this case takes advantage of a behaviourial concentration of the resource, which, in turn, is dependent on ecological conditions that bear no direct relationship to the oceanic environment.

TYPE 5. PRODUCTION AND STANDING STOCK DOMINATED BY SMALL CELLS

In oligotrophic oceanic waters, picoplankton typically account for > 50% of the chlorophyll *a* biomass (e.g. Herbland and Le Bouteiller, 1981; Platt *et al.*, 1983; Berman *et al.*, 1986; Legendre *et al.*, 1988), with values > 90% reported by Takahashi and Hori (1984). Given their high proportions in the biomass, the small cells account for a large proportion of the total primary production (typically ca. 50-60%; e.g. Paerl, 1977; Herbland and Le Bouteiller, 1981; Li *et al.*, 1983; Platt *et al.*, 1983). Azam *et al.* (1983) proposed that the small phototrophic cells are part of a "microbial food loop", which also comprises heterotrophic bacteria and protozoa. According to Cushing (1989), the microbial food web predominates in stratified waters because the exudates from phototrophic picoplankton (small eukaryotic algae and cyanobacteria) are not dispersed and can therefore be used by heterotrophic bacteria. In the scheme of Azam *et al.* (1983), the numbers of picoplankton cells are controlled by protozoa, including heterotrophic microflagellates which are in turn exploited by ciliates; this grazing activity is accompanied by excretion of ammonium, which is used by the photoautotrophs. Little material is exported from the nearly closed microbial food loop (e.g. Michaels and Silver, 1988) and this system is in a quasi-steady state (e.g. Cushing, 1989).

Even if the microbial food web is largely a recycling loop, export of production is nevertheless possible, through direct feeding by metazoans on small particles ("baleen whale" feeding; Fenchel, 1984), incorporation of small particles into marine snow and sinking, and their accumulation with particles of various sizes in hydrodynamic traps where they are subject to grazing by microphagous feeders (see Fig. 2, and also Goldman, 1988). Within the area, between 50°N and 50°S, where coccolithophore and foraminiferan production is significant, particulate calcium carbonate in the upper 500 m tends to peak between 10-15°C, with low concentrations being found in the warmest waters (Bishop, 1989). The export of carbon by calcareous tests should therefore be relatively low in warm oligotrophic waters, so that the amount of carbon exported by the microbial food web is likely to approximate the stoichiometric balance with the influx of allochthonous nitrogenous nutrients. Actual sequestration of carbon is however subjected to respiratory losses similar to those described above (Type 4), given the long and inefficient export pathways. In the oligotrophic ocean, the sources of allochthonous nutrients (Fig. 1) include the deep ocean (upwards eddy diffusion of NO_3^-) and the atmosphere (transport of NO_3^- and NH_3/NH_4^+; N_2 fixation). The atmospheric nitrogen sources probably contribute mainly to the microbial food web, while upwards diffusion may also fuel episodic blooms of large cells in the lower euphotic zone. These episodic blooms were hypothesized by Goldman (1988; see Type 1 above), who also suggested that phytoplankton production of Types 1 and 5 would operate quite independently of each other in oligotrophic waters (Fig. 8).

It has already been mentioned that the production cycle in the oligotrophic ocean is in quasi-steady state so that, according to Cushing (1989), the food chains are long and the organisms are dispersed. Such conditions do not favour sustained exploitation of fish stocks. Concerning the export and sequestration of carbon, the oligotrophic ocean is probably the system where the assumptions underlying the usual steady-state/stoichiometry model for export production (e.g. Eppley and Peterson, 1979) are best approximated. However, even in this case, the task of estimating new production is not trivial (e.g. Platt *et al.*, 1989b), and the

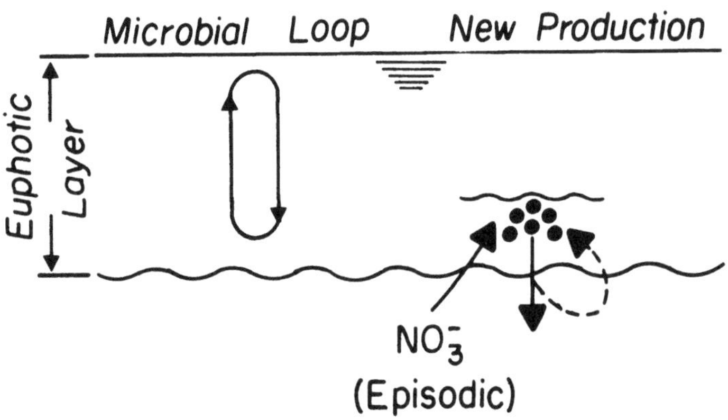

Figure 8. Schematic view of the extreme cases of the microbial food web and episodic blooms of large cells postulated by Goldman (1988) to take place in oligotrophic waters. The microbial web (Type 5 ecosystem) is portrayed as a recycling loop from which there is no export; it is dominated by ultraplankton and small protozoa, and occurs throughout the euphotic layer. The blooms of large cells (Type 1 ecosystem) are fuelled by episodic inputs, at the base of the euphotic layer, of nitrate and of an inoculum cell population (dashed line). The large phytoplankton are not readily grazed, and thus sink to deeper waters. Adapted from Goldman (1988).

generally accepted mass balance between new and export production is less than certain (Legendre and Gosselin, 1989).

CONCLUSIONS AND RECOMMENDATIONS

The five types of pelagic marine ecosystems discussed above correspond to different modes of phytoplankton production (as controlled by hydrodynamics) and different structures (as reflected in the standing stocks), which influence the pathways of carbon export and sequestration as well as the renewable marine resources. The discussion has shown that a significant component of the biologically driven flux of carbon dioxide into the oceans is determined from the level of individual cells up, and mediated by ecosystem structure. Relevant ecosystem properties are reflected in the relationship between the size-distributions of phytoplankton production and standing stock, which appears to be a fundamental characteristic of the structure of the overall pelagic ecosystem. The dynamics of marine ecosystems thus plays an important role in channelling biogenic carbon into export pathways leading to exploitable living resources or eventual long-term sequestration of carbon.

Given the role of cell and particle properties in this respect, it is obvious that individual cell and particle analysis could contribute to some of the most pressing issues in oceanography. In this context, the most useful analyses would concern the measurement of phytoplankton biomass per size class, the recognition of major taxonomic groups, and the estimation of production per size class and/or taxonomic group. In flow cytometry/image analysis, techniques for biomass and taxonomy are already available or under active development, but they are still largely lacking for production (e.g. Table 1 in Legendre and Yentsch, 1989). Within the context of the present scheme, the absence of direct measurement of particle size by most flow cytometers is certainly a major drawback. Concerning flow cytometry/image analysis, we therefore recommend the development of techniques for the analyses mentioned above, and their use within the framework of such interdisciplinary programs as JGOFS (Joint Global Ocean Flux Study) or GLOBEC. In addition, oceanographers using flow cytometry should stress to manufacturers the importance of direct measurements of particle size.

Concerning the development of models for the management of exploited stocks and the prediction of the global flux of carbon into the oceans, these could probably be significantly improved if they incorporated relevant characteristics of biogenic particle dynamics and of marine ecosystems. As discussed above, many of these characteristics could be derived from flow-cytometry and image analysis.

ACKNOWLEDGEMENTS

We thank Drs. T. Berman, P. Falkowski, L. Fortier, L. Gagné, P.M. Holligan and G. Rosenberg for comments or suggestions, and Ms. L. Roberge for assistance with bibliographic research. Financial support from Coopération France-Québec and Université Laval made possible a sabbatical leave for J.L.F. in Québec City, and grants from the Natural Sciences and Engineering Research Council of Canada to L.L. were instrumental in the completion of this work.

REFERENCES

Alldredge A (1984) The quantitative significance of gelatinous zooplankton as pelagic consumers. In: Fasham MJR (ed) Flows of energy and materials in marine ecosystems. Plenum, New York, 407-433

Atkins WRG (1945) Autotrophic flagellates as the major constituent of oceanic plankton. Nature 156:446-447

Azam FT, Fenchel T, Field JG, Meyer-Reil LA, Thingstad F (1983) The ecological role of water-column microbes in the sea. Mar Ecol Prog Ser 10:257-263

Barlow RG (1984) Dynamics of the decline of a phytoplankton bloom after an upwelling event. Mar Ecol Prog Ser 16:121-126

Bé AWH, Spero HJ, Anderson OR (1982) Effects of symbiont elimination and reinfection on the life processes of the planktonic foraminifer Globigerinoides sacculifer. Mar Biol 70:73-86

Beers JR (1966) Studies on the chemical composition of the major zooplankton groups in the Sargasso Sea off Bermuda. Limnol Oceanogr 11:520-528

Berger WH (1982) Deglacial CO_2 buildup: Constraints on the coral reef model. Paleogeog Paleoclim Paleoecol 40:235-253

Berger WH, Smetacek VS, Wefer G (1989) Ocean productivity and paleoproductivity. In: Berger WH, Smetacek VS, Wefer G (eds) Productivity of the ocean: present and past. Wiley and Sons, Chichester, 1-34

Berman T, Azov Y., Schneller A, Walline P, Townsend DW (1986) Extent, transparency and phytoplankton distribution of the neritic waters overlying the Israeli coastal shelf. Oceanol Acta 9:439-447

Berner RA, Honjo S (1981) Pelagic sedimentation of aragonite: its geochemical significance. Science 211:940-942

Bishop JKB (1989) Regional extremes in particulate matter composition and flux: effects on the chemistry of the ocean interior. In: Berger WH, Smetacek VS, Wefer G (eds) Productivity of the ocean: present and past. Wiley and Sons, Chichester, 117-137

Bishop SS, Yoder JA, Paffenhöfer GA (1980) Phytoplankton and nutrient variability along a cross-shelf transect off Savannah, Georgia, U.S.A. Est Coast Mar Sci 11:359-368

Boalch GT (1984) Algal blooms and their effects on fishing in the English Channel. Hydrobiologia 116/117:449-452

Boalch GT, Harbour DS, Butler EI (1978) Seasonal phytoplankton production in the western English Channel. J Mar Biol Ass UK 58:943-953

Brewer PG, Bruland KW, Eppley RW McCarthy JJ (1986) The Global Ocean Flux Study (GOFS): Status of the U.S. GOFS program. Eos 67:827-832

Capone DG (1983) Benthic nitrogen fixation. In: Carpenter EJ, Capone DG (eds) Nitrogen in the marine environment. Academic Press, New York, 105-137

Carpenter EJ (1983) Nitrogen fixation by marine *Oscillatoria* (Trichodesmium) in the world's ocean. In: Carpenter EJ, Capone DG (eds) Nitrogen in the marine environment. Academic Press, New York, 65-103

Codispoti LA (1989) Phosphorus vs. nitrogen limitation of new and export production. In: Berger WH, Smetacek VS, Wefer G (eds) Productivity of the ocean: present and past. Wiley and Sons, Chichester, 377-394

Cronin JR, Morris RJ (1982) The occurrence of high molecular weight humic material in recent organic-rich sediment from the Namibian shelf. Est Coast Shelf Sci 15:17-27

Curl H (1962) Analyses of carbon in marine plankton organisms. J Mar Res 20:181-188

Cushing DH (1988) The northerly wind. In: Rothschild BJ (ed) Toward a theory on biological-physical interactions in the world ocean. Kluwer, Dordrecht, 235-244

Cushing DH (1989) A difference in structure between ecosystems in strongly stratified waters and in those that are only weakly stratified. J Plankton Res 11:1-13

Cushing DH, Dickson R (1976) The biological response in the sea to climatic changes. Adv Mar Biol 14:1-122

Dugdale RC, Goering JJ (1967) Uptake of new and regenerated forms of nitrogen in primary production. Limnol Oceanogr 23:196-206

Dymond J, Lyle H (1985) Flux comparisons between sediments and sediment traps in the eastern tropical Pacific: Implications for atmospheric CO_2 variations during the Pleistocene. Limnol Oceanogr 30:699-712

Eppley RW (1989) New production: history, methods, problems. In: Berger WH, Smetacek VS, Wefer G (eds) Productivity of the ocean: present and past. Wiley and Sons, Chichester, 85-97

Eppley RW, Peterson BJ (1979) Particulate organic matter flux and planktonic new production in the upper ocean. Nature 282:677-680

Fenchel T (1984) Suspended marine bacteria as a food source. In: Fasham MJR (ed) Flows of energy and materials in marine ecosystems. Plenum, New York, 301-315

Gilmer RW, Harbison GR (1990) The diet of *Limacina helicina* (Gastropoda: Thecosomata) in Arctic waters in late summer. Eos 71:77 (abstract)

Goldman JC (1988) Spatial and temporal discontinuities of biological processes in pelagic surface waters. In: Rothschild BJ (ed) Toward a theory on biological-physical interactions in the world ocean. Kluwer, Dordrecht, 273-296

Gorsky G, Dallot S, Sardou J, Fenaux R, Carré C, Palazzoli I (1988) C and N composition of some northwestern Mediterranean plankton and micronekton species. J Exp Mar Biol Ecol 124:133-144

Gosselin S, Fortier L, Gagné JA (1989) Vulnerability of marine fish larvae to the toxic dinoflagellate *Protogonyaulax tamarensis*. Mar Ecol Prog Ser 57:1-10

Graf G, Bengtsson W, Diesner U, Schulz R, Theede H, (1982) Benthic response to a spring phytoplankton bloom: process and budget. Mar Biol 67:201-208

Graf G, Bengtsson W, Faubel A, Meyer-Reil LA, Schulz R, Theede H, Thiel H (1984) The importance of the spring phytoplankton bloom for the benthic system of Kiel Bight. Rapp P-V Réun Cons Int Explor Mer 183:136-143

Grall JR (1972a) Développement "printanier" de la diatomée *Rhizosolenia delicatula* près de Roscoff. Mar Biol 16:41-48

Grall JR (1972b) Recherches quantitatives sur la production primaire du phytoplancton dans les parages de Roscoff. Thèse Doc Sci Nat, Univ Paris 6

Haug A, Myklestad S, Sakshaug E (1973) Studies on the phytoplankton ecology of the Trondheimsfjord. I. The chemical composition of phytoplankton populations. J Exp Mar Biol Ecol 11:15-26

Heinbokel JF (1986) Occurrence of *Richelia intracellularis* (Cyanophyta) within the diatoms *Hemiaulus haukii* and *H. membranaceus* off Hawaii. J Phycol 22:399-403

Herbland A, Le Bouteiller A (1981) The size distribution of phytoplankton and particulate organic matter in the equatorial Atlantic Ocean: importance of ultraseston and consequences. J Plankton Res 3:659-673

Holligan PM, (1987) The physical environment of exceptional phytoplankton blooms in the northeast Atlantic. Rapp P-V Réun Cons Int Explor Mer 187:9-18

Holligan PM, Aiken J, Groom S (1988a) Bio-optical studies in the N.E. Atlantic in June, 1987. Eos 69:1117 (abstract)

Holligan PM, Ackleson SG, Balch WM (1988b) The 1988 coccolithophore bloom in the Gulf of Maine. Eos 69:1132 (abstract)

Holligan PM, Harbour DS (1977) The vertical distribution and succession of phytoplankton in the western English Channel in 1975 and 1976. J Mar Biol Ass UK 57:1075-1093

Holligan PM, Viollier M, Harbour DS, Camus P, Champagne-Philippe M (1983) Satellite and ship studies of coccolithophore production along a continental shelf edge. Nature 304:339-342

Honjo S (1980) Material fluxes and modes of sedimentation in the mesopelagic and bathypelagic zones. J Mar Res 38:53-97

Honjo S (1982) Seasonality and interaction of biogenic and lithogenic particulate flux at the Panama Basin. Science 218:883-884

Honjo S, Roman MR (1978) Marine copepod fecal pellets: production, preservation and sedimentation. J Mar Res 36:45-57

Iverson RL, Curl HC, O'Connors HB, Kirk D, Zakar K (1974) Summer phytoplankton blooms in Auke Bay, Alaska, driven by wind mixing of the water column. Limnol Oceanogr 19:271-278

Jørgensen BB, Erez J., Revsbech NP, Cohen Y (1985) Symbiotic photosynthesis in a planktonic foraminiferan, *Globigerinoides sacculifer* (Brady), studied with microelectrodes. Limnol Oceanogr 30:1253-1267

King KR, Hollibaugh JT, Azam F (1980) Predator-prey interactions between the larvacean *Oikopleura dioica* and bacterioplankton in enclosed water columns. Mar Biol 56:49-57

Klein B, Sournia A, (1987) A daily study of the diatom spring bloom at Roscoff (France) in 1985. II. Phytoplankton pigment composition studied by HPLC analysis. Mar Ecol Prog Ser 37:265-275

Lancelot C, Billen G (1985) Carbon-nitrogen relationships in nutrient metabolism of coastal marine ecosystems. Adv Aquat Microbiol 3:263-321

Lancelot C, Billen G, Sournia A, Weisse T, Colijn F, Veldhuis MJW, Davies A and Wassman P (1987) *Phaeocystis* blooms and nutrient enrichment in the continental coastal zones of the North Sea. Ambio 16:38-46

Lasker R (1988) Food chains and fisheries: an assessment after 20 years. In: Rothschild BJ (ed) Toward a theory on biological-physical interactions in the world ocean. Kluwer, Dordrecht, 173-182

Le Fèvre J (1986) Aspects of the biology of frontal systems. Adv Mar Biol 23:163-299

Le Fèvre J, Frontier S (1988) Influence of temporal characteristics of physical phenomena on plankton dynamics, as shown by North-West European marine ecosystems. In: Rothschild BJ (ed) Toward a theory on biological-physical interactions in the world ocean. Kluwer, Dordrecht, 245-272

Legendre L (1990) The significance of microalgal blooms for fisheries and for the export of particulate organic carbon in oceans. J Plankton Res 12:681-699

Legendre L, Demers S, Delesalle B, Harnois C (1988) Biomass and photosynthetic activity of phototrophic picoplankton in coral reef waters (Moorea Island, French Polynesia). Mar Ecol Prog Ser 47:153-160

Legendre L, Demers S, Lefaivre D (1986) Biological production at marine ergoclines. In: Nihoul JCJ (ed) Marine interfaces ecohydrodynamics. Elsevier, Amsterdam, 1-29

Legendre L, Gosselin M (1989) New production and export of organic matter to the deep ocean: consequences of some recent discoveries. Limnol Oceanogr 34:1374-1380

Legendre L, Le Fèvre J (1989) Hydrodynamic singularities as controls of recycled versus export production in oceans. In: Berger WH, Smetacek VS, Wefer G (eds) Productivity of the ocean: present and past. Wiley and Sons, Chichester, 49-63

Legendre L, Yentsch CM (1989) Overview of flow cytometry and image analysis in biological oceanography and limnology. Cytometry 10:501-510

Lewis MR, Harrison WG, Oakey NS, Hebert D, Platt T (1986) Vertical nitrate fluxes in the oligotrophic ocean. Science 234:870-873

Li WKW, Subba Rao DV, Harrison WG, Smith JC, Cullen JJ, Irwin B, Platt T (1983) Autotrophic picoplankton in the tropical ocean. Science 219:292-295

Lisitzin AP (1972) Sedimentation in the world ocean. Soc Econ Paleon Miner, Tulsa, OK, Spec Publ 17

Loeblich AR III (1970) The amphiesma or dinoflagellate cell covering. Proc North American Paleont Convention, Chicago, Sept 1969. Allen Press, Lawrence, Kansas, Vol. 2, Part G:867-929

Longhurst AR, Williams R (1979) Materials for plankton modelling: vertical distribution of Atlantic zooplankton in summer. J Plankton Res 1:1-28

Mague TH, Weare NM, Holme-Hansen O (1974) Nitrogen fixation in the North Pacific Ocean. Mar Biol 24:109-119

Mann KH (1982) Ecology of coastal waters. A systems approach. Studies in Ecology, Vol. 8. Univ of California Press, Berkeley

Margalef R (1978) Life-forms of phytoplankton as survival alternatives in an unstable environment. Oceanol Acta 1:493-309

Mazé R, Camus Y, Le Tareau JY (1986) Formation de gradients thermiques à la surface de l'océan, au-dessus d'un talus, par interaction entre les ondes internes et le mélange dû au vent. J Cons Int Explor Mer 42, 221-240

McAllister CD, Parsons TR, Strickland JDH (1960) Primary productivity and fertility at station "P" in the north-east Pacific Ocean. J Cons Int Explor Mer 25, 240-259

Michaels AF, Silver MW (1988) Primary production, sinking fluxes and the microbial food web. Deep-Sea Res 35:473-490

Miller CB, SUPER Group (1988) Lower trophic level production dynamics in the oceanic Subarctic Pacific Ocean. In: Nemoto T, Pearcy WG (eds) The biology of the subarctic Pacific, Proceedings of the Japan-United States of America seminar on the biology of micronekton of the Subarctic Pacific, Part I. Bull Ocean Res Inst Univ Tokyo 26:35-48

Morin P, Le Corre P, Le Fèvre J (1985) Assimilation and regeneration of nutrients off the west coast of Brittany. J Mar Biol Ass UK 65:677-695

Nelson DM, Smith WO (1986) Phytoplankton bloom dynamics of the western Ross Sea ice edge - II. Mesoscale cycling of nitrogen and silicon. Deep-Sea Res 33:1389-1412

Nevo Z, Sharon Z (1969) The cell wall of *Peridinium westii* a non-cellulosic glucan. Biochim Biophys Acta 173:161-175

Ohki K, Fujita Y (1988) Aerobic nitrogenase activity measured as acetylene reduction in the marine non-heterocystous cyanobacterium Trichodesmium spp. grown under artificial conditions. Mar Biol 98:111-114

Olson RJ (1980) Nitrate and ammonium uptake in Antarctic waters. Limnol Oceanogr 25:1064-1074

Omori M (1969) Weight and chemical composition of some important oceanic zooplankton in the North Pacific Ocean. Mar Biol 3:4-10

Paerl HW (1977) Nanoplankton vs netplankton photosynthetic and heterotrophic activities in Fijian waters of the South Pacific Ocean. Lau-Tonga Bull 117:211-221

Paerl HW (1988) Nuisance phytoplankton blooms in coastal, estuarine, and inland waters. Limnol Oceanogr 33:823-847

Parsons TR, Lalli CM (1988) Comparative oceanic ecology of the plankton communities of the subarctic Atlantic and Pacific oceans. Oceanogr Mar Biol Ann Rev 26:317-359

Peinert R, von Bodungen B, Smetacek VS (1989) Food web structure and loss rate. In: Berger WH, Smetacek VS, Wefer G (eds) Productivity of the ocean: present and past. Wiley and Sons, Chichester, 35-48

Pingree RD, Holligan PM, Mardell GT, Head RN (1976) The influence of physical stability on spring, summer and autumn phytoplankton blooms in the Celtic Sea. J Mar Biol Ass UK 56:845-873

Pingree RD, Mardell GT (1981) Slope turbulence, internal waves and phytoplankton growth at the Celtic Sea shelf-break. Phil Trans R Soc London A 302:663-682

Platt T. Calvert S, Denman K, Lewis M, Mann K, Vézina A, Wong CS (1989a) Canadian National Programme for the Joint Global Ocean Flux Study. Canadian Committee for JGOFS, Halifax

Platt T, Harrison WG, Lewis MR, Li WKW, Sathyendranath S, Smith RE, Vézina AF (1989b) Biological production of the oceans: the case for a consensus. Mar Ecol Prog Ser 52:77-88

Platt T, Subba Rao DV, Irwin B (1983) Photosynthesis of picoplankton in the oligotrophic ocean. Nature 301:702-704

Poutanen EL, Morris RJ (1983) The occurrence of high molecular weight humic compounds in the organic-rich sediments of the Peru continental shelf. Oceanol Acta 6:21-28

Ramanathan V, Cicerone RJ, Singh HB, Kiehl JT (1985) Trace gas trends and their potential role in climatic change. J Geophys Res 90:5547-5566

Redfield AC (1958) The biological control of chemical factors in the environment. Amer Scientist 46:205-222

Redfield AC, Ketchum BH, Richards FA (1963) The influence of organisms on the composition of sea water. In: Hill MN (ed) The sea. Interscience, New York, vol. 2, 26-77

Rougerie F, Wauthy B (1986) Le concept d'endo-upwelling dans le fonctionnement des atolls-oasis. Oceanol Acta 9:133-148

Rothschild BJ (1988) Dynamics of marine fish populations. Harvard Univ. Press, Cambridge, MA

Russell FS, Southward AJ, Boalch GT, Butler EI (1971) Changes in biological conditions in the English Channel off Plymouth during the last half century. Nature 234:468-470

Ryther JH (1969) Photosynthesis and fish production in the sea. Science 166:72-76

Sandström H, Elliott JA (1984) Internal tides and solitons on the Scotian shelf: A nutrient pump at work. J Geophys Res 89:6415-6426

Serpette A., Mazé R. (1989) Internal tides in the Bay of Biscay - A two-dimensional model. Cont Shelf Res 9:795-821

Sieburth JMcN, Smetacek V, Lenz J (1978) Pelagic ecosystem structure: heterotrophic compartments of the plankton and their relationship to plankton size fractions. Limnol Oceanogr 23:1256-1263

Silver MW, Gowing MM, Davoll PJ (1986) The association of photosynthetic picoplankton and ultraplankton with pelagic detritus through the water column (0-2000 m). In: Platt T, Li WKW (eds) Photosynthetic picoplankton. Can Bull Fish Aquat Sci 214:311-341

Simpson JH (1981) The shelf-sea fronts: Implications of their existence and behaviour. Phil Trans R Soc London A 302:531-546

Smetacek V, von Bodungen B, Knoppers B, Peinert R, Pollehne F, Stegmann P, Zeitschel B (1984) Seasonal stages characterizing the annual cycle of an inshore pelagic system. Rapp P-V Réun Cons Int Explor Mer 183:126-135

Smetacek V, Passow U (1990) Spring bloom initiation and Sverdrup's critical depth model. Limnol Oceanogr 35:228-233

Smith SL (1982) The northwestern Indian Ocean during the monsoons of 1979: distribution, abundance and feeding of zooplankton. Deep-Sea Res 29:1331-1353

Smith DJ, Eglinton G, Morris RJ (1983) Interfacial sediment and assessment of organic input from a highly productive water column. Nature 304:259-262

Smith WO, Nelson DM (1985) Phytoplankton bloom produced by a receding ice edge in the Ross Sea: spatial coherence with the density field. Science 227:163-166

Sournia A, Birrien JL, Douvillé JL, Klein B, Viollier M (1987) A daily study of the diatom spring bloom at Roscoff (France) in 1985. I. The spring bloom within the annual cycle. Est Coast Shelf Sci 25:355-367

Southward AJ (1974) Long term changes in abundance of eggs of the Cornish pilchard (*Sardina pilchardus* Walbaum) off Plymouth. J Mar Biol Ass UK 54:641-649

Southward AJ (1980) The western English Channel - an inconstant ecosystem? Nature 285:361-366

Southward AJ (1983) Fluctuations in the ecosystems of the Western Channel: a summary of studies in progress. Oceanol Acta, Proceedings 17th European Marine Biology Symposium (special issue):187-189

Suess E (1973) Interaction of organic compounds with calcium carbonate. II. Organo-carbonate association in recent sediments. Geochim Cosmochim Acta 37:2435-2447

Sugimura Y, Suzuki Y (1988) A high temperature catalytic oxidation method of non-volatile dissolved organic carbon in seawater by direct injection of liquid samples. Mar Chem 24:105-131

Sundquist ET (1985) Geological perspectives on carbon dioxide and the carbon cycle. In: Sundquist ET, Broecker WS (eds) The carbon cycle and atmospheric CO_2: natural variations archean to present. AGU Monograph 32, Amer Geophys Union, Washington DC, 5-59

Sverdrup HU (1953) On conditions for the vernal blooming of phytoplankton. J Cons Int Explor Mer 18:287-295

Sverdrup HU, Johnson MW, Fleming RH (1942) The oceans. Their physics, chemistry and general biology. Prentice-Hall, Englewood Cliffs, NJ

Takahashi K, Bé AWH (1984) Planktonic foraminifers: factors controlling sinking speeds. Deep-Sea Res 31:1477-1500

Takahashi M, Hori T (1984) Abundance of picoplankton in the subsurface chlorophyll maximum layer in subtropical and tropical waters. Mar Biol 79:177-186.

Takahashi M, Seibert DL, Thomas WH (1977) Occasional blooms of phytoplankton during summer in Saanich Inlet, B.C., Canada. Deep-Sea Res 24:775-780

Tett P (1987) The ecophysiology of exceptional blooms. Rapp P-V Réun Cons Int Explor Mer 187:47-60

Toggweiler JR (1989) Is the downward dissolved organic matter (DOM) flux important in carbon transport? In: Berger WH, Smetacek VS, Wefer G (eds) Productivity of the ocean: present and past, Wiley and Sons, Chichester, 65-83

Venrick EL (1974) The distribution and significance of *Richelia intracellularis* Schmidt in the North Pacific Central Gyre. Limnol Oceanogr 19:437-445

Volk T (1989) Sensitivity of climate and atmospheric CO_2 to deep-ocean and shallow-ocean carbonate burial. Nature 337:637-638

Volk T, Hoffert MI (1985) Ocean carbon pumps: Analysis of relative strengths and efficiencies in ocean driven CO_2 changes. In: Sundquist ET, Broecker WS (eds) The carbon cycle and atmospheric CO_2: natural variations archean to present. AGU Monograph 32, Amer Geophys Union, Washington DC, 99-110

Wafar M, Le Corre P, Birrien JL (1984) Seasonal changes of dissolved organic matter (C, N, P) in permanently well mixed temperate waters. Limnol Oceanogr 29:1127-1132

Walsh JJ (1989) How much shelf production reaches the deep sea? In: Berger WH, Smetacek VS, Wefer G (eds) Productivity of the ocean: present and past, Wiley and Sons, Chichester, 175-191

Ward BB, Zafiriou OC (1988) Nitrification and nitric oxide in the oxygen minimum of the eastern tropical North Pacific. Deep-Sea Res 35:1127-1142.

Webb KL, DuPaul WD, Wiebe W, Sottile W, Johannes RE (1975) Enewetak (Eniwetok) Atoll: Aspects of the nitrogen cycle on a coral reef. Limnol Oceanogr 20:198-210

White AW (1984) Paralytic shellfish toxins and finfish. In: Ragelis EP (ed) Seafood toxins. ACS Symposium series 262, Amer Chem Soc. Washington DC, 171-180

Wiebe WJ, Johannes RE, Webb KL (1975) Nitrogen fixation in a coral reef community. Science 188:257-259

Wilson DL, Smith WO, Nelson DM (1986) Phytoplankton dynamics of the western Ross Sea ice edge - I. Primary productivity and species-specific production. Deep-Sea Res 33:1375-1387

Wynne D, Patni NJ, Aaroson S, Berman T (1982) The relationship between nutrient status and chemical composition of Peridinium cinctum in Lake Kinneret. J Plankton Res 1:125-136

FROM THE OCEAN TO CELLS: COCCOLITHOPHORE OPTICS AND BIOGEOCHEMISTRY

P. M. Holligan[1] and W. M. Balch[2]

[1]Plymouth Marine Laboratory,
West Hoe,
Plymouth PL1 3DH,
United Kingdom.

[2]Rosenstiel School for Marine and
Atmospheric Science,
University of Miami,
Miami, Fl 33149-1098
United States

INTRODUCTION

The detection of extensive areas of highly reflective ocean waters by satellite visible-band radiometers, including the Landsat MSS (Gower *et al.*, 1980), the Nimbus-7 Coastal Zone Color Scanner (CZCS) (Holligan *et al.*, 1983) and the NOAA Advanced Very High Resolution Radiometer (AVHRR) (Groom and Holligan, 1987), and the identification of blooms of coccolithophores as the cause of the reflectance (Holligan *et al.*, 1983; Balch *et al.*, 1991) has led to renewed interest in the optical properties of this group of phytoplankton as well as in their ecology and biogeochemistry. The coccolithophores typically have complex life histories (Hibberd, 1980) which include a non-motile, planktonic phase characterised by external plates, or coccoliths, of calcium carbonate (Green, 1986; Westbroek *et al.*, 1989). It is the backscattering of light by the coccoliths attached to the cells or detached in the water that is detected by the satellite sensors. This phenomenon is well known to sailors and fishermen as "white water".

Studies of preserved phytoplankton samples, of sediment trap material and of bottom sediments have shown that *Emiliania huxleyi* (Lohm.) Hay et Mohler (Fig. 1) is the most widespread and abundant coccolithophore in the oceans today (Honjo, 1976). It is the only species that so far is known to produce high surface densities ($>10^{11}$ m^{-3}) of detached coccoliths and is generally the dominant coccolithophore in reflective waters of both coastal and oceanic situations. Although *E. huxleyi* is easily maintained in pure culture, as naked or calcifying cell strains, various studies of its growth and physiology (Klaveness and Paasche, 1979; Brand, 1982; Sikes and Wilbur, 1982; Westbroek *et al.*, 1983) have failed to provide

NATO ASI Series, Vol. G 27
Particle Analysis in Oceanography
Edited by S. Demers

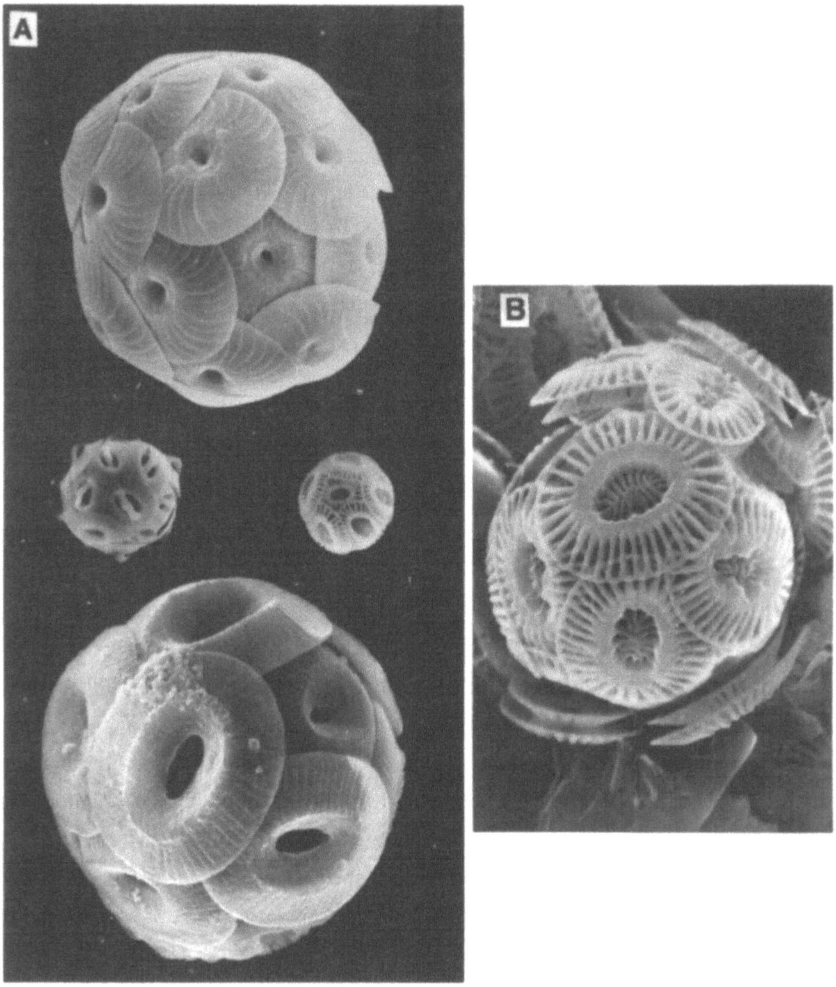

Figure 1. SEM pictures of the most abundant coccolithophores in the NE Atlantic Ocean. **A.** Top: *Cyclococcolithina leptopora*, Centre: *Gephyrocapsa 'caribbeanica'* and *Emiliania huxleyi*, Bottom: *Coccolithus pelagicus*. (All shown to the same scale. Larger cells are about 25μm in diameter). **B.** *E. huxleyi* cell from a natural population showing several layers of coccoliths before detachment.

an unequivocal physio-ecological explanation for coccolith formation. From a geological perspective, calcification in marine organisms is generally considered to be a biotic response to calcium toxicity which evolved some 600 million years ago at a time of rapidly increasing calcium concentrations in the sea (Degens and Ittekot, 1986). Biochemical investigations of algae indicate that precipitation of $CaCO_3$ is associated with increases in alkalinity and Ph during the photosynthetic assimilation of CO_2 and HCO_3- (Borowitzka, 1987). As suggested by the experiments of Baumann *et al.* (1978), calcification may be a means of getting rid of excess photosynthetic energy under conditions of nutrient limitation of cell growth. Various other hypotheses based on possible morphological functions of coccoliths, including light shading and cell sinking, have also been put forward.

The biogeochemical significance of the coccolithophores concerns, in particular, the global carbon and sulphur cycles. Calcium carbonate is the major form in which carbon is buried in the sea, with the most important groups of contributing organisms being the coccolithophores (Bramlette, 1958) and foraminifera in the oceans, and corals, benthic algae and molluscs in shallow waters. The present rate of marine carbonate-carbon burial is not well known (Milliman and Takahashi, 1990), but is probably at least twice that for organic carbon which is preserved mainly in shallow water, deltaic sediments (Berner, 1982). Coccolithophores also contribute to the organic carbon content of ocean sediments through the synthesis of refractory compounds such as long-chained ketones (Brassell *et al.*, 1986), and indirectly through the adsorption of organic matter onto calcite surfaces (Suess, 1973).

Over geological timescales, periods of high rates of carbonate burial (Chamberlin, 1898) and of abundant coccolithophores (e.g. Roth, 1987) are associated with warmer climates (and, presumably, elevated levels of atmospheric $CO_{2)}$. It has been postulated that increases in deep ocean carbonate burial by coccolithophores will tend to cause a relative warming of the earth (Volk, 1989) through effects on the geochemical carbonate-silicate cycle (Berner and Lasaga, 1989) which lead to an increase in levels of atmospheric CO_2. However, the potential influence of short term variations in coccolithophore productivity on atmospheric CO_2 and on global temperature, for example over glacial-interglacial cycles, is not well understood.

In the case of sulphur, coccolithophores are known to be a major oceanic source of dimethylsulphide (DMS) (Andreae, 1986; Turner *et al.*, 1988) which is thought to affect the formation of cloud condensation nuclei in the marine atmosphere (Charlson *et al.*, 1987). This problem is of climatological significance because of the important role of clouds in the regulation of global temperature (Slingo, 1990).

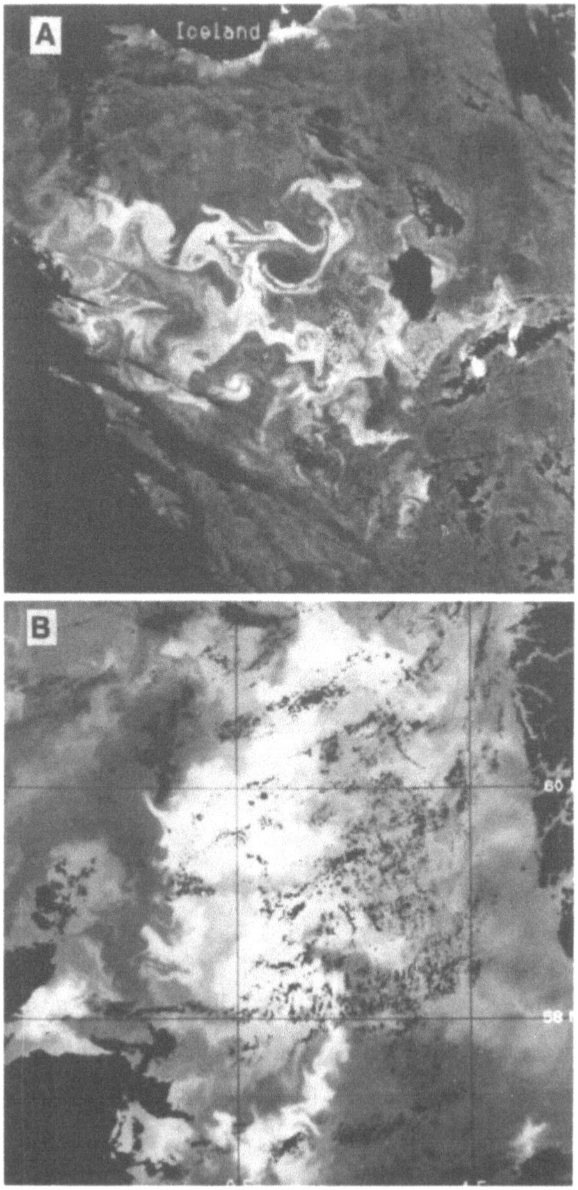

Figure 2. Examples of satellite images of the distributions of *Emiliania huxleyi* coccoliths. **A.** NE Atlantic (approximate area 55-64°N, 10-25°W), 14 June, 1985, NOAA-AVHRR channel 1. **B.** Northern North Sea (approximate area 56.5-61.5°N, 6°E-4°W), 12 July, 1982, CZCS channel 3. Image processing methods are described by Holligan *et al.*, 1989. Clouds and land are black.

In this paper the results of recent studies of coccolithophore blooms in the North Atlantic are summarised. The remote sensing and *in situ* observations on light scattering by coccoliths are considered in terms of providing new information about coccolithophore distributions and about the effects of this group of phytoplankton on the general optical and biogeochemical properties of the ocean-atmosphere system, rather than in relation to the complex optical processes themselves (see Gordon *et al.*, 1988). The possible causes of large scale changes in ocean phytoplankton ecology from diatom dominance (cold periods) to coccolithophore dominance (warm periods), and the implications of such changes in terms of feedback effects on the global environment (see Lovelock, 1986), are discussed. Finally, potential applications of new methods of individual particle analysis in studies of the ecology and biogeochemistry of coccolithophores are briefly evaluated.

COCCOLITHOPHORE OPTICS

Conditions of high (10-25%) water reflectance have now been observed frequently by satellite during spring and summer months for temperate and sub-polar oceans and shelf seas, including upwelling areas (e.g. Fukushima *et al.*, 1987; Dupouy and Demercq, 1987; GREPMA, 1988; Ackleson *et al.*, 1988) and, in all cases for which contemporary plankton data was obtained, *E. huxleyi* was abundant. Examples of *E. huxleyi* blooms in the North Sea and North Atlantic are illustrated in Fig. 2. The first attempt to relate satellite and plankton data quantitatively (Holligan *et al.*, 1983) demonstrated a correlation between reflectance and the log of cell density. However, numbers of detached coccoliths were higher by as much as two orders of magnitude and, on the grounds both of theoretical considerations based on particle size and shape and of subsequent optical studies (see below), it is now established that the free coccoliths as opposed to calcified cells are the dominant cause of light scatter.

Backscattering of light by coccoliths (Balch *et al.*, 1989) and other types of particles is an important potential source of error in algorithms for estimating chlorophyll using satellite ocean colour sensors such as the CZCS, so that knowledge of both absorption and scattering properties of coccolithophores is important for the quantitative analysis of such satellite data. Although CZCS reflectance spectra for coccolithophore blooms often exhibit maximum values in the blue waveband (443nm), observations on the development of individual blooms indicate that blue reflectance relative to green (520nm) or yellow (550nm) tends to be low initially, probably as a result of higher pigment concentrations (or high cell to coccolith ratios) during the growth phase of the populations (Groom and Holligan, 1987). In this context it should be noted that a major accessory pigment in coccolithophores, hexanoylfucoxanthin, shows peak absorption at 520nm (Haxo, 1985).

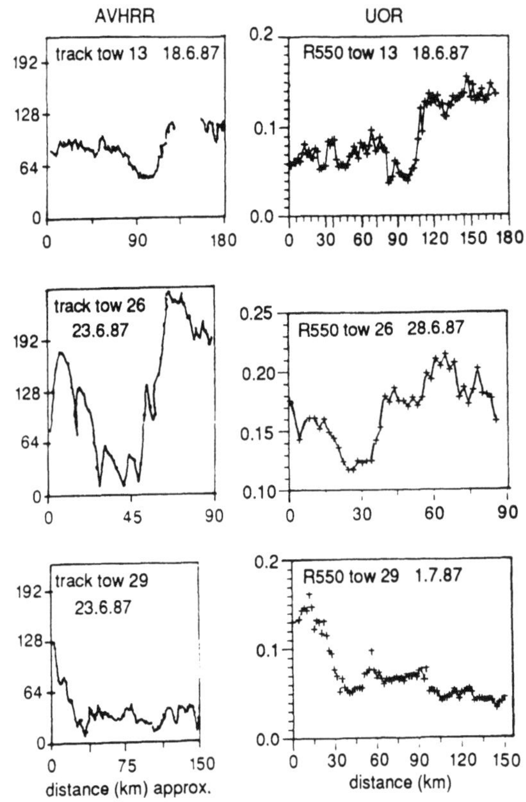

Figure 3. Comparison of along-track AVHRR radiances (arbitrary scale) and UOR reflectance at 550nm for three tows in the NE Atlantic, June 1987 (from Aiken and Bellan, 1989).

In general, increased backscatter by non-absorbing particles reduces the ratio of upwelling light at the wavebands used to determine chlorophyll (blue to yellow or green to yellow for the CZCS), so that phytoplankton biomass will tend to be underestimated. Bricaud and Morel (1986) measured high chlorophyll-specific scattering coefficients for suspensions of *E. huxleyi* cells compared to those for other species of phytoplankton, but the ratio of backscatter to total scatter was too low to explain the high satellite reflectances for natural blooms, supporting the earlier suggestion that the detached coccoliths cause most of the backscatter. Gordon *et al.*, (1988) developed a model for predicting the upwelled radiance at the sea surface as a function of light absorption and backscatter by phytoplankton pigments and coccoliths. However, validation of the model was not possible without values for the spectral backscattering properties of the coccoliths and other particulate matter in the water.

A study of coccolithophores in the North Atlantic in 1987 demonstrated a good qualitative correlation between satellite reflectance (580-680nm, NOAA-AVHRR, channel 1) and underwater reflectance measured with the Undulating Oceanographic Recorder (UOR) (Aiken and Bellan, 1990) as shown in Fig. 3. Quantitative analysis of the data was limited either by time differences >1 day between the satellite and ship measurements or by problems with locating ship sampling positions on the satellite images. However, measurements of coccolith abundance were found to be positively correlated both to reflectance measured with the UOR and to light scatter by non-chlorophyll particles derived from surface transmissometer (660nm) records (Fig. 4), so enabling the mapping of coccolith distributions at sea.

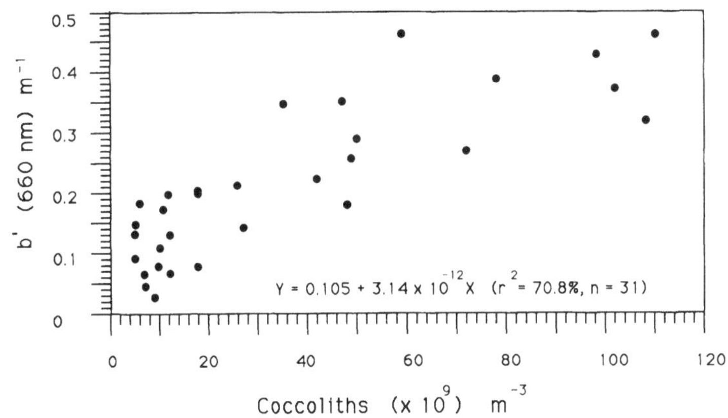

Figure 4. Plot of light scatter (b') at 660nm against coccolith abundance for the NE Atlantic, June 1987. The parameter b' was estimated from transmissometer data by first determining for non-coccolith waters a chlorophyll-specific absorption (a) + scattering (b) coefficient, and then for samples of coccolith-rich waters subtracting the product of this coefficient and the chlorophyll concentration from the beam attenuation (c) value to give an estimate of coccolith scatter.

It appears, therefore, that single channel satellite scanners for the 500-630nm waveband range (i.e. between the strong blue and red absorption bands for chlorophyll and water respectively) can be used to estimate surface coccolith abundance for ocean waters in which the coccoliths are the dominant cause of backscatter, provided that suitable calibration data is acquired, the sensors are not saturated at high reflectances, and appropriate atmospheric corrections are applied. Calibration factors are likely to depend on the size and form of the coccoliths which determine backscattering efficiency, and on the degree of light absorption by accessory pigments and other water constituents. The estimation of water column standing crop of calcite from satellite data will always depend on additional knowledge about the thickness of the coccolith-rich layer.

In order to apply the model of Gordon *et al.* (1988) for estimating chlorophyll from water leaving radiances in the presence of coccoliths, measurements of coccolith backscatter were made for water samples from the Gulf of Maine in 1989 using a light scattering photometer (Balch *et al.*, 1991). Total coccolith backscatter increased linearly for coccolith concentrations up to about 2.10^{11} m^{-3} with the slight decay at higher concentrations possibly related to multiple scattering conditions (Fig. 5). Coccolith backscatter accounted for about 80% of the total backscatter, and the coccolith-specific backscattering coefficients showed some wavelength dependence, being 1.4 times higher in blue light compared to green light.

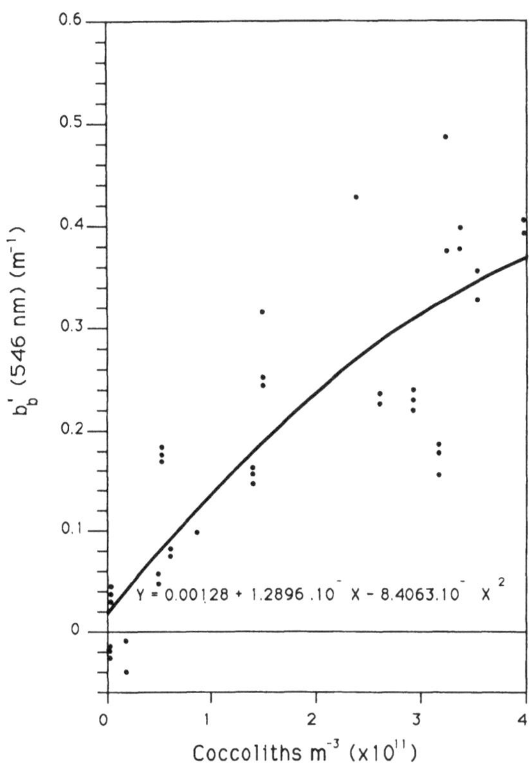

Figure 5. Plot of coccolith backscatter (b_b') at 546nm, measured with a Brice-Phoenix scatterometer, against coccolith abundance (from Balch *et al.*, 1991).

Confidence in the backscattering coefficient values is based is based on comparable estimates from the UOR reflectance data (Fig. 3) using the following approximate relationships (see Bricaud and Morel, 1986; Gordon *et al.*, 1988):

$$R = 0.35\ b_b/a+b_b$$

and

$$K = a + b_b$$

leading to

$$b_b = RK/0.35$$

where R is the sub-surface reflectance, K is the downwelling diffuse attenuation coefficient, and a and b_b are the absorption and backscattering coefficients at particular wavelengths. The constant 0.35 includes the term Q which is the ratio of upwelling radiance to upwelling irradiance. The ratio of backscatter to total scatter for coccoliths (b_b'/b' from Figs 4 and 5) is about 4% and falls within the range of values estimated by Bricaud and Morel (1986). Another important point is that the equation for R predicts, for high values of b_b, a maximum reflectance of about 35%. Somewhat higher values have been measured within coccolithophore blooms (Balch *et al.*, 1991), indicating multiple scattering conditions (Gordon *et al.*, 1988).

The models of water-leaving radiance (Gordon *et al.*, 1988) or reflectance (Balch *et al.*, 1991) still require a knowledge of backscatter by particles other than coccoliths which is difficult to derive independently for either natural water or culture samples. However, reasonable limits have been established for this component of backscatter relative to chlorophyll concentrations (e.g. Bricaud and Morel, 1986) so that it is unlikely to be a cause of serious errors in estimates both of surface coccolith abundance and of phytoplankton chlorophyll using future satellite ocean colour data.

SURFACE LAYER OPTICS

Backscattering of visible light by detached coccoliths causes a marked increase in the albedo of the sea surface and, therefore, a reduction in total light energy absorbed by the ocean. The penetration of solar radiation in the water column is also influenced by light scatter (coccoliths) and absorption (cells). Studies of absorption by water and by plant pigments within particle-rich layers have demonstrated that energy or heat trapping (Stavn, 1987) due to the increased scattering (absorption) path-length for light in the water and to the release of heat by photosynthetic organisms is likely to give rise to anomalous warming rates (Lewis *et al.*, 1983; Kirk, 1988). Such an effect will modify rates of vertical mixing, with important implications for the light environment and rates of growth of phytoplankton in surface layers (Lewis *et al.*, 1983).

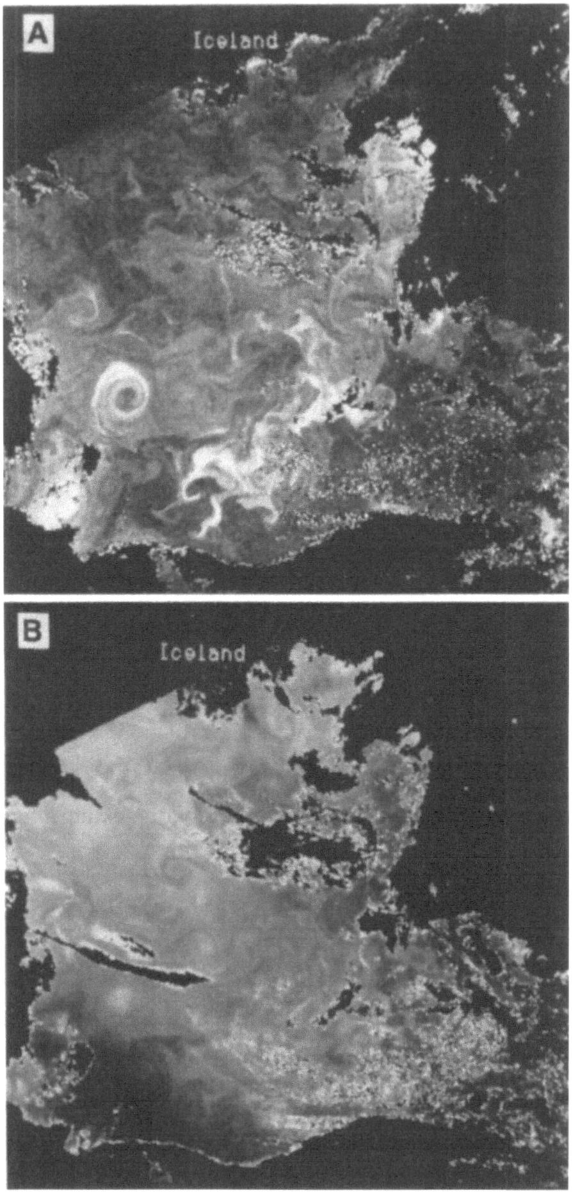

Figure 6. NOAA-AVHRR images of the NE Atlantic Ocean (approximate area 55-64°N, 10-25°W) for 7 July 1986. **A.** Channel 1 (visible) showing distribution of *E. huxleyi* coccoliths. **B.** Channel 4 (infra-red) showing sea surface temperature (warmer water in dark shades, clouds and land in black).

Satellite images show that the highest concentrations of coccoliths in oceanic situations are often along temperature gradients within eddy structures (Fig. 6). Such distributions may indicate local enhancement of coccolithophore growth and calcification due to the upward mixing of nutrients, or the surface outcropping of sub-surface coccolith-rich layers. On the continental shelf, however, regions of strong reflectance due to coccoliths tend to be associated with relatively warm surface water (Ackleson *et al.*, 1988) and it is possible to estimate from sequences of satellite images the enhancement of local heating rates attributable to coccolith light scatter. For the 1988 and 1989 *E. huxleyi* blooms in the Gulf of Maine, the coccoliths were associated with a rate of surface warming 0.05 to 0.08 °C d^{-1} greater than that for nearby regions outside the bloom (S.G. Ackleson pers.comm.). Experiments with cultures of naked and calcifying strains of *E. huxleyi* also show that the presence of coccoliths leads to significantly greater heating of the growth medium under equivalent conditions of illumination and external temperature (van Leeuwe, 1990).

UOR sections through a coccolithophore population illustrate how variations in reflectance (550nm) are accompanied by shifts in the depths of penetration for the 10 and 1% light levels at this wavelength (Fig. 7). Although quantitative comparisons of the light fields have yet to be carried out, it appears that the loss of light energy to the atmosphere due to near-surface backscattering is accompanied by an increase in overall light absorption within the coccolith layer. Sub-surface chlorophyll maxima are generally absent or only weakly developed within N. Atlantic coccolithophore blooms possibly due to light limitation of phytoplankton growth at the level of the seasonal thermocline. (Discrete chlorophyll determinations indicate that sub-surface maxima in chlorophyll fluorescence as seen in Fig. 7 are due largely to daytime solar inhibition of fluorescence near the surface.)

INORGANIC AND ORGANIC CARBON FLUXES

Laboratory studies with batch cultures of a calcifying strain of *E. huxleyi* isolated from the Gulf of Maine in 1988 have shown that carbon incorporation into calcite exceeds that into organic carbon during at least part of the growth cycle (Fig. 8). Field measurements have given similar results (unpubl. data), although the measured rates of calcification relative to photosynthesis are often low due in part to the presence of non-calcifying types of phytoplankton and in part to the highly localised distribution in space and time of actively calcifying cells (by the time blooms are detected as high reflectance features on satellite images coccolith formation has largely stopped). In the experiment illustrated in Fig. 8 the total production of new cells to coccoliths was in the proportion 1:53, assuming that each cell

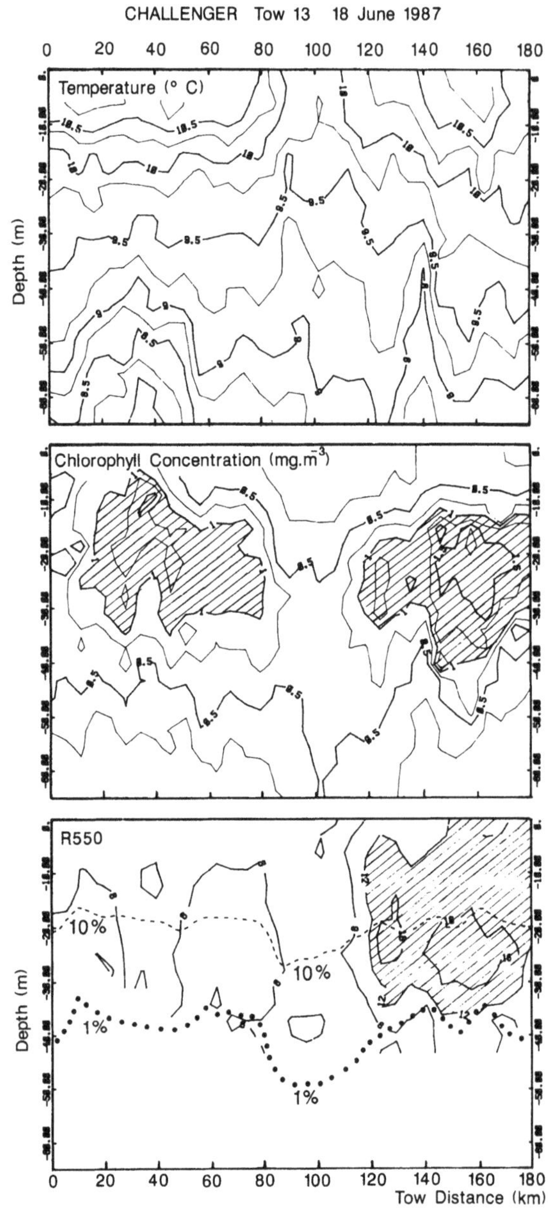

Figure 7. Undulating Oceanographic Recorder (UOR) sections for temperature (°C), chlorophyll a (mg m^{-3}), and reflectance at 550nm in the NE Atlantic, 18 June 1987. The tow was from 60°59'N, 13°57'W to 59°52'N, 09°59'W (see Aiken and Bellan, 1989). The depth of the 10 and 1% light levels (550nm) are shown in the bottom panel.

had 20 attached coccoliths. Independent estimates of carbon per cell and per coccolith indicate that the total fixation of carbon into organic matter and into calcite was about the same.

In cultures of *E. huxleyi* coccoliths are formed as rapidly as one per hour during the light period (Westbroek *et al.*, 1983; Fig. 8) so that the production rate might be as high as 10 cell^{-1} d^{-1} under natural conditions at temperate latitudes. However, without appropriate information on changes in cell abundance with time and on losses of coccoliths due to sinking,

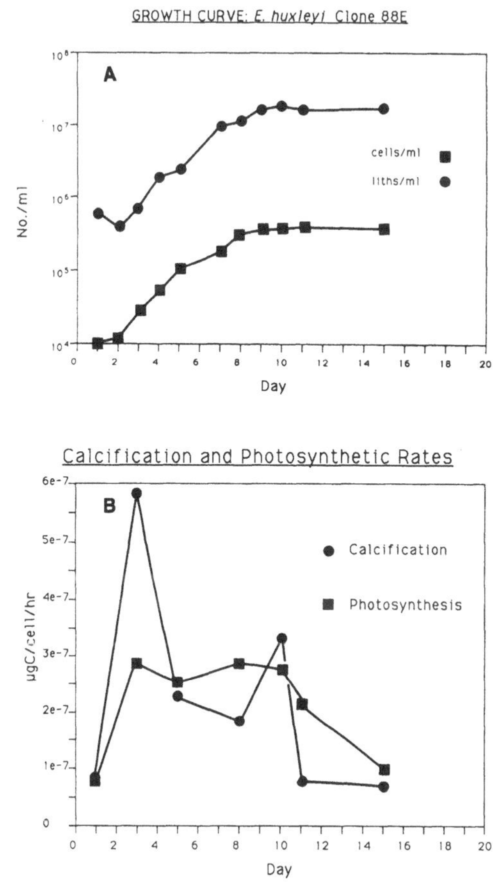

Figure 8. **A.** Changes in cell and coccolith densities during growth in f/50 medium of a batch culture of *E. huxleyi* (clone 88E from the Gulf of Maine). **B.** Changes in rates of photosynthesis and calcification for the same culture, as determined from assimilation rates for $^{14}CO_2$ in the light period. (From Balch *et al.*, 1991).

grazing, it is not possible to calculate total coccolith production. For illustrative purposes an estimate of total calcite productivity can be inferred from typical observed maximum densities of cells (3.10^9 m^{-3}) and coccoliths (10^{11} m^{-3}) if it is assumed that each cell has 20 coccoliths (Fig. 1B), that the thickness of the population layer is 20m and that an equal number of coccoliths to the maximum standing crop are lost by grazing and sinking during the development of the bloom. The calculated value is 64.10^{11} coccoliths m^{-2} which is equivalent to 13.0g $CaCO_3$ or 1.55g calcite-carbon m^{-2}, based on a volume for each *E. huxleyi* coccolith of 3/4 μm^3 (Westbroek *et al.*, 1983).

Such fluxes are compatible both with sediment trap data (Honjo, 1976; Cadee, 1985; Kempe and Jennerjahn, 1988; Hay *et al.*, 1990) and with information on the rate of accumulation of coccoliths in bottom sediments (see below) for coccolithophore-rich waters. However, more important is that the timescale of 3-4 weeks for bloom development inferred from ecological and remote sensing information gives an indication of the rates at which surface water optical properties and biogeochemical fluxes are likely to change as such blooms develop.

Holocene rates of accumulation of calcareous sediments in the NE Atlantic range from >20 cm ky^{-1} on sediment ridges to <3 cm ky^{-1} south of 45° N towards the subtropical gyre. The inorganic (calcite) and organic carbon contents are typically 7% (Fig. 10) and 0.3% (Romankevich, 1984) respectively, and mainly associated with the fine (< 63 μm diameter) material. For an intermediate deposition rate of 7 cm ky^{-1}, and assuming a sediment density of 0.76 g cc^{-1} and 80% of the calcite to be in the form of coccoliths (unpublished data), the accumulation rates of coccolith inorganic carbon and of total organic carbon are 2.98 and 0.16 g m^{-2} y^{-1} (a ratio of 19:1). The value for inorganic carbon is about twice that estimated above for an *E. huxleyi* bloom, but the coccoliths of other species (Fig. 1) are also abundant in the sediments. The significance of the inorganic to organic carbon ratio for coccolith-rich sediments lies in understanding the degree to which the two might be associated. Carbonate surfaces adsorb organic material (Suess, 1973; also see Degens and Ittekot, 1986), a process that enhances the final preservation of organic carbon in marine sediments (Gordon and Millero, 1985). Using the estimate of Suess that 1m^2 of carbonate surface holds 1.2 mg organic carbon, and a surface area value for each *E. huxleyi* coccolith of 15μm^2 (Westbroek *et al.*, 1983), a predicted inorganic to organic carbon ratio of about 13:1 is calculated for such coccoliths. Since this is lower than observed ratio for coccolith-rich sediments, most if not all the organic carbon in such sediments may be adsorbed onto the coccoliths. Some is derived directly from coccolithophores. including acidic polysaccharides that form the matrix of the coccoliths of *E. huxleyi* (Westbroek *et al.*, 1989) and long chained ketones in sediments that are known to be synthesised by this group of organisms (Brassell *et al.*, 1986; Prahl *et al.*, 1989).

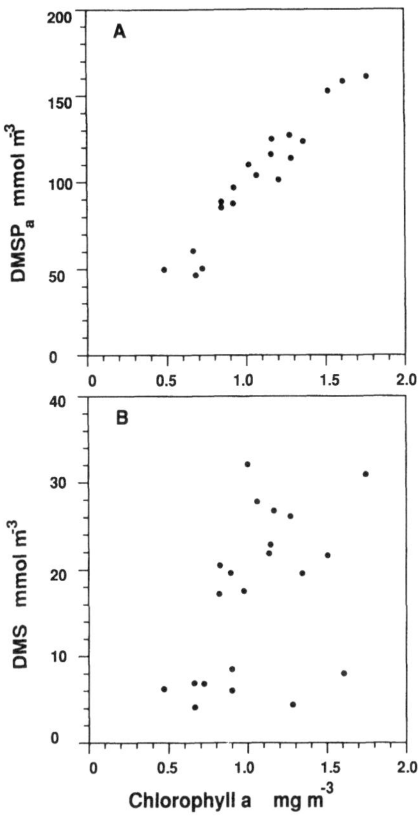

Figure 9. Paired measurements of A. algal DMSP and B. free DMS for coccolithophore-rich (>50% of total phytoplankton biomass) water samples from the North Atlantic, June 1987. (S. Turner and G. Malin, unpublished data).

However, the most important conclusion is that coccoliths, as the most abundant form of mineral particle in open ocean environments, have a significant influence on both the downward flux (as faecal pellets - see Honjo, 1976) and burial (by adsorption) of organic carbon.

There have been various studies of the effects of calcium carbonate formation in the sea on the level of CO_2 in the atmosphere over the timescales of the main geological cycles (Chamberlin, 1898; Volk, 1989) and of glacial-interglacial cycles (Berger, 1982; Dymond and Lyle, 1985). All conclude that an increase in the ratio of carbonate-carbon to organic carbon sedimentation will lead to an increase in atmospheric CO_2, due to the assimilation of bicarbonate:

$$Ca^{2+} + 2HCO_3^- = CaCO_3 + CO_2 + H_2O$$

The details of the mechanisms in terms of changes in the rates of exchange of CO_2 between the atmosphere, surface ocean and deep ocean are uncertain, and for this reason, studies of the effects of coccolithophore productivity on the state of the CO_2 system in surface ocean waters are needed.

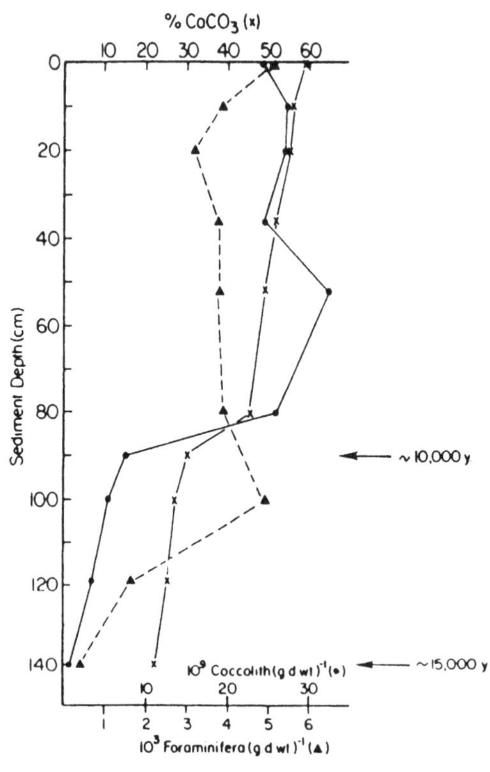

Figure 10. Vertical distributions of calcite (% dry weight), coccolith abundance (g^{-1}) and foraminifera (g^{-1}) in sediment core 850701 from the Rockall Trough, NE Atlantic (57°00'N, 12°01'W, water depth 2130m). Approximate dates before present are based on the occurrence of sinistral form of the foraminifera *Neogloboquadrina pachyderma*.

In recent review of the carbon physiology of coccolithophores Sikes and Fabry (1990) concluded that increases in the productivity of these organisms were unlikely to cause significant outgassing of CO_2 due mainly to the buffering capacity of seawater, but would maintain a downward flux of particulate carbon in the oceans. A more critical question concerns the difference in CO_2 exchange between ecosystems dominated by coccolithophores as opposed to non-calcifying phytoplankton (e.g. diatoms). A model of primary production in the North Atlantic (Taylor *et al.*, 1991) shows that, in the summer period of nutrient limited conditions, the presence of coccolithophores raises the partial pressure of CO_2 at the surface, thereby reducing the atmosphere-to-ocean flux of this gas while maintaining a higher downward flux of particulate (organic + inorganic) carbon due to the formation of calcite. This flux is balanced by the upward mixing of dissolved inorganic carbon from beneath the seasonal thermocline. The net result over a longer period of time is a change in the CO_2 chemistry of surface water (lower alkalinity and TCO_2) accompanied by higher atmospheric CO_2 which is the same conclusion reached by the geochemical mass budget models (Berger, 1982; Dymond and Lyle, 1985).

SULPHUR CYCLING

The recognition that the ocean is a strong source of biogenic sulphur gases which, on oxidation in the atmosphere, play an important role in the formation of cloud condensation nuclei (CCN) (Charlson *et al.*, 1987) and contribute to the acidity of rainfall has led to considerable efforts to characterise processes that may cause variations in the gas fluxes. It is now well known (Andreae, 1986) that the ocean emission is mainly in the form of dimethylsulphide (DMS) formed by the cleavage of dimethylsulphoniopropionate (DMSP), an important osmosolute in certain groups of phytoplankton (Dickson and Kirst, 1987). Measurements for natural water samples and cultures have demonstrated that coccolithophores contain large quantities of DMSP at internal concentrations consistent with an osmoregulatory function, and are associated with high levels of DMS in the water (Turner *et al.*, 1988; Keller *et al.*, 1989). However, the proportion of the total DMS flux from the oceans attributable to this group of phytoplankton remains uncertain.

Levels of algal (as opposed to free) DMSP measured in the surface waters of the NE Atlantic in summer 1987 were high and closely correlated with phytoplankton biomass within coccolithophore blooms (Fig. 9). By contrast, quantities of free DMS were extremely variable confirming that the processes controlling the conversion of DMSP to DMS are important determinants of the proportion of DMSP that reaches the atmosphere as DMS. These include grazing (Dacey and Wakeham, 1986), and microbial activity which affects both the degradation of DMSP and the consumption of DMS (Kiene and Bates, 1990); Another striking feature of the data in Fig. 9 is that the maximum DMS concentrations were about an order of magnitude greater than average values for surface ocean waters (Andreae, 1986) and represented as much as 20% of the algal DMSP. This high proportion suggest that the breakdown of DMSP to DMS is relatively rapid in coccolithophore blooms.

However, it is not possible to estimate global values for DMSP or DMS production by coccolithophores until there is comparable data on the productivity of this group of phytoplankton. Measurements of sulphur gases or CCN abundance in the atmosphere above coccolithophore populations have not been made, even though the large scale of the blooms (Fig. 2) and the high levels of DMS in the water (Fig. 9) represent ideal conditions for such an experiment. In the context of global budgets, a further point of interest is that the DMSP content of coccolithophores (estimated from the data in Fig. 9 assuming a carbon-to-chlorophyll ration of 40) represents about 15% of cellular organic carbon. This compound therefore appears also to have an important role in the marine carbon cycle given the abundance of coccolithophores and other DMSP-producing phytoplankton (Keller *et al.*, 1989) in the oceans.

DISCUSSION

The information summarised in this paper suggests that the coccolithophores have a significant effect on the radiation budget of the surface layers of the ocean and on the ocean-atmosphere cycles of carbon and sulphur. Also their distributions change with variations in climate, becoming more abundant and widespread in warmer periods. Such changes are most pronounced at mid-latitudes following displacements of the oceanic polar fronts (Ruddiman *et al.*, 1980; Ruddiman and McIntyre, 1981), as shown by data on coccolith abundance in sediment cores (Fig. 10) and on the distribution of $CaCO_3$ in the oceans (Bishop, 1989). Shifts of the polar fronts towards the poles during climatic amelioration lead to a marked increase in the areal extent of temperate and sub-polar waters and it is in this hydrographic regime that coccolithophores are most abundant today. At the same time the continental shelves have become flooded, creating shallow water environments in which *E. huxleyi* is abundant today and which are comparable to the major sites of coccolithophore deposition during the Jurassic and Cretaceous periods (Gallois, 1976).

The well-known seasonal succession in marine phytoplankton communities (Margalef, 1978) from diatoms to coccolithophores and then flagellates is observed consistently in a wide range of habitats. The appearance of coccolithophores is usually associated with the surface depletion of inorganic nutrients particularly nitrate, but possibly also trace nutrients such as iron (Martin, 1990) in the sub-polar regions, including the NE Atlantic, where nitrate levels remain high through the summer. Experimental work with cultures has demonstrated that coccolithophores continue to grow at very low iron concentrations (Kramer and Ryther, 1960; Brand *et al.*, 1983). Thus, as an extension to the hypothesis that the deposition of atmospheric dust (iron) over the oceans during the glacial period maintained higher rates of diatom productivity (Martin, 1990), it is suggested that low dust input during the interglacials favours the growth of coccolithophores, and the pelagic deposition of $CaCO_3$.

The overall effect on the global carbon cycle of such a change in ocean ecology remains uncertain, especially as the sedimentation of coccolithophores almost certainly enhances the downward flux of organic carbon. In general, it appears that an increase in coccolithophore productivity during interglacial periods would reinforce a lowering of surface alkalinity due to changes in ocean circulation which has been postulated by Broecker and Peng (1989) as a possible cause of higher atmospheric levels of CO_2. Simple models have demonstrated the potential effects of variation in carbon productivity due to corals (Berger, 1982) and coccolithophores (Dymond and Lyle 1985; Taylor *et al.*, 1991) on atmospheric CO_2, but such processes have yet to be incorporated into models which take account of changes in ocean circulation and in the lysocline depth for calcite.

The sulphur cycle presents a special problem in that ice core records show greater deposition biogenic sulphur during the glacial period (Legrand *et al.*, 1988) when diatoms (which are not an important source of DMS (Keller *et al.*, 1989)) are thought to have made a proportionately greater contribution to ocean productivity. However, interpretation of the ice core record must take account of the short life time (days) of biogenic sulphur in the atmosphere and of the possible scavenging role of the abundant dust (Petit *et al.*, 1990) in the glacial atmosphere. Thus, the observed changes in sulphur may reflect local variations in processes affecting sulphur deposition rather than global variations in phytoplankton ecology and productivity.

In considering the climatic implications of possible future changes in oceanic phytoplankton populations, in particular the effects of a potential increase in coccolithophore abundance at high latitudes associated with global warming due to anthropogenic CO_2, there are two major uncertainties. The first concerns the nature of interactions between the biogeochemical cycles of two or more critical elements and associated environmental impacts. For example, how might variations in the light climate of surface ocean caused by the effects DMS on cloud formation (Charlson *et al.*, 1987) affect the relative rates of photosynthesis and calcification and the exchange of CO_2 between atmosphere and ocean? The second concerns ecological predictions. How will changes in the surface ocean environment (temperature, seasonal stratification, ice cover etc.) influence the growth and distributions of the major phytoplankton groups such as diatoms and coccolithophores?

New techniques of individual particle analysis will contribute significantly to studies of the second problem particularly for types of phytoplankton, including coccolithophores, with complex optical properties and with life cycles that are difficult to investigate by standard microscopy. For *E. huxleyi* the satellite images (Figs. 2, 6) provide information on the spatial and temporal development of a single population (or set of populations), but the significance of the distributions and roles of the different life stages or of distinct genetic strains (Brand, 1982) for population dynamics of the species (including annual variations in abundance) is not known. Immunochemical techniques to identify non-calcifying cells of *E. huxleyi* have now been developed (Shapiro *et al.*, 1989) and, in conjunction with measurements of side scattering to detect intracellular coccolith formation, represent a powerful new approach to studying the dynamics of natural populations.

The process of calcification in coccolithophores is still poorly understood (Borowitzka, 1987; Sikes and Fabry, 1990) in terms both of biochemistry and of the effects of environmental factors (light, nutrients) on the partitioning of carbon and energy between organic matter synthesis and calcite formation. New single-cell methods of studying the uptake of carbon and calcium as well as intracellular variations in pH (e.g. Dixon *et al.*, 1989) will enable significant progress on these topics. Knowledge of the properties of coccoliths is important for investigations of sinking rates, ingestion and packaging by zooplankton (Cadee, 1986), the content of and binding capacity for organic carbon, and the isotopic ratios for carbon and oxygen in calcite. Methods to sort and count coccolithophores and to label specific associated organic compounds such as matrix polysaccharides (Westbroek *et al.*, 1989) are just beginning to be applied in plankton and sediment studies. A better understanding of the relationship between the stability of adsorbed organic carbon and the dissolution at depth of calcite under oxic and anoxic conditions is of particular relevance to the oceanic carbon cycle.

Finally, estimates of the conversion efficiency of DMSP in cells to DMS in the atmosphere will depend on new information about grazing rates of coccolithophores (see Dacey and Wakeham, 1986) and about the degradation of both DMSP and DMS by marine microbes (Kiene and Bates, 1990). Flow cytometric techniques have yet to be used in this context but represent a potentially powerful tool for studies of the role of the ocean biota in the sulphur cycle.

Changes in the distributions and productivity of coccolithophores will influence the general optical and biogeochemical properties of the oceans, and could lead to significant feedback effects on the global environment. Methods of single particle analysis are just beginning to be applied in research on these organisms and are likely to lead to a better understanding of how oceanic phytoplankton affect the climate.

ACKNOWLEDGEMENTS

We are most grateful to Dr. Jim Aiken, Dr. Gill Malin and Dr. Sue Turner, and Kay Kilpatrick for providing the data shown in Figures 7, 8 and 9 respectively . We also thank Stephen Groom for processing the satellite data, and Derek Harbour for many hours of identifying and counting coccolithophores.

REFERENCES

Ackleson SG, Balch WM, Holligan PM (1988) White waters of the Gulf of Maine. Oceanography 1 (2):18-22

Aiken J, Bellan I (1990) Optical oceanography: an assessment of a towed method. In: Herring PJ, Campbell AK, Whitfield M and Maddock L (eds) Light and life in the sea. Cambridge University Press pp 39-57

Andreae MO (1986) The ocean as a source of atmospheric sulfur compounds pp 331-362. In P Buat-Menard (ed). The Role of Air-sea Exchange in Geochemical Cycling. NATO ASI Series C Vol 185. D Reidel Publishing Company, Dordrecht

Balch WM, Eppley RW, Abbott MR, Reid FM (1989) Bias in satellite derived pigment measurements due to coccolithophores and dinoflagellates. J Plank Res 11:575-581

Balch WM, Holligan PM, Ackleson SG, Voss KJ (1991) Biological and optical properties of mesoscale coccolithophore blooms in the Gulf of Maine. Limnol Oceanogr (in press)

Baumann FG, Isenberg HD, Gennaro J (1978) The inverse relationship between nutrient nitrogen concentration and coccolith calcification in cultures of the coccolithophorid *Hymenomonas* sp J Protozool 25:253-256

Berger WH (1982) Deglacial CO_2 buildup: Constraints on the coral reef model. Palaeogeog Palaeoclim Palaeoecol 40:235-253

Berner RA (1982) Burial of organic carbon and pyrite sulfur in the modern ocean: Its geochemical and environmental significance. Am J Sci 282:415-473

Berner RA, Lasaga AC (1989) Modeling the geochemical carbon cycle. Sci Am 260:54-61

Bishop JKB (1989) Regional extremes in particulate matter composition and flux: Effects on the chemistry of the ocean interior. In Productivity of the Ocean: Past and Present. Eds Berger WH, Smetacek VS and Wefer G. pp 117-137. John Wiley & Sons Ltd

Borowitzka MA (1987) Calcification in algae: mechanisms and the role of metabolism. CRC Crit Rev Pl Sci 6:1-45

Bramlette MN (1958) Significance of coccolithophorids in calcium carbonate deposition. Bull Geol Soc Am 69:121-12.

Brand LE (1982) Genetic variability and spatial patterns of genetic differentiation in reproductive rates of the marine coccolithophores *Emiliania huxleyi* and *Gephyrocapsa oceanica*. Limnol Oceanogr 27:236-245

Brand LE, Sunda WG, Guillard RRL (1983) Limitation of marine phytoplankton reproductive rates by zinc, manganese and iron. Limnol Oceanogr 28:1182-1198

Brassell SC, Eglinton G, Marlowe IT, Pflaumann U and Sarntheim M (1986) Molecular stratigraphy: a new tool for climatic assessment. Nature 320:129-133

Bricaud A, Morel A (1986) Light attenuation and scattering by phytoplanktonic cells: a theoretical modelling. Appl Opt 25:571-580

Broecker WS, Peng TH (1989) The cause of the glacial to interglacial atmospheric CO_2 change: a polar alkalinity hypothesis. Global Biogeochemical Cycles 3:215-239

Cadee GC (1985) Macroaggregates of *Emiliania huxleyi* in sediment traps. Mar Ecol Prog Ser 24:193-196

Chamberlin TC (1898) The influence of great epochs of limestone formation upon the constitution of the atmosphere. J Geol 6:609-621

Charlson RJ, Lovelock JE, Andreae MO, Warren SG (1987) Oceanic phytoplankton, atmospheric sulphur, cloud albedo and climate. Nature 326:655-661

Dacey JWH, Wakeham SG (1986) Oceanic dimethylsulfide: Production during zooplankton grazing on phytoplankton. Science 233:1314-1316

Degens ET, Ittekkot V (1986) Ca2+-stress, biological response and particle aggregation in the aquatic habitat. Neth J Sea Res 20:109-116

Dickson DMJ, Kirst GO (1987) Osmotic adjustment in marine eukaryotic algae: The role of inorganic ions, quaternary ammonium, tertiary sulfonium and carbohydrate solutes. II. Prasinophytes and haptophytes. New Phytol 106:657-666

Dixon GK, Brownlee C, Merrett MJ (1989) Measurement of internal pH in the coccolithophore *Emiliania huxleyi* using 2' 7'-bis-(2-carboxyethyl)-5(and -6) carboxyfluoroscein acetoxymethylester and digital imaging microscopy. Planta 178:443-449

Dupouy C, Demercq H (1987) CZCS as an aid for understanding modalities of the plankton productivity during upwelling off Senegal. Adv Space Res 7:63-71

Dymond J, Lyle M (1985) Flux comparisons between sediments and sediment traps in the eastern tropical Pacific: Implications for atmospheric CO_2 variations during the Pleistocene. Limnol Oceanogr 30:699-712

Fukushima H, Hiramatsu K, Sugimori Y (1987) CZCS derived pigment concentration fields in a Japanese coastal area. Adv Space Res 7:79-82

Gallois RW (1976) Coccolith blooms in the Kimmeridge clay and the origin of North Sea oil. Nature 259:473-475

Gordon AS, Millero FJ (1985) Adsorption mediated decrease in the biodegradation rate of organic compounds. Microb Ecol 11:289-298

Gordon HR, Brown OB, Evans RH, Brown JW, Smith RC, Baker KS, Clark DK (1988) A semianalytical radiance model of ocean color. J Geophys Res 93:10909-10924

Gower JFR, Denman KL, Holyer RJ (1980) Phytoplankton patchiness indicates the fluctuation spectrum of mesoscale oceanic structure. Nature 288:157-159

Green JC (1986) Biomineralization in the algal class Prymnesiophyceae. In Leadbetter B and Riding R (eds) Biomineralization of Lower Plants and Animals. Clarendon Press Oxford pp 173-188

GREPMA (1988) Satellite (AVHRR/NOAA-9) and ship studies of a coccolithophorid bloom in the western English Channel. Mar Nature 1:1-1

Groom SB, Holligan PM (1987) Remote sensing of coccolithophore blooms. Adv Space Res 7:73-78

Haxo FT (1985) Photosynthetic action spectrum of the coccolithophorid *Emiliania huxleyi* (Haptophyceae): 19'Hexanoyloxyfucoxanthin as antenna pigment. J Phycol 21:282-287

Hay BJ, Honjo, S Kempe S, Ittekot VA, Degens ET, Konuk T, Izdar E (1990) Interannual variability in particle flux in the southwestern Black Sea. Deep Sea Res 37:911-928

Hibberd DJ (1980) Prymnesiophytes (=Haptophytes). In: Cox E (ed). Phytoflagellates Elsevier pp 273-317

Holligan PM, Aarup T, Groom SB (1989) The North Sea: Satellite colour atlas. Contin Shelf Res 9:667-765

Holligan PM, Viollier M, Harbour DS, Camus P, Champagne-Phillipe M (1983) Satellite and ship studies of coccolithophore production along a continental shelf edge. Nature 304:339-342

Honjo S (1976) Coccoliths: Production, transportation and sedimentation. Mar Micropaleont 1:65-79

Keller MD, Bellows WK, Guillard RRL (1989) Dimethyl sulphide production in marine phytoplankton. In Biogenic Sulphur in the Environment. Eds Saltzman ES and Cooper WJ. American Chemical Society Symposium Series 393, ACS Washington pp 167-182

Kempe S, Jennerjahn TC (1988) The vertical particle flux in the northern North Sea, its seasonality and composition. Mitt Geol-Palaont Inst Univ Hamburg 65:229-268

Kiene RP, Bates TS (1990) Biological removal of dimethyl sulphide from sea water, Nature 345:702-705

Kirk JTT (1988) Solar heating of water bodies as influenced by their optical properties. J Geophys Res 93:10897-10908

Klaveness D, Paasche E (1979) Physiology of coccolithophorids. In: Levandowski M and Hutner SH (eds) Biochemistry and Physiology of Protozoa (2nd edition) Vol 1 Academic Press

Kramer DJ, Ryther JH (1960) The iron requirement of some marine phytoplankton. Biol Bull 119:324

Legrand MR, Delmas RJ, Charlson RJ (1988) Climate forcing implications from Vostok ice-core sulphate data. Nature 334:418-420

Lewis MR, Cullen JJ, Platt T (1983) Phytoplankton and thermal structure in the upper ocean: Consequences of nonuniformity in chlorophyll profile. J Geophys Res 88:2565-2570

Lovelock JE (1986) Geophysiology: A new look at Earth Science. Bull Am Meteorol Soc 67:392-397

Margalef R (1978) Life-forms of phytoplankton as survival alternatives in an unstable environment. Oceanol Acta 1:493-509

Martin JH (1990) Glacial-interglacial CO_2 change: The iron hypothesis. Palaeoceanography 5:1-13

Milliman JD, Takahashi K (1990) Carbonate and opal production and accumulation in the ocean. In Global Surficial Geofluxes: Modern to Glacial. Eds Usselman TM, Hay W and Meybeck M. National Academy Press USA. In press

Petit JR, Mounier L, Jouzel J, Korotkevich YS, Kotlyakov VI, Lorius C (1990) Palaeoclimatological and chronological implications of the Vostok core dust record. Nature 443:56-58

Prahl FG, de Lange GJ, Lyle M, Sparrow MA (1989) Post depositional stability of long-chain alkenones under contrasting redox conditions. Nature 341:434-437

Romankevich EA (1984) Geochemistry of Organic Matter in the Ocean. Springer-Verlag 334 pp

Roth PH (1987) Mesozoic calcareous nannofossil evolution: Relation to paleoceanographic events. Paleoceanography, 2:601-611

Ruddiman WF, McIntyre A (1981) The North Atlantic during the last deglaciation. Palaeogeog Palaeoclimat Palaeoecol 35:145-214

Ruddiman WF, Molfino B, Esmay A, Pokras E (1980) Evidence bearing on the mechanism of rapid deglaciation. Clim Change 3:65-87

Shapiro LP, Campbell L, Haugen EM (1989) Immunochemical recognition of phytoplankton species. Mar Ecol Prog Ser 57:219-224

Sikes CS, Fabry VJ (1990) Photosynthesis and $CaCO_3$ deposition, coccolithophorids and the global carbon cycle. In photosynthetic Carbon Metabolism and Regulation of Atmospheric CO_2 and O_2. Eds NE Tolbert and J Preiss. In press

Sikes CS, Wilbur KM (1982) Functions of coccolith formation. Limnol Oceanogr 27:18-26

Slingo A (1990) Sensitivity of the Earth's radiation budget to changes in low clouds. Nature 343:49-51

Stavn RA (1987) The three-parameter model of the submarine light field: Radiant energy absorption and trapping in nepheloid layers recalculated. J Geophys Res 92:1934-1936

Suess E (1973) Interaction of organic compounds with calcium carbonate - II. Organo-carbonate association in recent sediments. Geochim Cosmochim Acta 37: 2435-2447

Taylor AH, Watson AJ, Ainsworth M, Robertson JE, Turner DR (1991) A modelling investigation of the role of phytoplankton in the balance of carbon at the surface of the North Atlantic. Global Biogeochem Cycles (in Press)

Turner SM, Malin G, Liss P, Harbour DS, Holligan PM (1988) The seasonal variations of dimethylsulphonioproprionate concentrations in near shore waters. Limnol Oceanogr 33:364-375

Van Leeuwe MA (1990) Heat generation in aquatic media containing scattering particles: theory, and measurements in *Emiliania hyxleyi* cultures (Coccolithophorids). Unpubl ms Groningen, The Netherlands

Volk T (1989) Sensitivity of climate and atmospheric CO_2 to deep-ocean and shallow-ocean carbonate burial. Nature 337:637-638

Westbroek P, Young PR, Linschooten K (1989) Coccolith production (biomineralization) in the marine alga *Emiliania huxleyi*. J Protozool 36:368-373

Westbroek P, de Jong EW, van der Wal P, Borman T de Vrind JPM, van Emburg PE, Bosch L (1983) Calcification in coccolithophoridae - wasteful or functional? Environment Biogeochem Ecol Bull 35:291-299

FROM CELLS TO THE OCEAN: SATELLITE OCEAN COLOR

M. R. Lewis[1] and J. J. Cullen[2,3]

[1]Department of Oceanography
Dalhousie University
Halifax, Nova Scotia
Canada B3H 4J1

[2]Bigelow Laboratory
West Boothbay Harbor, Maine
USA

ABSTRACT

Variations in the color of the ocean as seen from space are principally due to variations in the concentration and optical properties of biogenic materials, dissolved and particulate, in the upper ocean. From 1978 to 1986 the NIMBUS-7 Coastal Zone Color Scanner observed these variations over the global ocean; the wealth of data that has resulted is just now being appreciated. The resultant ability to observe the ocean from a biological perspective over synoptic scales revolutionized the field.

The satellite observations require algorithms to interpret the received signal in terms of meaningful geophysical quantities or processes. Most of the signal results from the atmosphere and corrections to permit analysis of the ocean signal is non-trivial. Assuming that this can be done with acceptable accuracy, it is still necessary to relate the observations of radiance leaving the surface of the ocean to more useful variables such as the concentration of chlorophyll in the sea surface or the primary productivity of the ocean. The link between the observations and the desired retrieval are the so-called bio-optical algorithms.

The relationship between water-leaving radiance and chlorophyll in the water column is not simple because most of the chlorophyll is contained within phytoplankton particles of varying dimension and with varying internal concentration of chlorophyll. The presence of ancillary and detrital pigments, in addition to chlorophyll, and the presence of non-chlorophyllous particles further complicate the issue. Two approaches to the bio-optical algorithms have been taken to resolve these problems. The first attempts to describe empirically the variability in particle size and composition in terms of coefficients of statistical relationships between water-leaving radiance and chlorophyll concentration. Primary production is described in a similar fashion. The second, not mutually exclusive, relies on first principles of radiative transfer and the physiology of phytoplankton coupled with a detailed understanding of the nature and size distribution of the particle population and their optical properties. One of the challenges of the latter path is to reproduce the water-leaving signal given the solar input and detailed knowledge of the particle population from analysis of individual cells. The inverse problem, infering the individual cell characteristics from the water-leaving radiance signal, is equally or more challenging.

Here, the bases for the remote measurement of ocean color from space, and the algorithms used in the estimation of chlorophyll concentration and primary productivity from these remote measurements will be discussed with a view towards interpretation based on the characteristics of the ensemble of individual particles in the upper ocean.

[3]Present Address: Department of Oceanography
Dalhousie University
Halifax, Nova Scotia
Canada B3H 4J1

NATO ASI Series, Vol. G 27
Particle Analysis in Oceanography
Edited by S. Demers

INTRODUCTION

The color of the ocean as seen from space results from variability in the optical properties of the upper ocean. Variations in the optical properties, absorption and scattering, modify the spectral and geometrical distribution of the underwater light field and likewise alter the spectral distribution of the water-leaving radiance and hence the color of the sea.

Knowledge and prediction of the variability in these oceanic optical properties are fundamental to both physical and biological oceanography. Absorption and scattering control the penetration and spectral composition of irradiance in the sea and this radiation not only fuels the the marine food chain but also modifies the local temperature field and influences upper ocean thermal dynamics.

The ability to view the global ocean from space has greatly expanded the scope of marine optics and has revolutionized oceanography. This is particularly true in the understanding of ocean biogeochemical cycles; the synoptic view of the variability in the global ocean's color during the 8 years that the Coastal Zone Color Scanner (CZCS) operated has drastically changed our understanding of biological oceanographic processes.

The optical properties of the upper ocean are determined by biogenic particles and dissolved material. Variations in the concentration of organic particles and dissolved compounds alter the optical properties; conversely, variations in the optical properties result in variations in photosynthesis which produces the organic material. Keys to an understanding of variability in ocean optics are both an understanding of upper ocean biogeochemical cycles and, as well, an understanding of the feedback by which optical variability results in biological variability. The satellite observations of ocean color provide a global view; however, they only resolve, or average over, spatial scales of order 1 km, nine orders of magnitude larger than the phytoplankton and bacterial cells which collectively are responsible for most of the variability in the color of the sea. The scale of the phytoplankton can be observed by microscopy or flow cytometry but synoptic measurements are not provided. To forge the linkage between the small scale biological processes and properties, which occur at the level of the individual cell, and the large scale forcing and ocean response in terms of ocean color, is one of the most difficult challenges currently facing biological oceanography. Resolving the optical properties of individual cells is one of the demonstrated strengths of flow cytometry; what is lacking is an appropriate coupling of the scales in order to permit a parameterization of satellite viewed ocean color in terms of the ensemble of individual phytoplankton cells in the upper ocean.

Here, we discuss the historical bases for observations of ocean color and its interpretation. We review some of the applications of remote sensing to computation of primary production at the global scale, and to computations of the variability in upper ocean heating rates associated with variations in the optical characteristics of the ocean. Finally, we discuss future observations of ocean color and some of the basic problems linking the scales of variability that must be addressed to interpret and apply these observations to current problems in oceanography.

BASES FOR OCEAN COLOR OBSERVATIONS

The Coastal Zone Color Scanner (CZCS), launched in 1978 and which operated until 1986, viewed radiance incident at the level of the spacecraft in 5 visible spectral bands centered at 443, 520, 550, 670 and 750 nm (e.g. Gordon *et al.*, 1980). These measurements did not derive solely from the ocean; as much as 95-99% of the radiance was derived from light scattered into the viewing angle of the sensor by the atmosphere,

$$L_t(\lambda,\theta,\theta_0,\phi) = L_{atm}(\lambda,\theta,\theta_0,\phi) + t(\lambda)L_w(\lambda) \quad (1)$$

where L_t is the radiance viewed at the level of spacecraft, L_{atm} is the radiance derived from the atmosphere, L_w is the radiance exiting the surface of the ocean and t is the atmospheric transmission, all of which are functions of wavelength, λ (Gordon and Morel, 1983). The dependencies indicated on the viewing angle of the spacecraft relative to the water (θ), on the Sun zenith angle relative to the water surface (θ_0) and on the azimuthal angle difference between the vertical planes of the spacecraft and Sun (ϕ) will be suppressed in what follows.

The atmospheric signal must be removed to enable interpretation of the water-leaving radiances. The correction depends, in a complex manner, on Rayleigh and aerosol scattering, on the relative transmission of atmospheric gases, especially ozone, and the atmospheric optical thickness. Given that L_{atm} is 95% of L_t, it is essential that the atmospheric correction be made with the greatest care. We will not discuss it further; a complete discussion can be found in Gordon (1978), Gordon and Morel (1983), Bricaud and Morel (1987), Gordon *et al.*, (1988), and André and Morel (1989).

It remains to interpret variations in L_w in terms of first the optical properties of the upper ocean, and second in terms of more biologically relevant quantities such as the chlorophyll concentration or primary production. Following Gordon *et al.* (1988), we define the normalized water-leaving radiance,

$$L_w|_n = [(1-\rho)(1-\rho')F_o R]/[m^2 Q(1-rR)] \tag{2}$$

where $L_w|_n$ is the water-leaving radiance that would occur under conditions where the sun is at zenith and there was no atmosphere. F_o is the solar flux, ρ and ρ' refer to components of the Fresnel reflectance at the sea-surface, R is the reflectance ($\equiv$ E_u/E_d where E_u and E_d are the upwelling and downwelling irradiances respectively), m is the index of refraction, Q is a parameter defining the distribution of radiance with respect to angle about the vertical ($\equiv$ E_u/L_u, where L_u is the upwelling radiance), and the term $(1-rR)$ takes account of internal reflectance at the sea surface by the upwelling radiance stream. To first order, the principal variables are R and Q; Eq. 2 can be rewritten,

$$L_w|_n \propto R/Q \tag{3}$$

Both R and Q are apparent optical properties; as opposed to inherent optical properties, they depend on the geometrical distribution of the underwater light field. Gordon (1986) has investigated their dependence on the inherent optical properties, the absorption coefficient, a, and the backscattering coefficient, b_b,

$$R/Q \propto b_b/(a+b_b) \tag{4}$$

or, to a reasonably good approximation,

$$R/Q = 0.11 b_b/K_d \tag{5}$$

where K_d is the attenuation coefficient for downward irradiance. This expression is clearly related to earlier efforts which focused on the reflectance, R,

$$R \approx 0.33 b_b / a \tag{6}$$

or,

$$R \approx 0.33 b_b / (\mu K_d) \tag{7}$$

where μ is defined as the average cosine of the radiance field (Gordon *et al.*, 1975; Morel and Prieur, 1977; Morel, 1988). In all of these formulations, $L_w|_n$ depends on backscatter and absorption, both of which vary with particle size, composition, and concentration.

BIO-OPTICAL ALGORITHMS

The remote sensing problem, assuming that the atmosphere can be adequately dealt with, is to predict the water-leaving radiances from an estimate of the inherent optical properties of the upper ocean. The inverse problem is more difficult, but of greater practical importance: given an estimate of the water-leaving radiances, to predict the inherent optical properties of the ocean. As can be seen in the following, the inherent optical properties are directly related to biological constituents of the sea, albeit in a complex manner. There are several approaches to the problem, ranging from determination of essentially empirical relationships between remotely-sensed water leaving radiances and reflectances, to fully analytical expressions linking the remotely-sensed parameters to the biological constituents of the upper ocean.

The inherent optical properties of absorption and scattering are additive; the absorptive and scattering properties of all constituents can be rigorously summed to yield the total value. In contrast, the attenuation coefficient, an apparent property, does not possess this attribute. In practice however, for reasonable sun angles and for the majority of the ocean where absorption dominates over scattering in the attenuation, it does possess 'quasi- inherent' properties and the constituent contributions to attenuation can be summed without substantial error (Baker and Smith, 1982; Kirk, 1983). For the two inherent properties of interest, and suppressing the wavelength dependence which is implicit,

$$a = a_w + a_p + a_d \tag{8}$$

where a_w is the absorption of pure seawater, a_p the absorption of particulate matter in the sea, and a_d is the absorption of dissolved organic matter, otherwise known as *yellow substance* (e.g. Bricaud *et al.*, 1983). A similar exercise for the backscattering yields,

$$b_b = (b_b)_w + (b_b)_p \tag{9}$$

where the subscripts refer likewise to the above. A difficulty with this approach is that the inherent optical properties are difficult to measure at sea; it is even more difficult to decompose the coefficients in terms of the constitutents unless statistical approaches are taken. The difficulty here is that apart from the water (taken as constant) the covariances between the quantities introduces additional terms in the statistical relationships; the problem is not well determined. More data exist for the apparent property K_d. Morel (1988) has compiled an extensive data base to estimate the component contribution to the attenuation coefficient for downward irradiance by statistical analysis; for wavelengths shorter than 575 nm, the explained variance in the log-log relationship between K_d and the concentration of chlorophyll plus phaeopigment, C, (a proxy for the particle concentration) is in excess of 80%. No attempt was made to subdivide the biological constituents beyond a bulk measurement of pigment. By the very nature of the goodness of fit, this results clearly demonstrates the substantial degree of covariance that exists between the numerous components that contribute to the attenuation of irradiance. The lack of linearity in the relation as evidenced by a monotonic increase in the first derivitive of the relationship between K_d and C also implies consistent changes that occur in the *chlorophyll-specific* attenuation coefficient. As the concentration of chlorophyll diminishes, the specific absorption coefficient increases. This *non-linear* biological effect presumably reflects relative increases of absorbing or scattering material relative to the chlorophyll-containing particles. It has been attributed to increases in dissolved organic matter (e.g. Morel, 1988) or detrital material (e.g. Yentsch, 1962), but may also reflect greater relative concentrations of small scattering particles such as bacteria or viruses (Morel, this volume).

A similar situation prevails with respect to the scattering coefficient. Morel (1980) and Gordon and Morel (1983) have shown that the total scattering coefficient at 550 nm likewise has a power dependency on the concentration of chlorophyll at sea,

$$b_{550} = 0.3C^{0.62} \tag{10}$$

where C is the concentration of pigment. The relationship between pigment and scattering is not nearly as good as that for K_d (Gordon and Morel, 1983). This uncertainty has roots in first the natural species to species variation in specific scattering (Bricaud *et al.*, 1983; Bricaud and Morel, 1986; Morel, 1987), the relative contributions of detrital to viable phytoplankton, and variations in the size structure of the marine particulate population. For comparison, Sathyendranath (1981) and Prieur and Sathyendranath (1981) compared the specific absorption coefficient for a range of phytoplankton cultures and natural populations respectively. The spectral distribution of the absorption coefficient was conservative; the amplitude of the curves varied by no more than a factor of 3-4. The specific scattering coeffient varied by more than a factor of 10 even for these cultures of viable phytoplankton. In the ocean, additional uncertainty in the pigment-scattering relationship is introduced by detrital particles and non-chlorophyllous biogenic particles.

At 550 nm, the scattering by particulate matter in the euphotic zone is generally much greater than the scattering of water. It remains to relate the backscattering coefficient to its dependency on wavelength and particle concentration (again through the pigment concentration). The backscattering probability is low, and varies inversely with wavelength. It is also depressed in the spectral regions where chlorophyll strongly absorbs. Based on these observations, purely empirical relationships between the concentration of chlorophyll, wavelength, and the particle backscattering coefficient have been developed by Gordon (1988) and Morel (1988). An alternative approach is under development whereby the particle size and refractive index distribution is estimated to base such a relationship more firmly in first principles (Morel, pers. commun.). Flow cytometry can be used to boster this approach and base it firmly on measurements. For example, flow cytometry has been used to estimate cell size and refractive index, and to relate taxonomic and physiological changes in these properties to the scattering properties of the phytoplankton population (Ackleson *et al.*, 1988 a & b). As pointed out recently by Kitchen and Zaneveld (1990), the skill in prediction of the scattering coefficient from the chlorophyll concentration is restricted to the upper ocean where there is good covariance between the constituent components; it fails deeper in the water column presumably since viable phytoplankton are not responsible for most of the scattering in the deeper ocean. In the end, the uncertainty in the relationship between the backscattering coefficient and chlorophyll concentration remains the largest impediment to further improvement in the remote sensing biooptical algorithms.

The remote sensing problem reduces to inverting the following,

$$L_w|_n(C) \propto b_b(C)/(a(C)+b_b(C)) \tag{11}$$

An improvement can be had by taking the ratio of $L_w|_n$ at two wavelengths; this reduces the uncertainty principally through the cancellation of variation in the backscattering coefficient. An analogous model can be developed for the retreival of the attenuation coefficient at wavelengths of interest (Austin and Petzold, 1981). The relationships developed for use with the CZCS archive are empirical representations of these relationships between upwelled radiance and the concentration of pigment in the upper ocean, for example (Gordon 1988),

$$C = 1.15(L_w(443)|_n / L_w(560)|_n)^{-1.42} \quad C<1 mg\, m^{-3} \tag{12}$$

and

$$C = 3.64(L_w(500)|_n / L_w(560)|_n)^{-2.62} \quad C>1 mg\, m^{-3} \tag{13}$$

where the coefficient of determination is greater than 0.95 and the relative error is approximately 20% for Eq. 12, and 30% for Eq. 13.

PRIMARY PRODUCTION ALGORITHMS

For many of the current problems in understanding global biogeochemical cycles, a knowledge of the large-scale distribution of chlorophyll is only a first step. At present, much effort is being devoted to developing algorithms that would permit prediction of the rate of primary production from satellite estimates of surface pigment concentration, and various ancillary variables such as sea-surface temperature and incident irradiance (Eppley *et al.*, 1985; Platt 1986; Balch *et al.*, 1989; Platt and Lewis, 1987; Platt *et al.*, 1988). As in the satellite determination of chlorophyll, these efforts span the range from purely empirical relationships to estimates constructed from principles of optics and algal physiology. Given an exact estimation of chlorophyll, sea-surface radiation and the attenuation coefficient (not possible from space observations), it appears that some of these relationships may account for as much as 75% of the variability in primary production on a global scale; when the uncertainties in the chlorophyll concentration, solar flux and attenuation is propagated through, the error explained is much less, perhaps 20-30% (see Kuring *et al.*, 1990). Nonetheless, this represents a significant improvement over current ship-based estimates where the bias due to undersampling at the large scale is enormous.

Alternatively, one could in principle eliminate the intermediate step in the calculation of chlorophyll, and attempt to estimate the light absorbed by viable photosynthetic organisms directly (Perry, pers. commun). Given an estimate of the quantum efficiency of photosynthesis, the photosynthetic rate could be directly determined. Problems with this approach lie in ambiguity introduced through the backscattering coefficient as in the estimation of chlorophyll, in estimating the portion of the ocean particle absorption that is due to photosynthetically competent pigments rather than detritus or photoprotective pigments, and achieving robust estimates of quantum efficiency.

Another alternative of great current interest is the prediction of the photosynthetic rate from a remote measurement of the upwelled radiance associated with red fluoresence which results from absorption of solar energy by photosynthetic pigments (e.g. Gower and Borstad, 1981). The fluoresced radiance seen by a remote sensor is emitted from the upper ocean by biogenic particles, it is less dependent on the backscattering coefficient, and on the concentration of dissolved organic matter. The estimation of photosynthetic rate depends on a predictable ratio of the quantum yield for fluorescence to that of photosynthesis defined on the basis of constant irradiance and on the bulk phytoplankton assemblage. It is not clear at present that this condition is met to a desired degree over the ocean provinces of interest and remains an active area of research.

ESTIMATES OF SURFACE HEATING RATES

The remote measurement of water-leaving radiances not only provides insight into the biogeochemical processes of the upper ocean; it also provides an estimate of the clarity of the surface ocean to visible radiation. This is important in the evolution of the upper ocean thermal structure since the distribution of the absorption of solar radiation with respect to depth is a key determinant of the heating rate (e.g. Denman, 1973; Lewis *et al.*, 1983; Lewis, 1987; Ackleson *et al.*, 1988; Lewis *et al.*, 1990). Areas of the ocean where high concentrations of phytoplankton exist permit relatively little solar energy penetration with the result that most of the solar energy is absorbed and heats the upper few meters. Clearer, open ocean conditions result in as much as 40% of the energy contained in visible frequencies pentrating to heat waters deeper than 30 m. The development of thermal structure, such as the upper mixed layer, is strongly affected on time-scales ranging from the diurnal to the seasonal and longer. On interannual scales, Lewis *et al.* (1990) using CZCS derived estimates of the attenuation coefficient for the Pacific Ocean equatorial basin, demonstrated the importance of the penetrating irradiance flux in the heat budget of the tropical ocean and the potential variations induced in air-sea heat exchange by variations in the pigment content of the upper ocean. For the heating problem, the principal variable of interest is the attenuation coefficient where more retreival skill exists than is possible with pigment concentration.

OPTICAL MEASUREMENTS FROM ALTERNATE PLATFORMS

Measurement of the color of the sea, and upper ocean optical properties, is not restricted to satellite observations. For many applications, other platforms are preferable. For example, for many parts of the world's oceans, including some of the most productive, clouds obscure the satellite's view for much of the time. Ships, buoys and aircraft can to more or less a degree, alleviate this, albeit with a loss in the synoptic coverage that the satellite affords. Higher spatial and temporal resolution is also possible and for ships, buoys and low altitude aircraft, the error associated with uncertainties in the atmospheric correction is to large extent, eliminated. Details of the vertical structure of the optical properties in the ocean can be determined. Power, mass and data rate are also not such great constraints; higher spectral resolution and consequent improvements via advanced spectroscopic analysis is attainable. Use of these platforms, integrated with the large-scale view provided by satellite coverage, can give great insight into some of the outstanding problems of the day in oceanography (e.g. Brown *et al.*, 1985).

Coupling of ship and buoy measurements with those from the satellite will provide the most effective means to merge the scales of variability in the optical properties of the ocean. The requirement, which at present remains a goal, is a parameterization of the small scales of variability, preferably in terms of remote observations, in a form that will be useful in the development of mesoscale and synoptic scale models of the oceanic optical fields.

FUTURE DIRECTIONS AND COUPLING TO FLOW CYTOMETRY

Several nations have plans for ocean color sensors in the near future. These planned future satellite ocean color missions will have better sensor performance in terms of digitization resolution and signal to noise ratios. This will permit improvements in the bio-optical algorithms as well. Still, considerable uncertainty will remain in attempts to relate the measured upwelled radiance to surface chlorophyll concentrations, as well as future estimates of primary production and upper ocean heating rates. As seen above, the principle uncertainty in the biooptical algorithms lie in the parameterization of the backscattering coefficient. This is potentially an area where flow cytometry could be of great importance; modifications to current flow cytometers that would permit the analysis of the size distribution of particles in the sea, as well as the size-dependent backscattering coefficient, would permit a much more rigorous parameterization which will be required to span the 9 orders of magnitude between the satellite resolution and the size of particles ultimately responsible for the color of the sea. Essential to this effort will be absolute calibration of the instruments, to permit both retrievals in absolute units and for comparibility between instruments. Future satellite ocean color missions will require cooperation on an international scale in the effort to develop algorithms that adequately parameterize the small scale variability in order to be useful in reducing current uncertainties in global biogeochemical cycles and physical climate. Analysis of the optical and physiological properties of single cells will be a key part of this effort.

REFERENCES

Ackleson S, Balch WM, Holligan PM (1988) White waters of the Gulf of Maine. Oceangr 1:18-22

Ackleson SG, Spinrad RW (1988) Size and refractive index of individual marine particulates: a flow cytometric approach Appl Optics 27:1270-1277

Ackleson SG, Spinrad RW, Yentsch CM, Brown J, Korjeff-Bellows W (1988) Phytoplankton optical properties: flow cytometric examinations of dilution-induced effects. Appl Optics 27:1262-1269

André JM, Morel A (1989) Simulated effects of barometric pressure and ozone content upon the estimate of marine phytoplankton from space. J Geophys Res 94:1029-1037

Austin RW, Petzold TJ (1981) Remote sensing of the diffuse attenuation coefficient of sea water using the coastal zone color scanner. in Oceangraphy from Space, JRF Gower (ed.) pp 239-256. Plenum NY

Balch WM, Abbott MR, Eppley RW (1989) Remote sensing of primary production-I. A comparison of empirical and semi-analytical algorithms. Deep Sea Res 36:281-295

Baker K, Smith RC (1979) quasi-inherent characteristics of the diffuse attenuation coefficient for irradiance. SPIE 208:60-63.

Bricaud A, Morel A (1986) Light attenuation and scattering by phytoplanktonic cells: a theoretical modeling. Applied Optics 25:571-580

Bricaud A, Morel A, Prieur L (1983) Optical efficiency factors of some phytoplankters. Limnol and Oceanogr 28:816-832

Bricaud A, Morel A (1987) Atmospheric corrections and interpretation of marine radiances in CZCS imagery: Use of a reflectance model. Oceanol Acta 7:33-50

Brown OB, Evans RH, Brown JW, Gordon HR, Smith RC, Baker KS. Phytoplankton blooming off the U.S. East Coast: A satellite description. Science 229:163-167

Denman KL (1973) A time-dependent model of the upper ocean. J Phys Oceangr 3:173-184.

Eppley RW, Stewart E, Abbott MR, Heyman U (1985) Estimating ocean primary production from satellite chlorophyll. Introduction to regional differences and statisitics for the Southern California Bight. J Plankton Res 7:57-70.

Gordon HR (1978) Removal of atmospheric effects from satellite imagery of the oceans. Appl Opt 17:1631-1636.

Gordon HR (1986) Ocean color remote sensing: Influence of the particle phase function and the solar zenith angle. EOS Trans AGU 14:1055

Gordon HR (1988) Ocean color remote sensing systems: Radiometric requirements. SPIE 924:151-167.

Gordon HR, Morel AY (1983) Remote Assessment of Ocean Color for Interpretation of Satellite Visible Imagery: A Review. Springer-Verlag, NY 114 pp

Gordon HR, Brown J, Evans RH (1988) Exact Rayleigh scattering calculations for use with the Nimbus-7 Coastal Zone Color Scanner. Appl Opt 27:862-871

Gordon HR, Brown OB, Jacobs MM (1975) Computed relationships between the inherent and apparent optical properties of a flat homogeneous ocean. Appl Opt 14:417-427

Gordon HR, Clark DK, Mueller JL, Hovis WA (1980) Phytoplankton pigments derived from the Nimbus 7 CZCS: Initial comparisons with surface measurements. Science 210:63-66

Gordon HR, Brown OB, Evans RH, Brown JW, Smith RC, Baker KS, Clark DK (1988) A semianalytic radiance model of ocean color. J Geophys Res 93:10909-10924

Gower JFR, Borstad G (1981) Use of the in-vivo fluorescence line at 685 nm for remote sensing surveys of surface chlorophyll a. In, J.F.R. Gower (ed.) Oceanography from Space. Plenum Press, NY pp 281-294

Kirk JTO (1983) Light and photosynthesis in aquatic ecosystems. Cambridge University Press, NY 401pp.

Kitchen JC, Zaneveld JRV (1990) On the noncorrelation of the vertical structure of light scattering and chlorophyll a in Case I waters. J Geophys Res 95:20237-20246

Kuring N, Lewis MR, Platt T, O'Reilly JF (1990) Satellite-derived estimates of primary production in the Northwestern Atlantic. Cont Shelf Res 10:461-484

Lewis MR, Cullen JJ, Platt T. Phytoplankton and thermal structure in the upper ocean: consequences of nonuniformity in chlorophyll profile. J Geophys Res 88:2565-2570

Lewis MR (1987) Phytoplankton and thermal structure in the tropical ocean. Oceanol Acta SP:91-95.

Lewis MR, Carr ME, Feldman GC, Esaias W, McClain C (1990) Influence of penetrating solar radiation on the heat budget of the equatorial Pacific Ocean. Nature 347:543-545

Morel A (1980) In water and remote measurement of ocean color. Boundary Layer Meteorol 18:177-201

Morel A (1987) Chlorophyll-specific scattering coefficient of phytoplankton. A simplified theoretical approach. Deep Sea Res 34:1093-1105

Morel A (1988) Optical modeling of the upper ocean in relations to its biogenous matter content (Case I waters. J Geophys Res 93:10747-10768

Morel A, Prieur L (1977) Analysis of variations in ocean color. Limnol Oceangr 22:709-722.

Platt T (1986) Primary production of the ocean water column as a function of surface light intensity: algorithms for remote sensing. Deep-Sea Res 33:149-163

Platt T, Lewis MR (1987) Estimation of phytoplankton production by remote sensing. Adv Space Res 7:131-135

Platt T, Sathyendranath S, Caverhill CM, Lewis MR (1988) Ocean primary production and available light: further algorithms for remote sensing. Deep Sea Res 35:855-879

Prieur L, Sathyendranath S (1981) An optical classification of coastal and oceanic waters based on the specific absorption curves of phytoplankton pigments, dissolved organic matter and other particulate materials. Limnol Oceangr 26:671-689

Sathyendranath S (1981) Influence des substances en solution et en suspension dans les eaux de mer sur l'absorption et la reflectance. Modelisation et applications a la teledetection. PhD Thesis, 3rd cycle Univ Pierre et Marie Curie, Paris 123 pp

Yentsch CS (1962) Measurement of visible light absorption by particulate matter in the ocean. Limnol Oceangr 7:207-217

INDIVIDUAL AND BULK ANALYSIS OF THE OPTICAL PROPERTIES OF MARINE PARTICULATES: EXAMPLES OF MERGING THESE TWO SCALES OF ANALYSIS[1]

R. Iturriaga[2], A. Morel[3], C. Roesler[4] and D. Stramski[2]

[2]Department of Biological Sciences
University of Southern California
Los Angeles, California
USA

[3]Laboratoire de Physique et Chime Marine
B.P. 806230
Villefranche-sur-Mer
France

[4]Oceanography WB-10
University of Washington
Seattle, WA 91895
USA

In recent years cytofluorometric and microphotometric techniques have significantly contributed to the field of biological and optical oceanography. Such determinations have enabled us to better understand the composition of some natural phytoplankton communities. We can now discriminate between *Synechococcus* clone types in the ocean, sort and analyze minute prochlorophytes, follow microalgae cell cycles, and determine the spectral absorption properties of individual cells as well as detrital particulates of field samples. Analysis at the individual level constitutes another valuable tool for the study of phytoplankton cells and detrital particulates in aquatic systems.

Until recently most absorption, scattering and fluorescence determinations were performed on bulk measurements. When dealing with algal populations or other microorganisms grown in culture, such bulk measurements can be interpreted in terms of optical properties at the level of a single cell, i.e. efficiency factors for absorption or scattering which represent a "mean" cell of the population (Morel and Bricaud, 1986). This requires that cell abundance and size

---

[1]Comments from the open session on the optical properties of marine particles

NATO ASI Series, Vol. G 27
Particle Analysis in Oceanography
Edited by S. Demers

distribution be determined. When dealing with mixed populations that exhibit a continuous size distribution as in natural environments, bulk measurements cannot be interpreted in the same way. The advent of individual measurements has created new ways of studying the intersample and intrasample variability of particle absorption and fluorescence properties; however, the gap between individual and bulk determination needs to be addressed in order to assess the impact of environmental forcing on the variability of natural assemblages of cells and other particulates in the ocean.

These informal talks and discussions centred on the problems and limitations of analyzing marine particulates at the bulk and individual cell levels, providing a broad perspective of the elements that must be considered for this type of analysis.

Consensus pivoted on the relevance of determining spectral absorption and possibly scattering, because this type of determination is fundamental to understanding the propagation of submarine light, including the fraction returned toward atmosphere ("ocean color" signature). In addition, the spectral absorption in relation to pigment content and packaging is fundamental to understanding the photophysiology of light adaptation and its subsequent effect on the photosynthetic capabilities and the fluorescence signature of phytoplankton cells.

Because field samples are too dilute for direct spectrophotometric determinations, the most common approach for characterizing the bulk absorption by oceanic particles has been the concentration of samples on glass fiber filters. Such a sample filter is then scanned in a spectrophotometer against a reference blank filter, and care must be taken to ensure a constant reproducible water saturation of the filter before and after filtration. All agree that this technique is, as yet, the most practical for field experiments. A high quality determination of particulate absorption on a filter is a difficult task, however. The major concern of investigators is the need to make a correction for the pathlength amplification effect ß (e.g. Mitchell and Kiefer, 1984, 1988). The error associated with this correction may be large and is one of the major sources of uncertainty in the derived values of absorption (Bricaud and Stramski, 1990). The situation is complicated by the fact that the ß factor is dependent not only on the amount of material retained on the filter (i.e. the optical density measured on the filter), but also on the composition of particulate assemblage. Although the correction of ß has been widely applied, there has been little uniformity with respect to correction methodology. We wish to encourage theoretical studies as well as extensive laboratory work designed specifically to acquire a better understanding of the ß effect. Standardization and optimization of measuring techniques are also desirable for the purpose of reliable intercomparison; the geometrical configuration of the measuring set-up is to be controlled and relevant information provided.

There is also evidence that other methodological difficulties related to imperfect retention of particles and determination of baseline occur when the filter pad technique is applied. This may introduce errors into the magnitude and spectral shape of the optical density measured on the filter before any corrections for the ß effect are made. Based on his experience with the filter pad technique, Dariusz Stramski suggests a number of recommendations which should be followed if best quality results are to be obtained.

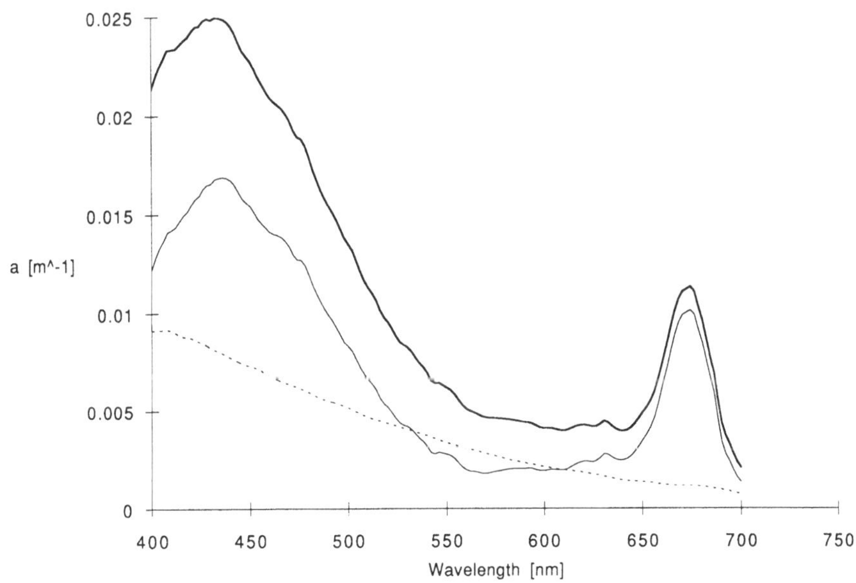

Figure 1. Microphotometrically determined particulate (heavy solid), phytoplankton (solid) and detrital (dash) absorption spectra from Biowatt station 04-03 20 m (Iturriaga and Siegel, 1989).

First, consecutive filtration and the additional measurement of absorption by particles that passed through the GF/F filter but were retained on a second pass through the same type of filter will substantially improve results in oligotrophic waters dominated by small-sized particles.

Second, special caution must be exercised to correct for the baseline and minimize baseline shifts. Because the optical density of blank filters, even from a specified lot of filters, can vary by as much as 0.05, the baseline should be measured separately for each filter prior to filtration. Also, it is not sufficient to rely on the automatic baseline correction which is built in to most of the commercial spectrophotometers. The spectrum of the baseline should be saved to a file, and then subtracted from the sample spectrum when processing the data. This procedure requires the use of a spectrophotometer on board the ship which is consistent with another recommendation that favors the immediate analysis after sampling rather then long-term storage of samples and post-cruise analysis. In addition, the filters should be positioned exactly in the same way for the baseline and sample scans (a small marker on the edge of the filter is helpful).

Third, it may be advantageous to measure the optical density on the filter against air rather than against the blank filter whenever a geometrical configuration of the instrument allows such a measurement. In this case, the reference beam is kept exactly constant, so the possible shifts of the baseline associated with the blank filter (e.g. changes in water saturation) are avoided.

Finally, the commonly used null correction of absorption spectra at 750 nm does not appear to be justified for total particulates. A number of measurements made at sea with careful determinations of baseline for each filter separately (D. Stramski, oligotrophic Pacific, unpubl. data), showed that the absorption by total particulates at 750 nm is significantly different from zero. Only one of 35 measurements, which had been corrected for baseline, exhibited negative value of optical density at 750 nm, the remaining were positive. The significance of this observation is supported by the fact that the optical density on a filter at 750 nm differs insignificantly from zero when measured with the same care and precision with healthy phytoplankton cultures free from other particles.

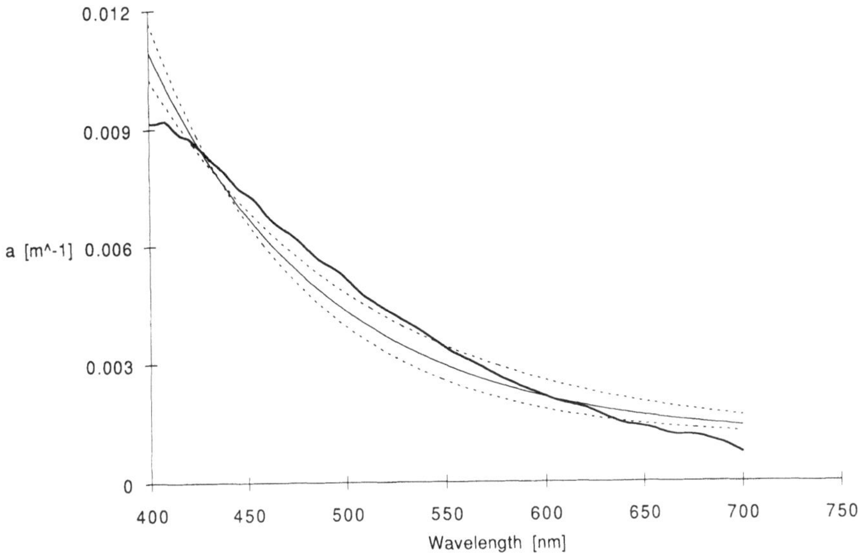

Figure 2. Microphotometrically determined (heavy solid) and modeled (c_2 = 0.011 $\pm$ 0.002; solid $\pm$ dash) detrital absorption spectra. The percent difference between the modeled and measured spectrally integrated absorption coefficients is 3.1 %.

Recent development and successful application of microphotometric techniques for measuring absorption by individual particles has provided a new means for addressing the problem of marine light absorption (e.g. Iturriaga and Siegel, 1988). Such measurements appear to be relatively accurate although they are time consuming. The important advantage is that each particle, for which the absorption is measured, can be unambiguously identified because the technique is based upon the use of a microscope. The limitation of microphotometry lies, however, in the fact that a significant contribution to absorption may be associated with small-sized microorganisms (< 2 μm), which are inaccessible to this analysis.

Recently, several models were developed to partition the total particulate absorption into the phytoplankton and detrital components (Morrow *et al.*, 1989; Roesler *et al.*, 1989; Bidigare *et al.*, 1990; Bricaud and Stramski, 1990). Filter pad determinations of total particulate absorption served as a database in this development. The partitioning algorithms involve certain assumptions which may limit their applicability. Further studies are thus needed to test and compare these algorithms using more and better data. In addition, the microphotometric analysis of field samples has proven useful in characterizing the variability in absorption efficiency between individual particles for both viable phytoplankton and non-living detrital material.

An alternate way to partition the total particulate absorption into the phytoplankton and detrital components is to make a supplementary spectrophotometric measurement on a filter pad after extraction treatment in organic solvent (Kishino *et al.*, 1985). Assuming that all pigments associated with viable phytoplankton are removed by extraction, the phytoplankton component is obtained as a difference between the values that had been measured before and after extraction. The drawback of such a method is that non-extractable algal pigments (phycobilins, cytochromes) and bleached cells contribute to the estimate of detrital absorption. In addition, some pigments present in detritus are extracted; hence they contribute to the estimate of absorption by viable phytoplankton.

Given the fact that none of the present techniques for measuring the absorption of field samples is ideal, they must rather be regarded as complementary. A comparison of microphotometric and bulk filter pad data is desirable. Such initial comparisons show that the spectral shape and magnitude of absorption efficiency derived from both methods agree favourably for laboratory cultures of phytoplankton and natural samples (Iturriaga and Siegel, 1988).Another interesting exercise is to apply a partitioning model for the total particulate absorption based upon bulk spectrophotometric analysis and compare the results with the data representing individual phytoplankton and detrital particles. An example is provided by Collin Roesler using the model described in Roesler *et al.* (1989) and absorption data collected in the Sargasso Sea by Iturriaga and Siegel (1989). Total particulate absorption spectrum, $a_p(\lambda)$, is the sum of the absorption spectra of phytoplankton and detritus, $a_{ph}(\lambda)$ and $a_d(\lambda)$, respectively. Thus, phytoplankton absorption spectrum is given by:

$$a_{ph}(\lambda)-a_p(\lambda)-a_d(\lambda) \quad (1)$$

Based upon analysis of spectral variability of detrital absorption (Roesler *et al.*, 1989), detrital absorption spectra can be assumed to approximate an exponential curve:

$$a_d(\lambda)-c_1\exp[-c_2(\lambda-400)] \quad (2)$$

where C_1 is the magnitude of the function at 400 nm and C_2 is the coefficient defining the slope of the exponential (0.011 $\pm$ 0.002). In order to determine C_1, the magnitude of the curve has to be determined at a wavelength of significant detrital absorption. This method has been outlined in Roesler *et al.* (1989). Once detrital absorption is determined, phytoplankton absorption can be estimated from equation (1).

As already mentioned, the model was applied to data from the Sargasso Sea (Iturriaga and Siegel, 1989). Figure 1 shows detrital and phytoplankton spectral absorption coefficients from Biowatt station 04-03 (20m) determined from microphotometric analysis of individual

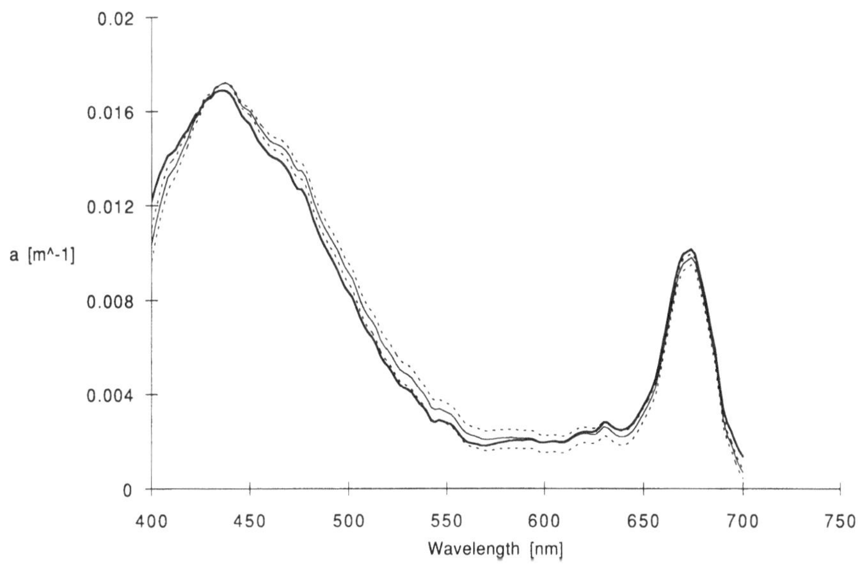

Figure 3. Microphotometrically determined (heavy solid) and modeled (c_2 = 0.011 $\pm$ 0.002; solid $\pm$ dash) detrital absorption spectra. The percent difference between the modeled and measured spectrally integrated absorption coefficients is 1.8 %.

particles. The sum of these two spectra is the total particulate absorption spectrum on which the model was run. Model estimated detrital absorption (using C_2 = 0.011 $\pm$ 0.002) is shown in Figure 2 with the microphotometrically measured detrital absorption. Using these modeled detrital spectra, phytoplankton absorption spectra are computed and shown with the microphotometrically determined spectra (Fig. 3).

The agreement between the modeled and measured component absorption spectra is quite good. Especially gratifying is the fact that the measured spectra are derived from analysis of individual particles while the modeled spectra are based upon a bulk analysis model of component absorption. Thus, the gap between individual and bulk analysis can be bridged with relatively small errors.

Many questions can be raised regarding future priorities for assessing the basic determinations of spectral absorption by cells and detrital components in the ocean. Besides new technology and data processing to come in the future, an interactive communication between researchers will certainly prove to be highly productive. Emphasis should be placed on reinforcing theoretical and experimental studies concerning the optics of single cells including the smallest ones to take a full advantage of the capabilities offered by, or to be developed for, cytometric instrumentation.

It seems timely to envisage the constitution of an optical data bank. High quality absorption spectra gathered in various provinces of the world ocean could form one part of such a bank. A catalog of optical signatures (absorption, scattering, fluorescence and possibly backscattering) of the common species, grown in culture under various conditions, could form the other part. An example of such a contribution is given in Bricaud *et al.* (1988). The optical catalog should be associated with a catalog of the identified (HPLC) pigments which is now under development for taxonomic applications.

ACKNOWLEDGEMENTS

We thank the members, in particular Clarice Yentsch, of the Organizing Committee of the NATO Advanced Study Institute for encouraging the open session on the optical properties of marine particulates. We also express our gratitude to all participants of this discussion for their contribution.

REFERENCES

Bidigare R, Ondrusek ME, Morrow JH, Kiefer DA (1990) *In vivo* absorption properties of algal pigments, p. 290-302. In Ocean Optics X, Proc. SPIE 1302, Bellingham

Bricaud A, Bedhomme AL, Morel A (1988) Optical properties of diverse phytoplanktonic species: experimental results and theoretical interpretation. J Plankton Res 10:851-873

Bricaud A, Stramski D (1990) Spectral absorption coefficient of living phytoplankton and nonalgal biogenous matter: A comparison between the Peru upwelling area and the Sargasso Sea. Limnol Oceanogr 35:562-582

Iturriaga R, Siegel D (1988) Microphotometric distinction of phytoplankton and detrital absorption properties, p. 277-287. In Ocean Optics IX, Proc. SPIE 925, Bellingham

Iturriaga R, Siegel D (1989) Microphotometric characterization of phytoplankton and detrital absorption properties in the Sargasso Sea. Limnol Oceanogr 34:1706-1726

Kishino M, Takahashi M, Okami N, Ichimura S (1985) Estimation of the spectral absorption coefficients of phytoplankton in the sea. Bull Mar Sci 37:634-642

Mitchell BG, Kiefer DA (1984) Determination of absorption and fluorescence excitation spectra for phytoplankton, p. 157-169. In O. Holm-Hansen *et al.* [eds.]. Marine phytoplankton and productivity. Springer

Mitchell BG, Kiefer DA (1988) Chlorophyll *a* specific absorption and fluorescence excitation spectra for light-limited phytoplankton. Deep-Sea Res 35:639-663

Morel A, Bricaud A (1986) Inherent optical properties of algal cells including picoplankton: theoretical and experimental results, p. 521-559. In T. Platt and W.K.W. Li [eds.]. Can Bull Fish Aquat Sci 214

Morrow JH, Chamberlin WS, Kiefer DA (1989) A two-component description of spectral absorption by marine particles. Limnol Oceanogr 34:1500-1509

Roesler C, Perry MJ, Carder K (1989) Modelling *in situ* phytoplankton absorption from total absorption spectra in productive inland marine waters. Limnol Oceanogr 34:1510-1523

Flow cytometry in oceanography

ADVANCES IN OCEANOGRAPHY THROUGH FLOW CYTOMETRY

R. J. Olson[1], E. R. Zettler[1], S. W. Chisholm[2], J. A. Dusenberry[2]

[1]Biology Department
Woods Hole Oceanographic Institution
Woods Hole, MA 02543 USA

[2]Ralph M. Parsons Laboratory
48-425 Massachusetts Institute of Technology
Cambridge, MA 02139 USA

INTRODUCTION

As forecast by Boyd (1978) well over a decade ago, the analysis of individual plankton cells by flow cytometry has become a powerful tool in oceanography, and it has given us a new window through which to look at the plankton community. Here we illustrate some of the ways in which flow cytometry has already advanced the science of oceanography (with emphasis on direct field applications) and point out some of the directions in which further advances can be expected. We will also address some of the limitations of existing technology and the efforts needed to overcome these limitations.

Current applications of flow cytometry to oceanography fall into two general categories: ecological and physical (Fig. 1). The first has as its goal understanding a particular group of organisms or relationships between organisms, while the second represents attempts to understand physical properties of the ocean through consideration of the optical properties of its particles. There is considerable overlap between these two categories, and the same data sets can be applied to questions in both; for example, measurements of diel changes in the light scatter signal from cells can be used to estimate their growth rates and to interpret bulk beam attenuation patterns.

NATO ASI Series, Vol. G 27
Particle Analysis in Oceanography
Edited by S. Demers

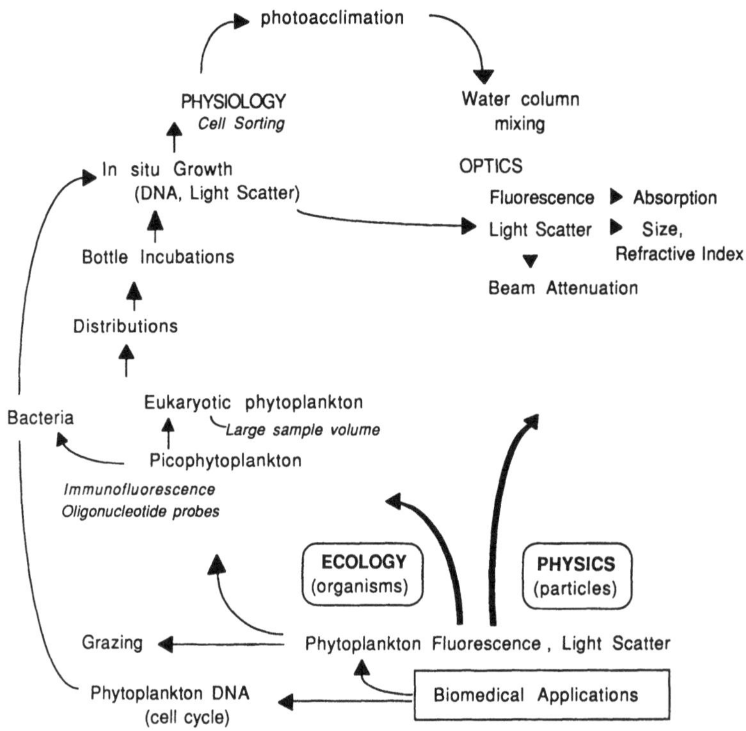

Figure 1. Outline of applications of flow cytometry to oceanography, attempting to reflect the evolution of these applications and relationships between various research efforts. The major split is between ecological questions, which are primarily concerned with the organisms themselves, and physical questions, which are primarily concerned with the optical effects of particles, regardless of their origin or composition. Italics indicate areas expected to figure prominently in future advances.

Before covering progress in the areas described in Fig. 1 in detail, the general features of the instrumentation should be described for the uninitiated. More complete treatments can be found in Shapiro (1988) and Melamed *et al.* (1990). A central feature of flow cytometry is that several measurements are made on an individual particle during its flow through a sensing region (Fig. 2). This region is defined by a focussed light beam of a given wavelength (although instruments that can also make electronic particle volume measurements do so in a separate location). Particles passing through this light beam give rise to two kinds of optical signals: scattered light and fluorescence. The amount of light scattered from the beam by a particle passing through it and the angular distribution of the scattering are determined by the particle's size, shape and refractive index, although to a first approximation light scattering is often regarded as proportional to cell size.

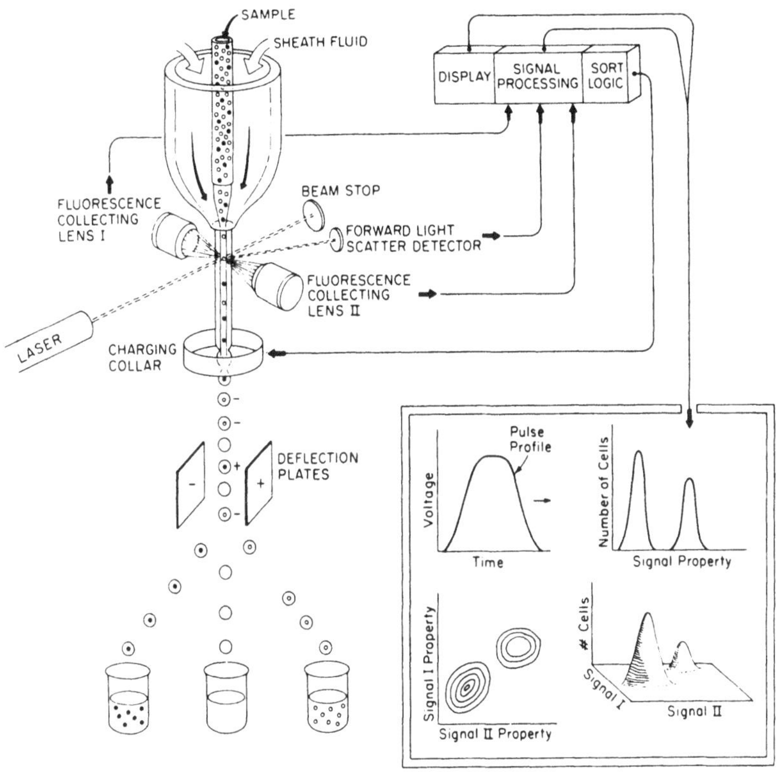

Figure 2. Schema of a typical flow cytometer/cell sorter. Sample is injected into the center of a particle-free, flowing sheath fluid at a rate that ensures particles pass through the laser beam one at a time. Analog pulse profiles of light scatter and fluorescence generated by each particle are digitized and accumulated by a computer. This data can then be displayed and analyzed as multiparameter plots. Particles of interest can be physically sorted from the rest of the sample by breaking the stream into precisely positioned droplets with a piezoelectric oscillator and electrostatically deflecting the charged droplets containing these particles into a collection vessel. From Chisholm *et al.* 1986.

The light beam can also excite fluorescence of various wavelengths, depending on the presence in the cell of autofluorescent pigments such as chlorophyll (red fluorescence) or phycoerythrin (orange), or of added fluorescent dyes such as DAPI (blue) or fluorescein (green). Optical filters are used to separate fluorescence signals according to their wavelengths, so that several constituents can be measured simultaneously. The intensity of fluorescence is related to the amount of each constituent.

Pulse peak heights or integrated signals from each detector are typically stored as a list of multiparameter data for each cell, and visualized in one-, two-, or three-dimensional plots in which populations of similar cells form characteristic flow cytometric "signatures." For example, marine *Synechococcus* cells can be distinguished from most other phytoplankton because they have orange fluorescence from phycoerythrin (PE) as well as the red fluorescence from chlorophyll (CHL) that all phytoplankton exhibit (Fig. 3A). Prochlorophyte picoplankton (Chisholm *et al.* 1988) can be distinguished from larger eukaryotic phytoplankton by virtue of their lower light scatter and red fluorescence values (Fig. 3B). *Synechococcus* cells, which are intermediate in size between these groups, do not appear in Fig. 3B because all cells with orange fluorescence have been eliminated from this particular view of the data. This is known as "gating," and is a common procedure in flow cytometric data analysis. The identity of cells in a given population can be confirmed by sorting (physical isolation of cells with specific combinations of signals) for subsequent microscopic examination. Sorting is a feature available on some commercial instruments.

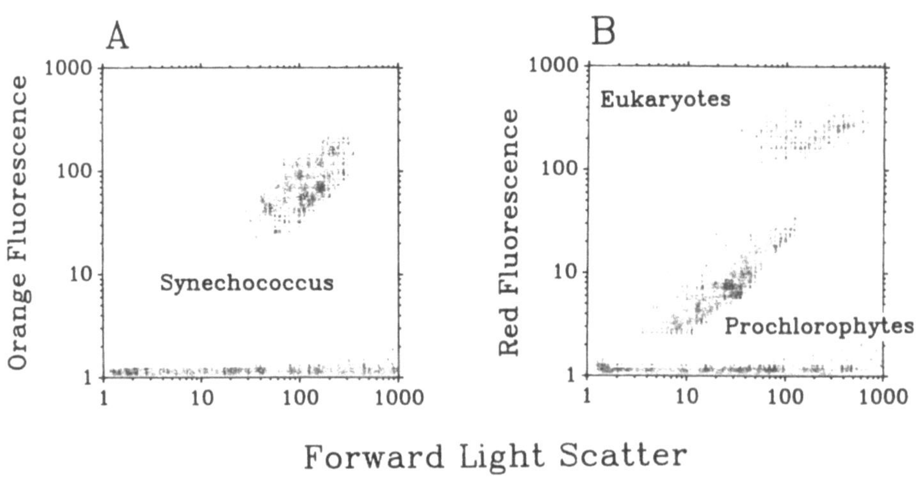

Figure 3. Examples of flow cytometric signatures of (A) *Synechococcus* and (B) prochlorophytes and larger eukaryotic phytoplankton. Forward light scatter (related to size) is plotted against either orange (530-630 nm, phycoerythrin) or red (660-700 nm, chlorophyll) fluorescence, on 3-decade log scales. In (B), only cells without orange fluorescence have been plotted, thus "gating out" *Synechococcus*. From Olson *et al.* 1990b.

It is important to note that the amount of light scattered or emitted as fluorescence by an individual cell is very small, so signal sizes must be maximized by making the exciting light beam very intense. This is typically done by focussing the beam to a very small spot, sometimes only a few-fold larger than the cells of interest. This in turn necessitates confining the sample flow stream to a very small diameter in the center of the flow channel and places an upper limit on the sample volume that can be analyzed in a given time. This is a significant constraint for many of the applications we will discuss.

APPLICATIONS TO ECOLOGY

Experimental Studies

Cell Cycle Analyses: One of the first applications of flow cytometry to the study of marine organisms was a direct outgrowth of biomedical studies of the cell cycle: DNA distributions in cultures of phytoplankton grown under different conditions were measured to investigate the influence of environmental factors on the regulation of cell growth and division (e.g. Chisholm *et al.* 1986; Olson *et al.* 1986; Vaulot *et al.* 1986). For this work it proved necessary to remove the phytoplankters' autofluorescent pigments to get good signals from the fluorescent DNA stains used. This was done either by photo-oxidation (Yentsch *et al.* 1983) or by methanol treatment (Olson *et al.* 1983), making the phytoplankters quite analogous to the non-pigmented cells for which the techniques had been designed. These laboratory studies have improved our understanding of cell cycle regulation in different phytoplankton groups but they did not initially lend themselves to field applications. Only through the use of microspectrofluorometry (Chang and Carpenter 1988) could de-chlorophyllized cells belonging to different species be distinguished from one another well enough to resolve the DNA histograms in field samples (Chang 1989).

Recently, however, Boucher *et al.* (in press) have shown that when natural populations are stained with DAPI, DNA histograms from the picoplankton can be resolved without extracting the chlorophyll from the cells. Prochlorophytes in particular could be recognized from their flow cytometric signatures and their cell cycle position could be determined. Moreover, the distributions of DNA per cell reveal relative growth rates for cells growing under different conditions, and after suitable calibration (McDuff and Chisholm 1982; Carpenter and Chang 1988) may provide absolute growth rates. This technique is especially attractive because it can be used with natural samples and does not require bottle incubations.

Grazing: Another early application of flow cytometry to oceanographic questions was the analysis of grazing selectivity by planktivorous herbivores. Traditionally, grazing rates have been estimated indirectly by observing the loss of food particles from the experimental volume or directly by observing food particle accumulation in the grazers (Peters 1984). Both approaches have been used in conjunction with flow cytometric analysis. Although in some cases flow cytometry is simply a way of automating cell enumeration, which is usually done microscopically, in other cases it allows one to discriminate between cells and particles which would be difficult to distinguish under the microscope.

By presenting grazers with a selection of cultured prey species that can be discriminated by flow cytometry, relative grazing preferences have been demonstrated quite convincingly in laboratory studies. For example, Shumway *et al.* (1985) were able to show that a filter feeder

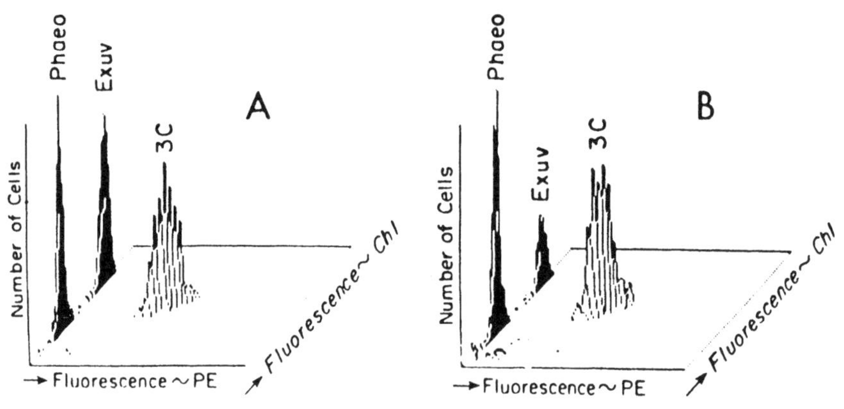

Figure 4. Results from a grazing experiment in which oysters (*Ostrea edulis*) were fed a mixture of a diatom (Phaeo), a dinoflagellate (Exuv) and a cryptophyte (3C). The number of cells is plotted against phycoerythrin (PE) and chlorophyll (Chl) fluorescence. Flow cytometric analysis allowed discrimination among food items in the mixture and showed that while at time 0 (panel A) the numbers of each type of algal cells were approximately equal, after 60 minutes of grazing (panel B) the dinoflagellate (Exuv) had been selectively removed. From Shumway *et al.* 1985.

(*Ostrea edulis*) preferentially removed dinoflagellate cells from a mixture of three species (a diatom, a dinoflagellate, and a phycoerythrin-containing cryptophyte) (Fig. 4). Several other studies have shown grazing preferences by a variety of filter feeders (Cucci *et al.* 1985; Stoecker *et al.* 1986; Shumway and Cucci 1987; Burkill *et al.* 1990).

Flow cytometry has been utilized in these kinds of grazing studies in the field as well. Newell *et al.* (1989) observed a reduction in the number of red-fluorescing particles along a transect across a mussel bed (relative to particles without red fluorescence), which could be interpreted as selective removal of chlorophyll-containing phytoplankton by the mussels. Field studies

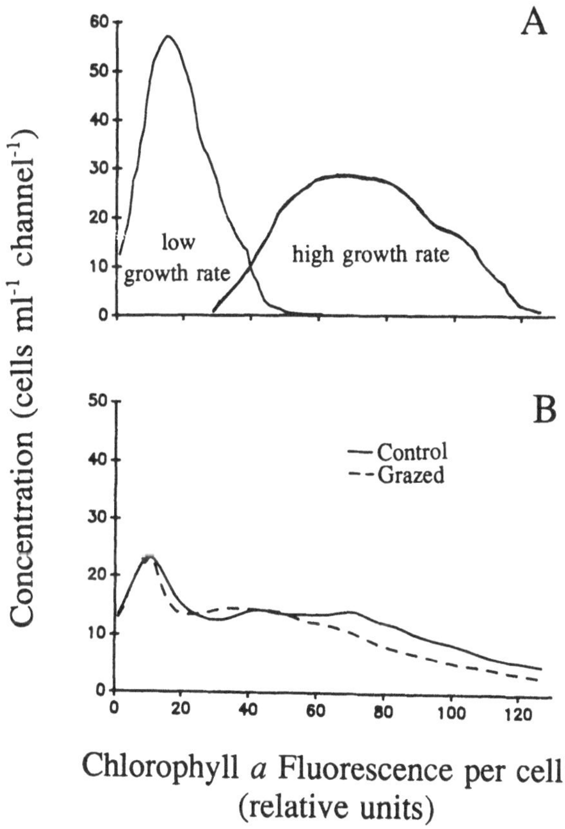

Figure 5. (A) Using flow cytometry, *Thalassiosira weissflogii* grown at low and high growth rates could be distinguished in a mixture by their relative chlorophyll fluorescence, even though they were the same size (size data not shown). **(B)** When a mixture was fed to the copepod *Acartia tonsa*, high chlorophyll cells were ingested out of proportion to their abundance. After Cowles *et al.* 1988.

such as this, however, are more difficult to interpret because several processes are likely to co-occur (i.e., removal and production of particles), and because the heterogeneity of natural particle assemblages may make discrimination among particle classes difficult.

Flow cytometry has also allowed us to demonstrate food selection by grazers on the basis of food "quality" within a single species of phytoplankter (Cowles *et al.* 1988). For this experiment, diatoms were grown in N-limited chemostats at high and low growth rates. Even though both kinds of cells were the same size, they could be distinguished from each other by their chlorophyll fluorescence: the fast-growing cells had more chlorophyll than the slow-growing, severely N-limited ones (Fig. 5A). When a mixture of the two types was grazed by copepods, more of the fast-growing cells were removed (Fig. 5B), indicating that the grazers could sense some difference at the single-cell level.

Direct detection of particles ingested by individual grazers is an area where flow cytometry promises to be a particularly useful tool in terms of both quantitation and speed. In a flow cytometric study of grazing by the ciliate *Tetrahymena pyriformis*, it has been shown that cell fluorescence from ingested polystyrene beads increases linearly as a function of bead ingestion over a range from 1 to 150 beads (Lavin *et al.* 1990). Similarly, phycoerythrin-containing *Synechococcus* cells have been used to measure feeding by the chlorophyll-containing chrysophyte *Ochromonas* (Cucci *et al.* 1989). Grazers that ingest *Synechococcus* cells become distinctively labelled by their increased orange fluorescence. Non-fluorescent particles can also be stained for use in direct grazing studies; for example, grazing on fluorescently labelled bacteria is currently measured using epifluorescence microscopy (Sherr *et al.* 1987) and should be easily detected and quantified using flow cytometry.

Gerritsen *et al.* (1987) compared flow cytometric and microscopic analyses of grazing. They found that the flow cytometer was faster and required less sample preparation, but pointed out that background fluorescence could in some cases bias the results obtained by flow cytometry. Background fluorescence is a problem that must be dealt with in many flow cytometric studies. With the aid of a microscope, the human eye can easily distinguish the subtle differences in color and shape between the discrete ingested food particles within the grazer and the dim overall background fluorescence of detritus or the grazer itself, while the flow cytometer "sees" only total fluorescence from a particle within relatively broad spectral bands determined by the optical filters.

Field Studies

The relatively low concentrations of the larger phytoplankton in natural ocean waters, and conversely, the relatively high concentrations of picoplankters, have defined the applications of commercially available flow cytometers for field studies to date. Typical sample flow rates are a fraction of a milliliter per minute, which limits efficient analysis to those types of cells whose abundance is in accord with reasonable sample analysis times. The picoplankters *Synechococcus* and prochlorophytes fit this category and thus have been most extensively studied using flow cytometric methods. In fact, the latter were effectively discovered using flow cytometry (Chisholm *et al.* 1988). As is discussed below, larger and less abundant cells can be studied using this technology and with some minor technological advances we will be able to analyze them as readily as the smaller cells.

Synechococcus: Synechococcus has been a prime target of study using flow cytometry since the first open ocean seawater samples were analyzed using this method (Yentsch *et al.* 1983; Olson *et al.* 1985). These cells have very distinctive signatures because they are very small (about 1 μm) and contain phycoerythrin, which fluoresces brightly in the orange (570 nm) region of the spectrum. They are found in surface waters of all the warm oceans at concentrations high enough (10^3-10^5 ml^{-1}) to analyze rapidly with standard commercial flow cytometers.

Synechococcus was discovered using epifluorescence microscopy (Waterbury *et al.* 1979; Johnson and Sieburth 1979), and the general distribution of these cells has been fairly well established using this technique (Waterbury *et al.* 1986; Glover *et al.* 1986; Glover 1985; Iturriaga and Marra 1988). Flow cytometry, in addition to making more detailed sampling possible, has enabled us to quantify the optical characteristics of the cells, such as fluorescence intensity and cell size, and to do more detailed studies of the factors that regulate the distribution and abundance of various pigment types (Olson *et al.* 1988, 1990a).

Our work in this area was stimulated by observations made during surveys of the distribution of *Synechococcus*, in which we sometimes noticed two populations of cells, having relatively "bright" and "dim" fluorescence (Fig. 6A). When we saw these bimodal flow cytometric signatures, we wondered whether they were the result of recent mixing of deep and shallow populations that had adapted to different light intensities, or whether they were actually different kinds of *Synechococcus*. The cell sorting capability of the flow cytometer allowed us to resolve this question: we isolated cells of each type into culture, grew them under identical conditions for many generations, and then re-analyzed them (Fig. 6B and 6C). The cells originally sorted from the bright population were still bright, and the cells from the dim population were still dim: thus we concluded they must be different strains.

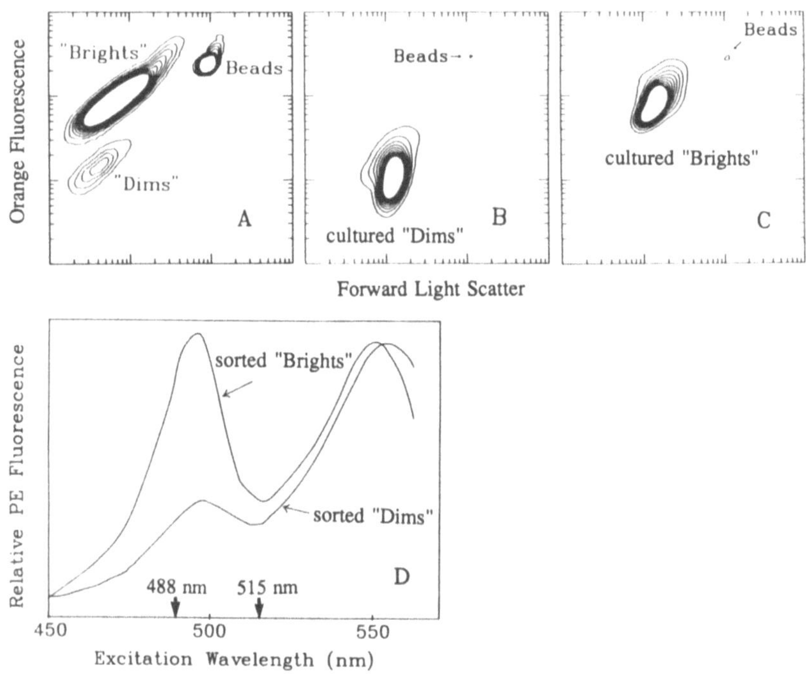

Figure 6. (A) Surface sample from the Gulf Stream showing a dual population of bright and dim *Synechococcus*. A flow cytometer was used to sort 100 cells from each population into culture medium. After more than 12 generations of growth under identical conditions the cultures retained their *in situ* fluorescence characteristics (B, C), indicating they were different strains. (D) Fluorescence excitation spectra of natural populations of "bright" and "dim" *Synechococcus* which had been sorted at sea and frozen for later analysis. Fluorescence emission was measured at 580 nm. The location of the two wavelengths used for dual beam flow cytometry (488 and 515 nm) are indicated, and show how strains with high phycourobilin will excite much better at 488 nm than low phycourobilin strains. After Olson *et al.* (1988).

Cell sorting also enabled us to determine something about what made the strains different: by sorting several million cells of each kind from a natural sample, we were able to obtain

fluorescence excitation spectra (Fig. 6D) that indicated that the pigments from the two kinds of cells were strikingly different. The fluorescence of the "bright" cell type was excited by blue light much more effectively than that of the "dim" type, while both cell types were also excited by green light. These results explain why we observed bright and dim fluorescence from the two populations: since we were using a blue laser beam (488 nm) to excite fluorescence, the cells with blue-absorbing pigment absorbed more light and hence emitted brighter fluorescence than the "dim" cells (Fig. 6D).

These fluorescence excitation patterns, it turns out, reflect differing proportions of two chromophores in the phycoerythrin of these cells: phycoerythrobilin (PEB) absorbs maximally at 550 nm, and a second kind of chromophore, phycourobilin (PUB), absorbs at about 490 nm (Waterbury *et al.* 1986). It is noteworthy that PUB, the blue-absorbing chromophore, was described in 1985 as a new and unusual chromophore (Ong 1984), only rarely encountered in cultured strains. Since many oceanic organisms are not easily cultured, we wondered how prevalent these cells were in the oceans as opposed to in laboratory collections; we used another flow cytometric technique to address this question.

Because the fluorescence intensity of each cell depends not only on its pigment type but also the amount of pigment (which can change dramatically depending on photoacclimation or nutrient status), we can't simply use "brightness" as an index for pigment type (except perhaps in the relatively small number of cases where we see both bright and dim cells together). However, by using dual beam excitation in the flow cytometer, one can a obtain crude, 2-point excitation "spectrum" from each cell. If we pass the cell through first a blue beam and then a green beam, we can measure two sets of fluorescence signals, and their ratio will tell us the relative absorption properties of the cell's fluorescent pigment. In this case, we use the 488 and 515 nm laser lines from the argon ion laser: 488 nm is absorbed much more effectively by PUB than by PEB, and 515 nm is absorbed about equally by both pigments (Fig. 6D).

Using this method of analysis, we surveyed *Synechococcus* in the North Atlantic and Pacific Oceans to better understand what regulates the distribution of the various pigment types (Olson *et al.* 1990a). Although we found a large range in the relative PUB content of natural populations, we could see two main groups. The large majority of cells in the open oceans were of the blue-absorbing (high-PUB) type. Low-PUB populations were found only in coastal waters, and these were limited to relatively shallow depths. In most cases the low PUB cells were part of dual populations, so that high-PUB cells were present at all but a few stations.

This distribution of pigment types is consistent with the optical characteristics of the oceanic euphotic zone, in that the cells best adapted to absorb blue light were found in the open ocean (where blue light penetrates deepest), while the cells with PE adapted for green-light absorption were limited to shallow, coastal waters where green light is available. It seems likely that these findings will have an impact on the interpretation of remotely sensed ocean color measurements, but this will depend on the light absorption by these numerous but very small cells relative to the total absorption.

Campbell and Iturriaga (1988), using single-cell fluorescence excitation spectra obtained by microspectrofluorometry, also concluded that *Synechococcus* in the Sargasso Sea had high PUB:PEB ratios, and Wood *et al.* (1985) were able to discriminate between pigment types with epifluorescence microscopy by examining each cell with two different filter sets. The rapidity and objectivity of flow cytometry, however, allows us to put these observations on a quantitative basis for large areas of the world ocean in a way that other methods cannot.

Prochlorophyte picoplankton: Flow cytometry has played an even more pivotal role in our understanding of the picoplankton group currently designated "prochlorophytes," awaiting detailed characterization. These cells were first observed in samples from the deep euphotic zone, although we recognized that their apparent absence from most surface samples was due to the detection limits of our instrument (Chisholm *et al.* 1988). More recently we have been able to measure even the dim cells at the surface by using more sensitive instruments (Olson *et al.* 1990b) (Fig. 7). The flow cytometric signatures of these cells were first recognized as extremely numerous populations with very dim red fluorescence, no orange fluorescence, and light scatter signals even smaller than *Synechococcus*. Even at depths where the flow

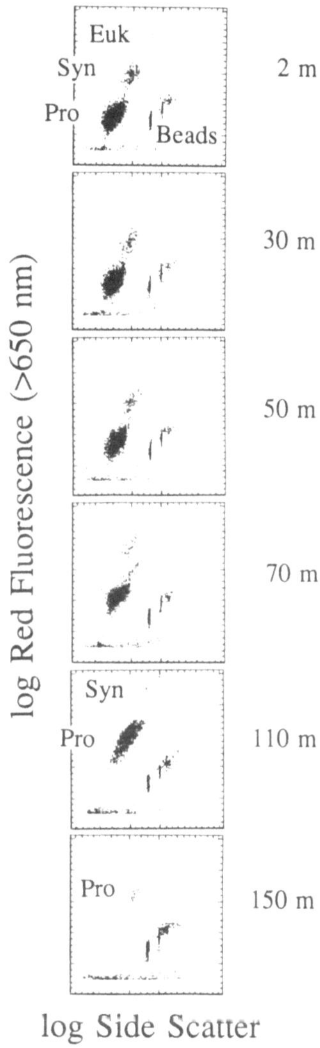

Figure 7. Depth profile showing *Synechococcus*, prochlorophytes, and larger eukaryotic phytoplankton from a station in the Sargasso Sea north of Bermuda (35°15'N, 66°33'W), October 1989. Side scatter and red (Chl) fluorescence are plotted from samples analyzed using a Becton Dickinson FACScan. Uniform fluorescent beads (0.57 μm diameter, Polysciences) were added as internal standards; these are visible as both single beads and doublet peaks. The FACScan is sensitive enough to detect the prochlorophytes in surface waters as well as deep in the euphotic zone at most stations. On occasion, when the surface cells have very little pigment, the Coulter EPICS (modified for high sensitivity) is required to separate the cells completely from the noise (Olson *et al.* 1990b).

cytometer could easily detect these cells, they could not be readily enumerated by epifluorescence microscopy because upon illumination their fluorescence faded within a few seconds (flow cytometric measurements are carried out in a few microseconds). The cells are very abundant (up to 10^5 cells ml^{-1}) and appear at depths where the *Synechococcus* are no longer able to grow. We have found them at nearly all areas of the oceans where we see *Synechococcus* (Chisholm *et al.* 1988; Olson *et al.* 1990b), and their widespread distribution has been more recently documented by others: they have now been observed in the Mediterranean (Vaulot *et al.* 1990), the equatorial Pacific (sample courtesy of N. Price), off Hawaii (unpublished data from samples provided by D. Vaulot), off Southern California (Chisholm *et al.* 1988), in Monterey Bay (unpublished data from samples provided by K. Buck and F. Chavez, MBARI) as well as in the North Atlantic (Chisholm *et al.* 1988; Neveux *et al.* 1989; Li and Wood 1988; Veldhuis and Kraay 1990).

Because culturing these cells eluded us for several years, our initial characterization of them had to be done using natural samples (Chisholm *et al.* 1988). The cells were much less effectively excited by the 515 nm laser line than by the 488 nm line, suggesting the absence of carotenoid accessory pigments which absorb efficiently at 515 nm and which would be expected in non-chlorophytes. Since we knew the cells did not contain phycobiliproteins, we suspected they might contain chlorophyll *b*. The first evidence that these cells were prokaryotes came from transmission electron micrographs of a bulk water sample from 120 m in the Sargasso Sea, where flow cytometry indicated the prochlorophytes were 100-fold more abundant than *Synechococcus* (Chisholm *et al.* 1988). Using flow cytometric analysis to assess the community composition, we were sometimes able to obtain samples highly enriched with prochlorophytes, either through filter fractionation (Fig. 8) or cell sorting, for further analysis. This was possible in part simply because these cells can grow at greater depths than *Synechococcus*. HPLC analysis of the pigments of the prochlorophyte-enriched samples indicated that the cells contained both chl *a* and chl *b*, zeaxanthin, and α-carotene; this, combined with their prokaryotic ultrastructure, is the reason for their designation as prochlorophytes, after Lewin's (1981) original description.

The pigment analyses also revealed that the chlorophylls in these cells were "red-shifted" (Soret absorption is shifted 8-10 nm toward longer wavelengths compared to normal

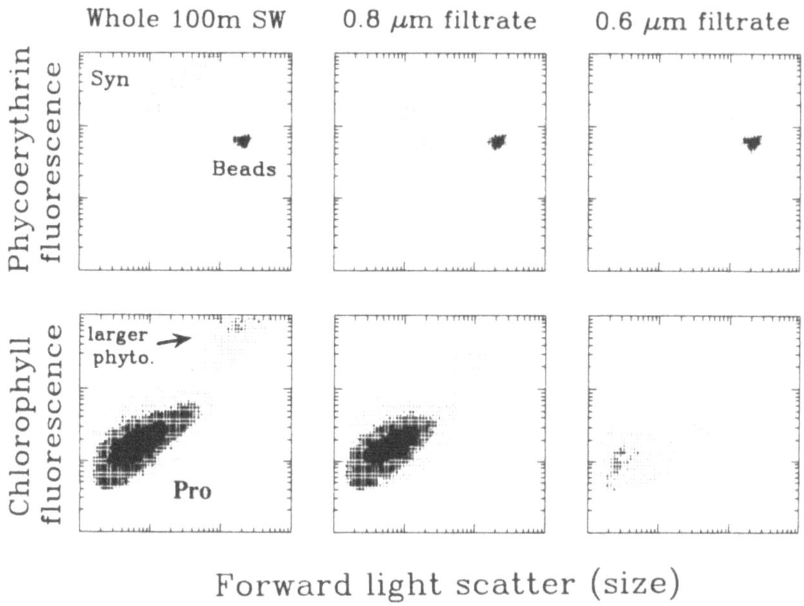

Figure 8. Results of a filter fractionation of water from 100 m in the Caribbean Sea. Water was gravity filtered through different pore-size Nuclepore filters and the filtrates analyzed on the flow cytometer. Forward Light Scatter (FLS) is plotted against Phycoerythrine (PE) and Chlorophyll (CHL) on three decade log scales, with *Synechococcus* gated out of the CHL plots. Most prochlorophytes passed through the 0.8 μm filter, while very few *Synechococcus* and larger phytoplankton did.

chlorophylls), and are therefore almost certainly the same as the unusual chlorophylls documented by others (Gieskes and Kraay 1983, 1986; Vernet 1983) in bulk water samples analyzed by HPLC (see below). Through NMR/MS analysis of pigments extracted from cultures of the cells, it has recently been demonstrated that the red-shifted chlorophylls are divinyl chlorophylls (Goericke and Repeta 1991), which are metabolic intermediates of chlorophyll synthesis in most plants, and are normally present only in trace amounts. Small amounts of a chl-*c*-like pigment have also been found in the cultured cells, as well as α-carotene. Thus these organisms contain a pigment complement which is unique and of evolutionary interest.

The study of prochlorophyte picoplankton exemplifies how individual cell analysis enabled us to better utilize provocative but incomplete information from bulk water measurements. Several sets of HPLC measurements had revealed "unusual" (red-shifted) chlorophyll pigments at high concentrations (sometimes up to half the total chlorophyll) in tropical and subtropical oceans (Gieskes *et al.* 1978, 1988; Gieskes and Kraay 1983, 1986; Vernet 1983; Goericke 1990), and size fractionation experiments (Gieskes and Kraay 1983) indicated the unusual pigment was in the < 1 μm fraction. These pigments are almost certainly the same divinyl chlorophylls that we find in the prochlorophytes, but without the means to see both the cells and their pigments together it was not possible to complete the story. Similarly, what appears to be the same cell as our prochlorophyte was previously reported in bulk water samples examined by transmission electron microscopy (Johnson and Sieburth 1979) as a type of *Synechococcus*. Without the clue given by fluorescence measurements of the living cells, no other interpretation would be suggested. Finally, flow cytometric analysis made possible a rapid and detailed study of the ecology of this organism, which appears to occupy a distinctly different niche from that of its *Synechococcus* relatives (Olson *et al.* 1990b).

Flow cytometric studies of the picoplankton have been facilitated by the recent development of a relatively effective preservation method for these cells, involving fixation with glutaraldehyde and storage in liquid nitrogen (Vaulot *et al.* 1989). Although even this technique is not quantitative for all species or for all parameters measured, it has already allowed detailed studies involving many more samples (e.g., see Fig. 20) than were possible to process when "real time" sample analysis was necessary.

Eukaryotic phytoplankton: The larger eukaryotic phytoplankton have not been studied in the same detail as the picoplankters. This is partly because of sampling limitations and partly because it is difficult to discriminate among the majority of the cell types with flow cytometry, at least compared to traditional microscopic methods. For example, Li and Wood (1988) demonstrated that by using flow cytometric analyses of chlorophyll fluorescence and Coulter volume they could distinguish two kinds of ultraplanktonic (2-5 μm) eukaryotes, whereas analysis of the same samples with epifluorescence microscopy suggested that a dozen or more categories could be distinguished based on the size and shape of the chloroplasts. (Note that flow cytometry of picoplankton also gives us only a relatively crude separation of cell types, but this limitation is less troubling here because the lack of morphological clues in these small cells makes light microscopy even more limiting).

The chlorophyll fluorescence common to all phytoplankton cells is usually roughly proportional to cell size, and the signatures of different populations often overlap. Even in cases where discrete eukaryotic populations can be visualized on 2-parameter plots, they don't have the kind of distinctive characteristics that we see for *Synechococcus* and prochlorophytes; it is not easy to look at a signature and say "those are diatoms" or "those are dinoflagellates" without sorting each sample for confirmation. One exception is the phycoerythrin-containing cryptophyte group. These cells are easy to recognize, but are often not an important component of the phytoplankton, especially in the open ocean. Even in this case one could not be sure that the PE fluorescence represented cryptophytes and not *Synechococcus* cells ingested by grazers.

Flow cytometry allows us to monitor in detail changes in the phytoplankton community, even if we can't identify the members of the community. This facet of flow cytometry can be a valuable adjunct to traditional bottle incubation experiments to measure such oceanographic staples as productivity and photosynthesis vs. irradiance curves. For example, Li (1989) showed that during incubations ranging from 4 h to 2 d, the properties of the phytoplankton changed in terms of both cell numbers and the characteristics of the cells. There were significant differences among phytoplankton categories in how the cells responded to the range of light intensities in the experiment: the smaller eukaryotes were more sensitive to high light intensity than were the larger cells, and *Synechococcus* cell numbers, in contrast to the other phytoplankton, were not affected during the time of the experiments (Fig. 9). These kinds of information are of vital importance in interpreting the results of bottle incubations.

The analyses of eukaryotes discussed thus far concern the "ultraplankton" size class (2-5 μm), where cell numbers are relatively high (10^3 ml^{-1} or higher). Flow cytometry has also been used successfully to study coastal phytoplankton blooms, in which single species can be dominant and abundant. For example, a fish-killing bloom of chrysophytes off the coast of Norway was mapped (Borsheim *et al.* 1989), and the optics of coccolithophore blooms in the Gulf of Maine have been studied using shipboard flow cytometry (P. Holligan, B. Balch, pers. comm.). When we begin looking at the rarer cells in the nanoplankton and larger classes in non-bloom situations, however, flow cytometers run into a sampling problem. This was pointed out by Perry and Porter (1989) in a study of chlorophyll fluorescence as an indicator of phytoplankton light absorption. They noted that in the Sargasso Sea, a very few large cells (which were not well represented in the 0.2 ml samples analyzed by their flow cytometer)

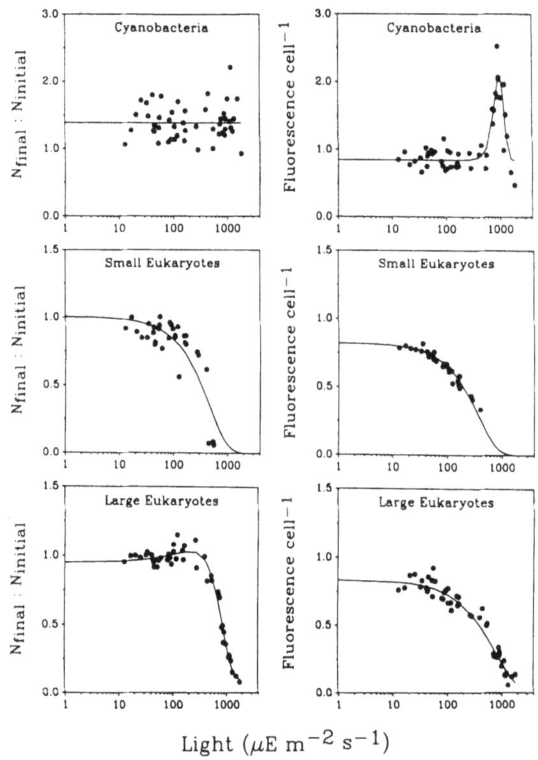

Figure 9. Cell concentration (left) and fluorescence (right) of cyanobacteria, small eukaryotes, and large eukaryotes from 100 m in the Sargasso Sea, analyzed with flow cytometry after a four hour deck incubation at a range of light intensities. All data is normalized to the initial values, so a value of one indicates no change. From Li (1989).

accounted for much of the total fluorescence. One solution to the problem of sampling the larger, rare cells would be preconcentration by filtration or other means, but as yet no concentration technique has been shown to be quantitative and practical. Another solution is to use a larger flow stream and excitation beam in the flow cytometer, to increase the sample throughput. This has been implemented in the Optical Plankton Analyzer (Dubelaar *et al.* 1989), and allows sampling rates of up to several ml min^{-1}. With this flow system it is possible to study very large subjects, such as colonies and filaments of freshwater blue-greens, which are important indicators in field studies of eutrophication (Fig. 10). In this particular application, a combination of chlorophyll fluorescence and time-of-flight through the sensing region was used to discriminate filaments, single cells, and colonies.

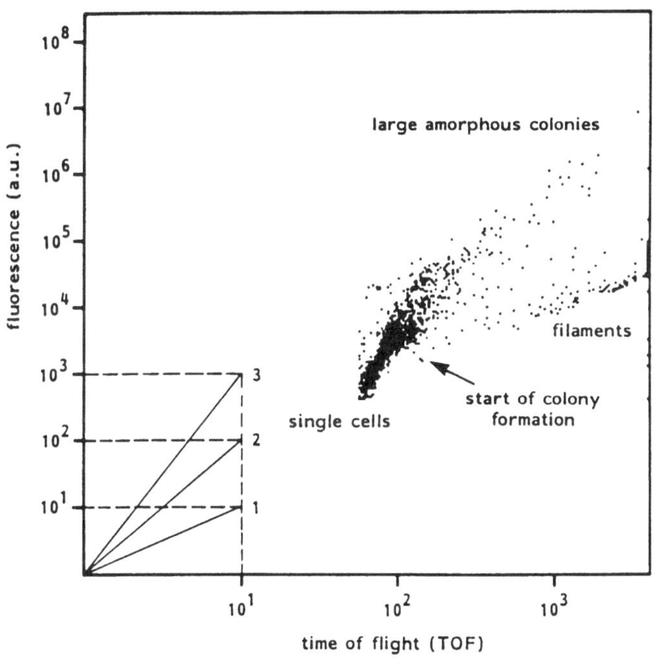

Figure 10. Chlorophyll fluorescence (arbitrary units) plotted against time-of-flight for a sample of river water run on the Optical Plankton Analyser (OPA). Predicted slopes of 1, 2, and 3 for one-, two-, or three-dimensionally shaped particles are indicated for reference. Linear filaments lie on a different slope than spherical cells and colonies. After Dubelaar *et al.* 1989.

We have modified our commercial EPICS flow cytometer in an effort to look at the larger (and less abundant) phytoplankton in open ocean samples. With an analysis rate of about 10 ml min^{-1}, we were able to examine one component of the phytoplankton, the coccolithophores, in the Sargasso Sea where they were present at only about 25 cells ml^{-1} (Fig. 11). Coccolithophores were discriminated from other phytoplankton in their size class by taking advantage of the crystalline nature of the calcium carbonate coccoliths covering these cells;

the polarized laser light is depolarized as it is scattered by these structures (Olson *et al.* 1989). Flow cytometric analysis revealed that not only were these cells most numerous deep in the euphotic zone, but also that their depolarized scatter increased with depth, suggesting that the deep cells were relatively more heavily calcified than those growing near the surface.

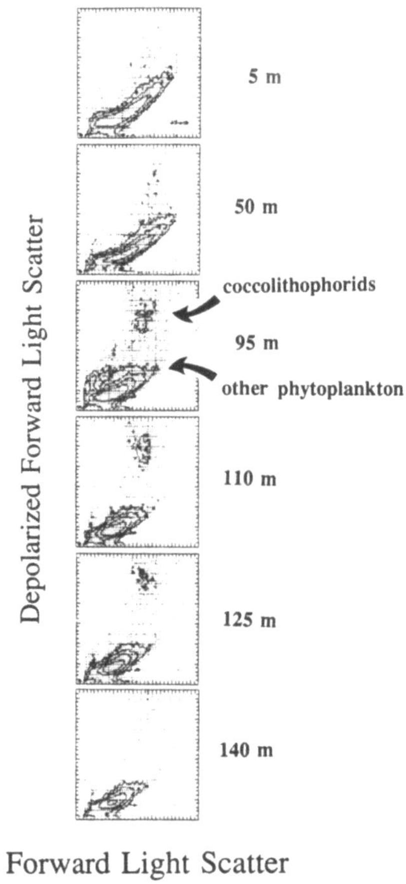

Figure 11. Analysis of the larger phytoplankton (approximately 2-20 μm) on an EPICS V flow cytometer modified to run samples at 10 ml min^{-1}. Depolarized forward light scatter is plotted against forward light scatter (on 3-decade log scales) for all the chlorophyll-containing cells in samples from a depth profile off Bermuda in May 1989. The population of cells which depolarizes more light (which was not distinguishable from other phytoplankton based on autofluorescence alone) was sorted and found to be coccolithophorids.

Note that the depolarized scatter of this population increases with depth, suggesting increased numbers of coccoliths per cell. Microscopic examination of sorted cells indeed revealed multiple layers of coccoliths on some deep cells.

Coccolithophores have been receiving increased attention of late because their coccoliths can drastically alter water reflectance properties during blooms (Holligan *et al.* 1983; Ackleson *et al.* 1988), and the production of dimethyl sulfide by these cells may be important in affecting weather patterns (Charlson *et al.* 1987). Furthermore, their $CaCO_3$ precipitating capabilities could be an important consideration in the global carbon budget. Flow cytometric analysis could greatly facilitate the study of their ecology.

Optics

Particles play an important role in determining the bulk optical properties of seawater, and measurements of the optical properties of individual particles by flow cytometry are of obvious value in this context (Ackleson and Robins 1990). Progress has been made in interpreting flow cytometric measurements of both chlorophyll fluorescence and light scattering by phytoplankton cells with this in mind.

Fluorescence: Chlorophyll fluorescence measurements of bulk water samples are widely used in biological oceanography for studying phytoplankton biomass and physiology. The unique place of flow cytometry in the suite of *in vivo* fluorescence techniques is now being appreciated (Cullen *et al.* 1988), and a biophysical framework within which we can interpret these measurements of individual cells is being constructed (Neale *et al.* 1989; see chapter by T. Owens, this volume). Because fluorescence is highly dependent on instrument characteristics such as time scale and excitation light intensity, and since flow cytometers used in oceanography vary widely in these characteristics, empirical evaluation of flow cytometric fluorescence signals has been a necessary first step.

Sosik *et al.* (1989) analyzed flow cytometric fluorescence measurements from three species of phytoplankton and found that variations in fluorescence/chlorophyll *a* could be accounted for by considering the accessory pigments absorbing the light in each species (Fig. 12). The dinoflagellate in this study (*Amphidinium carteri*) contained more accessory pigments per chlorophyll *a* than either of the other species examined (a diatom and a coccolithophore), so that more of the excitation light was absorbed and passed on to chlorophyll *a*, which then fluoresced. We also suggested that pigment ratios could account for differences in results

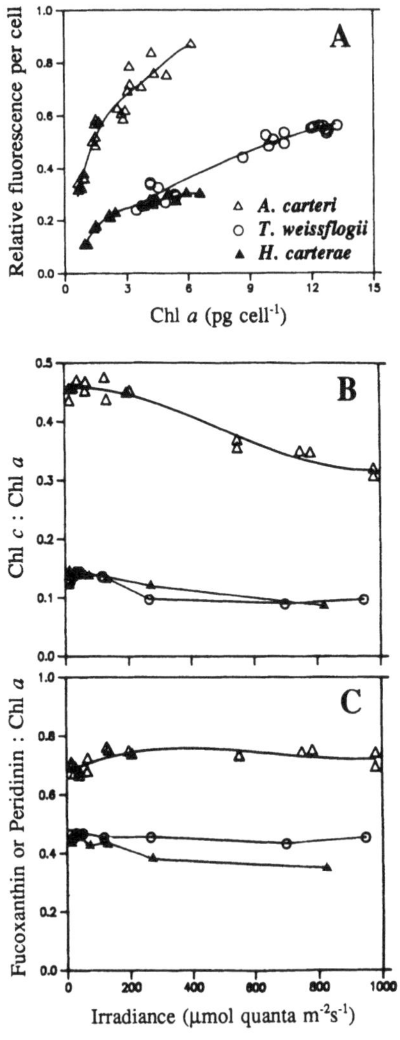

Figure 12. (A) *In vivo* chlorophyll fluorescence per cell measured on an EPICS V flow cytometer vs. extracted Chl *a* per cell. Variations in Chl *a* were achieved by growing batch cultures of three phytoplankton cultures (*Amphidinium carteri*, *Thalassiosira weissflogii*, and *Hymenomonas carterae*) at a range of light intensities. (B) and (C) Molar accessory pigment ratios, as measured by HPLC, for the same phytoplankton species. After Sosik *et al.* 1990.

between the two flow cytometers tested, a laser based EPICS with excitation at 488 nm (primarily absorbed by accessory pigments), and the Hg-arc lamp based "Cytomutt" which excites primarily at 436 nm (absorbed by chlorophyll *a*). The ratio of chlorophyll *c* : chlorophyll *a* increased dramatically in *Amphidinium carteri* grown at the lower end of the range of light intensities, and this trend was reflected in the flow cytometric measurements: 488-nm excited fluorescence/chlorophyll *a* increased more over this range than did that excited by 436 nm.

Perry and Porter (1989) have further shown that fluorescence measurements made by flow cytometry are directly related to bulk absorption measurements made in a spectrophotometer and normalized to a per-cell basis. The cross-section absorption coefficients for individual phytoplankters of a wide variety of species and growth irradiance conditions can be determined quite accurately from flow cytometric measurements of fluorescence (Fig. 13).

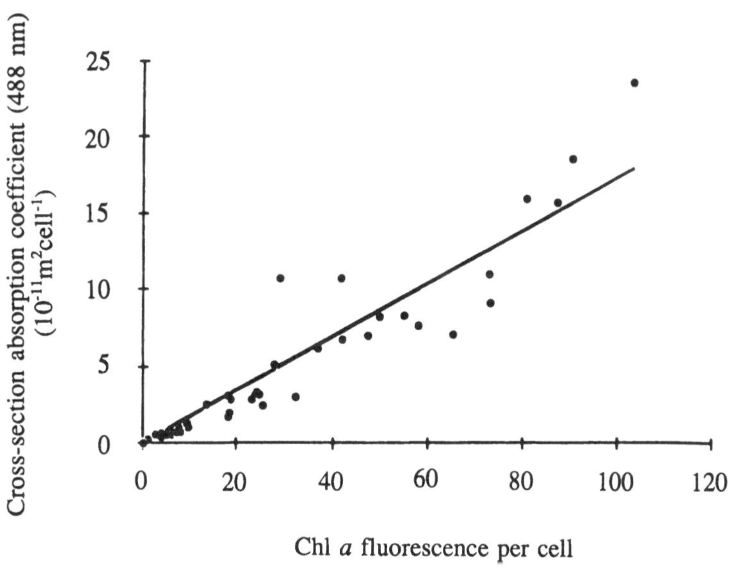

Figure 13. Regression of the cross-section absorption coefficient, measured spectrophotometrically, on the geometric mean Chl *a* fluorescence per cell (measured with a flow cytometer at 488 nm) for a variety of phytoplankton species and growth irradiances. After Perry and Porter 1989.

The flow cytometric measurements in both these studies were made only at single wavelengths in the blue, so it is necessary to make assumptions about absorption of cells at other wavelengths. Perry and Porter (1989) estimated the error caused by extrapolating results from 488 nm to the whole spectrum to be relatively small (on the order of 20%), based on measurements of a variety of cultured phytoplankton species. The use of additional excitation wavelengths might reduce this source of error, as well as allow some information to be gained about the accessory pigment composition of the individual cells (e.g., Olson *et al.* 1989).

Measurements of single-cell fluorescence should be useful in modeling photosynthetic rates in the sea and in studies of bulk water absorption. Effects of nutrient limitation and absorption by photoprotective pigments in cells near the surface remain to be investigated.

Light Scatter: The combination of forward and side scatter signals normally measured in flow cytometry can yield useful information about the size and refractive index of individual particles (see chapter by G. Salzman, this volume). Using polystyrene particles (spherical, homogenous particles with constant refractive index), Ackleson and Spinrad (1988) have shown that measured forward and side scatter values agreed well with differential scattering cross-sections computed from Mie theory. The ratio of forward to side scatter can be used as an indicator of refractive index (Spinrad and Brown 1986), as demonstrated by analyzing droplets of oils with different refractive indices. Ackleson and Spinrad (1988) combined these approaches by plotting forward against side scatter, and comparing the data for measurements of several oil suspensions against Mie computation results assuming either constant particle size and varying refractive index or vice versa (Fig. 14). The measured data for each suspension of oil droplets (which covered a range of sizes) fell on a line parallel to the computed curves for constant refractive index but not to those for constant size. This approach has been used to evaluate the size and refractive index of both cultures (Ackleson and Spinrad 1988) and natural populations of phytoplankton (Ackleson *et al.* 1988), with the calculated results close to expected values. This is encouraging, considering that the calculations assumed spherical cells with constant and homogenous refractive index, which are not true for phytoplankton cells with cell membranes and inclusions such as nuclei and chloroplasts.

Most studies using flow cytometry to measure cell size have utilized either direct measurements of Coulter volume (e.g., Yentsch *et al.* 1986; Li and Wood 1988; Demers *et al.* 1989), or empirically derived relationships between cell size and forward light scatter (FLS). The relatively tight correlations obtained in empirical calibrations (Robertson and Button 1988; Olson *et al.* 1988) suggest that for many cell types, deviations from sphericity and homogeneity may not be of major importance, and that quantitative interpretations of light scatter from field studies may in fact be reasonable. It is worth noting that for phytoplankters that are far from spherical (e.g., pennate diatoms that are long and narrow), distinctive angular light scattering patterns have been observed (Olson *et al.* 1989; Buonnacorsi and Cunningham 1990) that may be useful as discriminating factors, especially if an independent measure of size such as time-of-flight is also available.

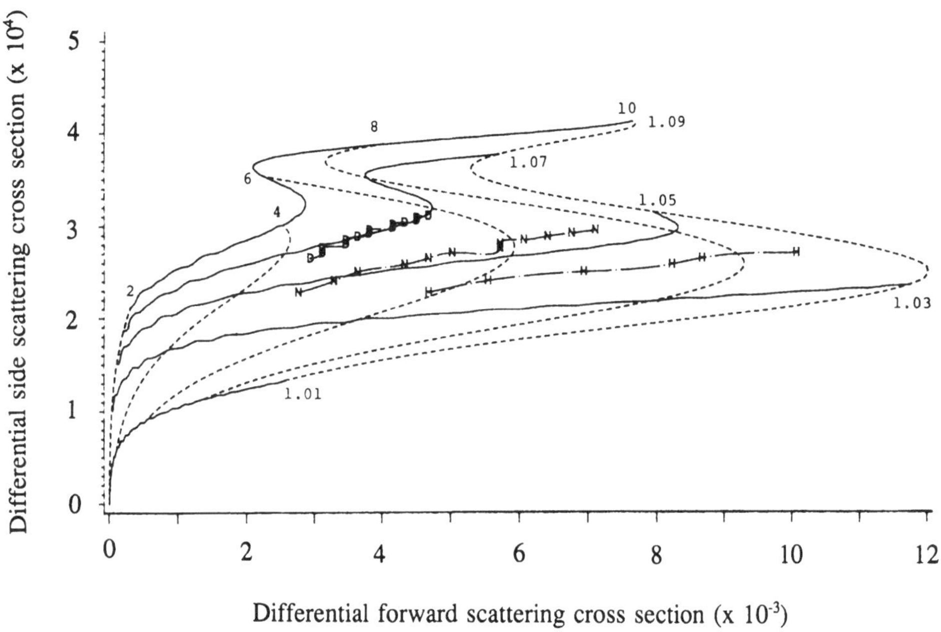

Figure 14. Flow cytometric measurements of differential scattering cross section from three oil suspensions: heptane (H, refractive index = 1.0435); nonane (N, refractive index = 1.0567); and dodecane (D, refractive index = 1.0689). The background grid represents Mie scatter computations where solid lines indicate contours of constant refractive index and dashed lines indicate contours of constant droplet diameter (μm). After Ackleson and Spinrad (1988).

As studies of individual cell optics were taking place in the laboratory, oceanographers were noticing distinctive diel patterns in a bulk water inherent optical property, beam attenuation. Beam attenuation increases during the day and decreases at night. This pattern appears to be well correlated with primary production (Siegel *et al.* 1989; Cullen *et al.* 1991). Since beam attenuation is quick and easy to measure, its potential usefulness is enormous in this context. It is not clear, however, whether the diel patterns in beam attenuation are a result of changes in cell numbers or in cell sizes (or both), and what the relative contributions of various sized particles are to the bulk signal. Flow cytometric analysis of individual cell properties has shed some light on the matter.

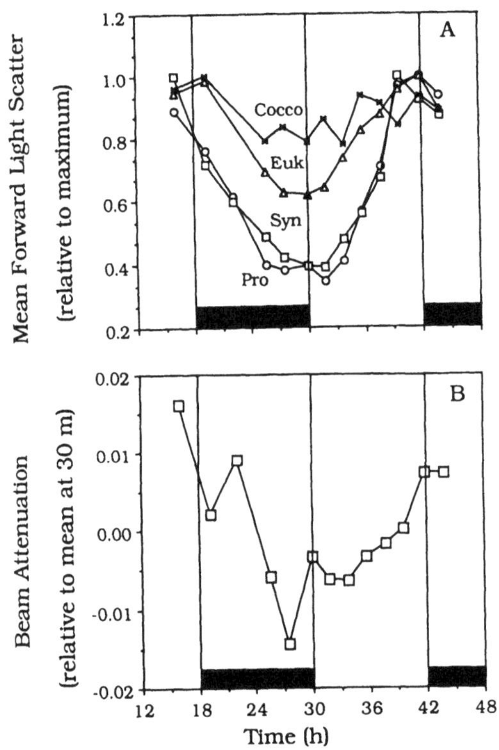

Figure 15. Diel patterns in (A) relative mean forward light scatter of phytoplankton groups as measured by flow cytometry, and (B) beam attenuation, at a station in the Sargasso Sea (October 1989). Results from 30 m are shown: *Synechococcus* (Syn) and prochlorophytes (Pro) were measured using a Becton Dickinson FACScan, and the eukaryotic ultraplankton (Euk) and coccolithophores (Cocco) were measured using a Coulter EPICS V modified to run 50 ml samples. Note that forward light scatter per cell for all the phytoplankton groups analyzed parallels changes in beam attenuation over the diel cycle.

Measurements of light scattering by individual phytoplankters at sea do in fact, show diel patterns in phase with those in beam attenuation (Fig. 15) for both the picoplankton (Olson *et al.* 1990a) and the larger phytoplankton. This evidence, while suggestive, does not rule out the possibility that the beam attenuation patterns are caused by changes in cell concentration,

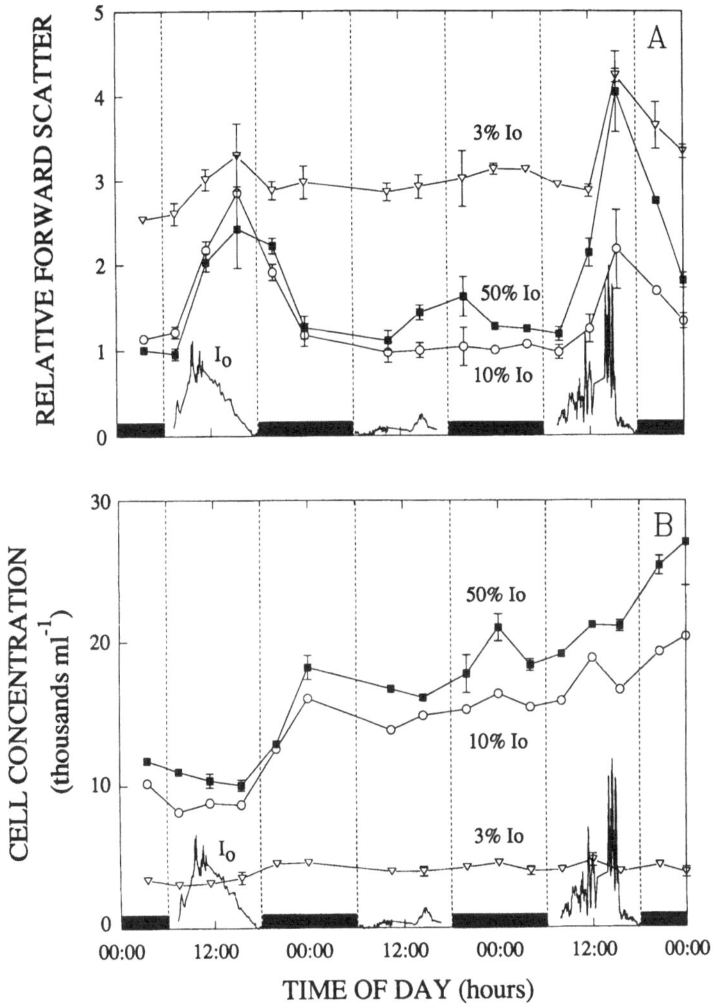

Figure 16. Diel patterns in *Synechococcus* forward light scatter (A) and cell concentration (B) from a bottle incubation experiment in the Sargasso Sea in October 1989. Water samples from three depths representing 3%, 10%, and 50% of the surface light (77, 48, and 15 m, respectively) were incubated under simulated *in situ* conditions on deck, and sampled every four hours for flow cytometric analysis. Relative incident light level (I_0) throughout the incubation is shown at the bottom of each panel. Note that *Synechococcus* FLS per cell and cell concentrations increased more during sunny days than during the cloudy second day, and that the amplitude of the diel changes is a function of light attenuation. From Dusenberry *et al.* (1990).

since we do not know the relative contributions of the large and small cells to beam attenuation. In the diel sampling experiment of Fig. 15, the picoplankton cell abundance varied in the opposite direction from beam attenuation (data not shown), but the data for the larger phytoplankton (which are difficult to sample quantitatively) were too noisy to detect any pattern. It is worth noting that most eukaryotic phytoplankton species divide at night (Chisholm 1981), so we would not expect the concentration of the larger cells to increase during the day, when beam attenuation increases.

A simple explanation for the FLS patterns we observed is that they reflect cell growth during the day as the cells photosynthesize; cell division and respiration at night would then cause the mean FLS per cell to decrease. In a bottle incubation experiment in which *Synechococcus* was monitored (Fig. 16), this explanation is supported by the cell concentration data: the cells increased most in numbers after sunny days 1 and 3, and the FLS increases during the sunny days were also larger than during the intervening cloudy day.

An alternate explanation for the observed patterns in light scatter could be that physiological changes affect the cells' scattering properties without actually changing their biomass. There is some evidence for this in the form of observations of abrupt changes in light scattering on a short time scale when light intensity is changed or a culture is diluted with fresh medium (Ackleson *et al.* 1988). The mechanisms underlying these changes, however, remain to be resolved.

Physiology

Cell growth as reflected in cell size or cell cycle measurements represents an integration of many physiological processes such as photosynthesis, photoacclimation, and nutrition. Several of these processes have been studied using flow cytometric analysis. These can be as simple as measurements of autofluorescence and cell size. For example, Sakshaug *et al.* (1987) used flow cytometry to monitor subpopulations in cultures of the diatom *Thalassiosira pseudonana* (Fig. 17). Upon transfer to very low light, cultures developed bimodal fluorescence distributions, with the appearance of small, low-fluorescence cells. The average fluorescence per cell of the whole population decreased with time in low light, but this was due to an increase in the proportion of low-fluorescence cells; the average fluorescence of the cells in each of the two populations was stable. The low-fluorescence cells were viable, since sorted cells returned to optimum light conditions grew normally. This provides a nice illustration of how simply measuring average properties of cells can be misleading regarding the actual population dynamics.

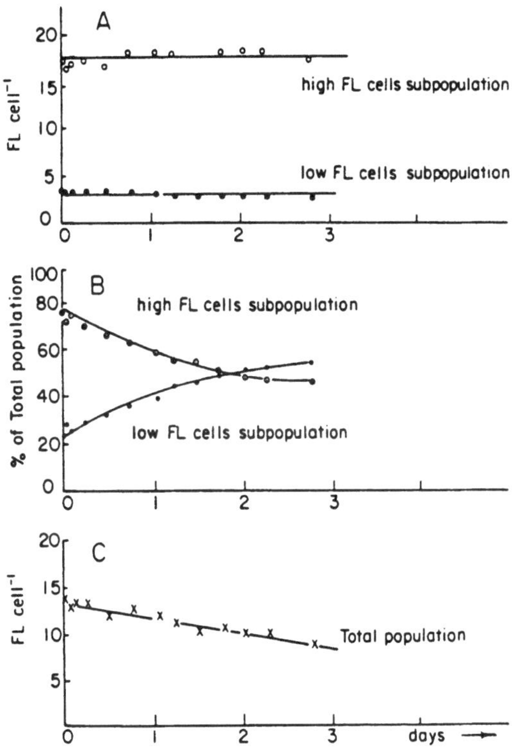

Figure 17. Time course experiment of *Thalassiosira pseudonana* transferred to low light (17 μEin m^{-2} s^{-1}) prior to the experiment. (A) Average fluorescence per cell (relative units) within each subpopulation; (B) relative frequency of each subpopulation; (C) resulting average fluorescence per cell of the total population. From Sakshaug *et al.* (1987).

Differences in cell fluorescence in the field are often equivocal; both light and nutrient limitation can affect chlorophyll and fluorescence levels. Demers *et al.* (1989), using a flow cytometer to look at all chlorophyll-containing particles, observed that phytoplankton from surface water samples had low fluorescence per biovolume compared to cells from the subsurface chlorophyll maximum. To evaluate whether this result arose from high light growth conditions or from nutrient depletion, incubations of both surface and deep samples were carried out at the same intermediate light level, but with different nutrient additions. The results suggested that the original surface samples were nutrient limited, while the samples from the chlorophyll maximum were not (Fig. 18).

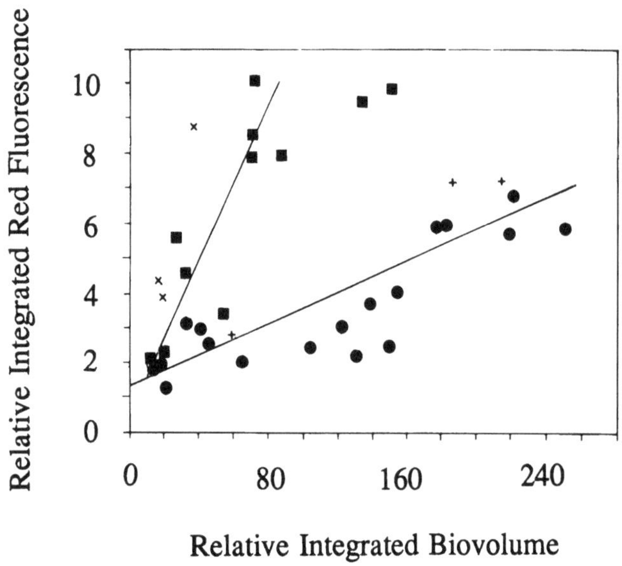

Figure 18. Relationship between integrated red fluorescence and biovolume of chlorophyll-containing cells, as estimated with a flow cytometer equipped with a Coulter-type volume analyzer. The data lie along one of two slopes: initial samples from the surface (+'s) and samples from both the surface and the subsurface chlorophyll maximum that were incubated without added N or P (circles) have relatively low Chl fluorescence per biovolume; initial samples from the subsurface chlorophyll maximum (x's) and all samples incubated with added N and P (squares) have higher Chl per biovolume. After Demers *et al.* (1989).

Knowledge of cell activity is more useful than that of biomass alone. Biomass, as determined by counting and sizing cells with chlorophyll fluorescence, can be misleading because the activity of the cells can vary and all cells may not even be viable. Fluorescein diacetate (FDA) is a fluorescent probe which in combination with flow cytometry offers a quick and relatively easy way to assess cell viability without the use of radioactive isotopes. FDA is lipophilic (thus able to cross the plasma membrane) and non-fluorescent until it is cleaved by intracellular esterases. The resulting fluorescein molecule is fluorescent and hydrophilic (thus it cannot easily pass back through the cell membrane).

Dorsey *et al.* (1989) used flow cytometry to quantify the uptake of FDA by a variety of phytoplankton cultures. All phytoplankton investigated (five classes of eukaryotes and one prokaryotic *Synechococcus*) accumulated FDA and exhibited fluorescein fluorescence which peaked about eight minutes after the addition. FDA activity and photosynthetic activity both correlated with cell volume, as would be expected. Furthermore, in dark release experiments, both the percentage of cells exhibiting FDA activity and the mean fluorescein fluorescence of

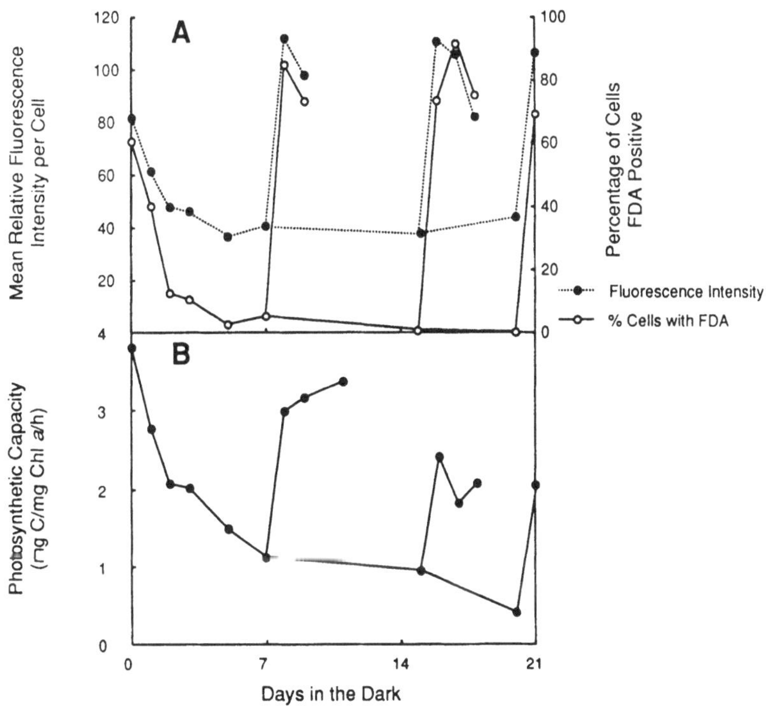

Figure 19. A comparison of FDA activity (A) with photosynthetic capacity (B) in a culture of the prasinophyte Ω48-23 during a dark release experiment. Aliquots from a culture placed in the dark at time 0 were returned to light conditions after 7, 14, and 21 days. Note the similarity between FDA activity and ^{14}C derived photosynthetic capacity of cells released from the dark. From Dorsey *et al.* 1989.

these cells showed responses similar to that of the photosynthetic capacity as determined by ^{14}C uptake experiments (Fig. 19). The green fluorescein fluorescence does not interfere with red chlorophyll fluorescence, so both biomass (from chlorophyll fluorescence) and viability (from FDA activity) can be determined simultaneously on a cell-by-cell basis. The authors point out that FDA could also be useful for examining the metabolic activity of heterotrophic cells, but caution that we still do not understand all of the physiological properties that influence FDA activity.

Fluorescein diacetate is just one example of fluorescent probes being developed for a myriad of molecules within cells (Haugland 1989). As methods are adapted from epifluorescence microscopy to flow cytometry, these probes will make it possible to monitor the levels of other molecules within individual phytoplankton cells. Each new probe allows us to investigate how another cell component or property changes in response to environmental stresses and provides more hints on how cells deal with changing growth conditions.

Future Research Directions

The proliferation of both custom-built and commercial flow cytometers, combined with improvements in technology, continue to expand the scope of individual cell analyses in oceanography. Improvements in our ability to recognize and discriminate among different particles, to make sensitive and reproducible measurements, and to analyze the large amounts of data acquired are growing continually. Discussed below are some of the areas in which these improvements might lead to further advances in the near future.

Physical Processes: The biological responses of phytoplankton cells to changes in their environment can be used to provide information on physical processes such as water column mixing and nutrient supply. If we know how fast cells can respond to changes in light levels, we can get an idea from the depth distribution of their properties how fast they have been moved from one depth to another. A variety of bulk water parameters, such as photosynthesis vs. irradiance characteristics (Falkowski 1983) and carbon:chlorophyll ratio (Cullen and Lewis 1988), have been proposed as potential candidates for investigating vertical mixing. Flow cytometric analyses offer the potential for added specificity since discrete components of the community can be followed, which reduces the potential for interference by changes in the community composition. We have observed, for example, that the shape of the depth distribution of the mean fluorescence of prochlorophytes at a station in the Sargasso Sea changed over a diel cycle (Fig. 20), from a homogeneous distribution in the upper water column (i.e., the cells were being mixed faster than they could photoacclimate) to one in which the surface cells were dimmer and the deeper cells brighter. The timing of this change

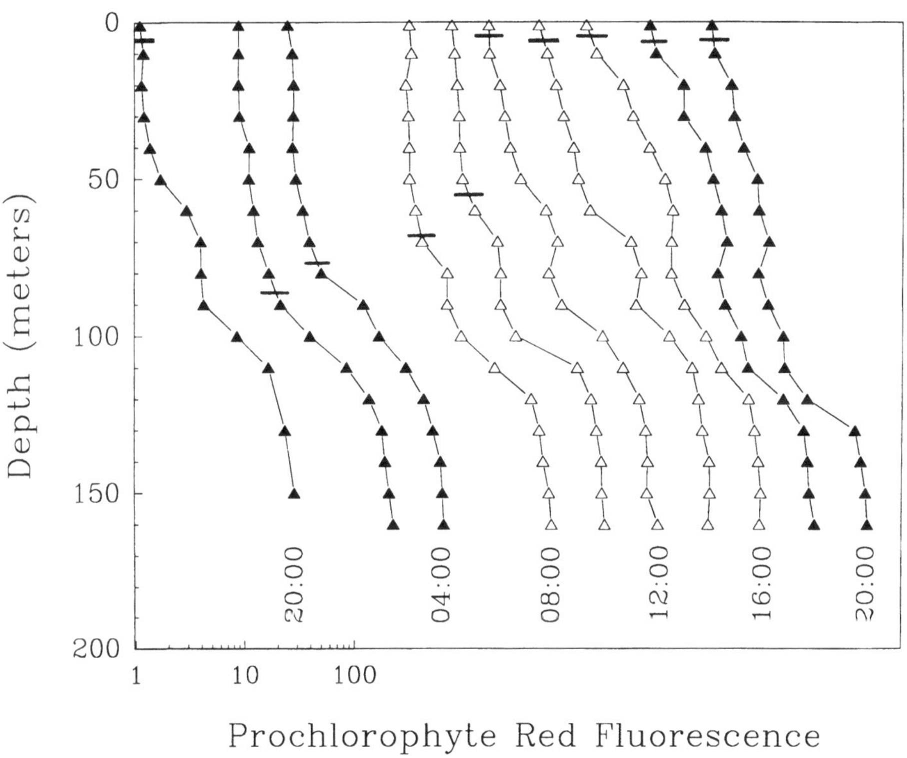

Figure 20. Mean fluorescence intensity of prochlorophytes over a diel cycle at a station in the Sargasso Sea in October 1989. Profiles taken during the dark are indicated with filled symbols. The fluorescence intensity scale is shown for the first profile only; subsequent profiles are offset by a factor proportional to the time between samples. The depth of the mixed layer (defined as the depth at which temperature was 0.02°C less than the surface temperature) is indicated by horizontal bars; it shoaled dramatically during the day due to surface heating. Note the development of a gradient in fluorescence with depth after the mixed layer shoaled, apparently because the cells were able to photoacclimate when they were no longer being rapidly mixed. From Dusenberry *et al.* (1990).

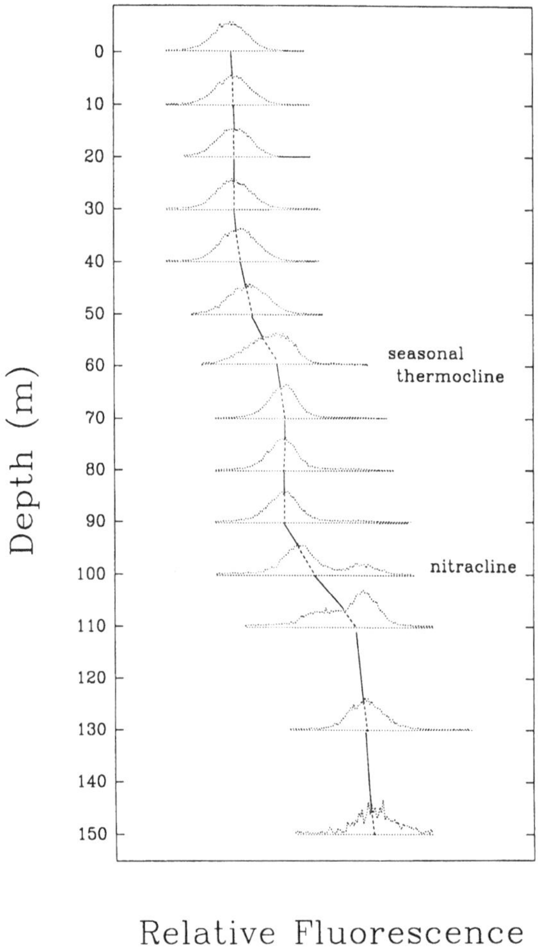

Figure 21. Depth profile of distributions of individual prochlorophyte cell fluorescence (log scale) from the first profile in Figure 20. Note the correspondence between changes in the distributions and water column physical features, e.g. the increased skewness of the distribution at the depth of the seasonal thermocline and the appearance of bimodal distributions at the depth of the nitracline. From Dusenberry *et al.* 1990.

coincided with a shoaling of the mixed layer caused by surface heating, suggesting that cessation of mixing allowed the cells at various depths to photoacclimate to their ambient light levels (Dusenberry *et al.* 1990).

Since the flow cytometer makes measurements on individual cells, we also have access to information contained in the distribution of properties *within* each population. In the example illustrated above, the distributions as well as the means in fluorescence per cell show some dramatic changes with depth, including skewing of the distribution near the seasonal thermocline (Fig. 21). This skewness is predicted to occur at a reflecting boundary layer using a single cell mixing model (Dusenberry *et al.* 1990). We have also seen bimodal distributions in the prochlorophytes near the nitracline, which we hypothesize to reflect a transition from nutrient-limited to light-limited growth. The record of each individual cell's trajectory through its environment represented by these measurements may provide insights into complex processes, such as mixing, that are not possible in any other way.

Bacteria: High sensitivity flow cytometry has been used to characterize the heterotrophic as well as the phototrophic prokaryotes in seawater (Robertson and Button 1989; Button and Robertson 1989; Frankel *et al.* 1990). Since most bacteria have no autofluorescent pigments, this has been done using light scattering and the fluorescent DNA stain DAPI. In some samples, flow cytometric analysis has revealed distinct subpopulations among the bacteria (Button 1990), and there is potential for growth rate information in the DNA distributions obtained. Cell cycle regulation has recently been studied in both heterotrophic marine bacteria (Henderson 1990) and cyanobacteria (Armbrust *et al.* 1989; Binder and Chisholm 1990) with flow cytometry in the laboratory, thus we are beginning to develop the models necessary to interpret field data in terms of growth rates. At present, even simple enumeration of bacteria in seawater samples is typically carried out by epifluorescence microscopy, but the advent of taxa-specific fluorescent probes (see below) should make flow cytometry an increasingly useful tool in marine microbiology.

Probes: Advances in molecular biology are being applied with increasing frequency to oceanographic questions (see chapter by DeLong, this volume). Our knowledge of the marine bacterial community, for example, is very sketchy because most of the cells in a given water sample cannot yet be cultured and the cells cannot easily be distinguished by morphology. Molecular biological probes, however, can allow us to study specific cell types as well as total bacterial biomass, and thus to ask more detailed questions about the distribution and function of different components of the community. Fluorescently labelled antibodies can be used to identify and enumerate specific organisms, or even specific molecules within cells. It is not necessary to isolate the antigen to generate antibodies to a specific organism; the entire organism can be used (Campbell *et al.* 1983), although extensive testing to determine the specificity of the antibody obtained is then required.

Perhaps the best example of the use of fluorescently-labelled antibodies to marine bacteria has been the study of nitrifiers, a restricted class of chemoautotrophs responsible for the regeneration of nitrate from reduced forms of nitrogen (e.g., Ward *et al.* 1982). Studies

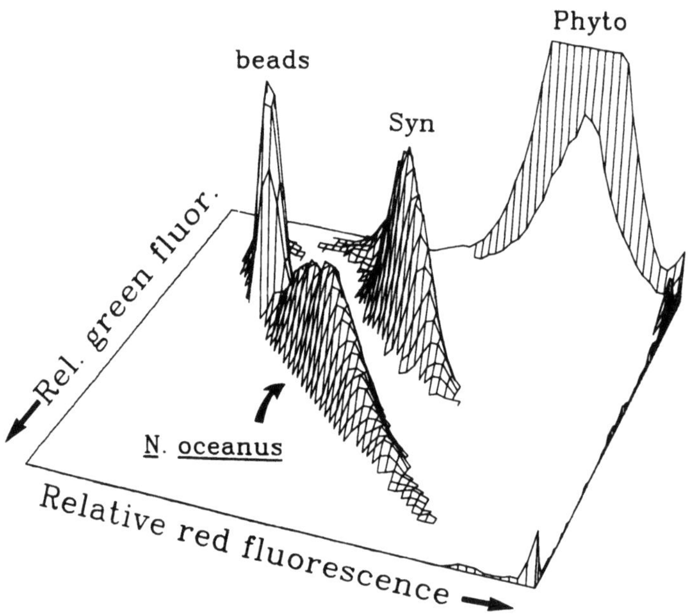

Figure 22. Cultured *Nitrosococcus* cells diluted with Vineyard Sound (Massachusetts) water containing ambient plankton and stained with a species-specific primary antibody (Ward and Carlucci 1985) and a secondary antibody tagged with FITC. The origin of the graph is in the upper left corner. The X and Y axes are 3-decade log scales and represent relative "green" (515-630 nm) and red (660-700 nm) fluorescence. The Z axis is linear and represents number of cells. The tagged *N. oceanus* are easily separated from beads (added as an internal standard), the autofluorescent *Synechococcus* ("Syn," which have both "green"phycoerythrin fluorescence and red chlorophyll fluorescence), and eukaryotic phytoplankton ("Phyto," with chlorophyll alone). From Ward (1990).

utilizing epifluorescence microscopy of filter-concentrated samples showed that these chemoautotrophic organisms are present in seawater in much greater abundance than was reviously estimated by culturing techniques. These cells are still quite rare, (typically 10 cell ml^{-1}), which would make them difficult to sample with present flow cytometers. However, it has been shown that if present in sufficient concentration, immunofluorescently-labelled bacteria could be detected in the presence of autofluorescent phytoplankton by flow cytometry (Ward 1990), as demonstrated by diluting cultured cells of the ammonia oxidizer *Nitrosococcus* into whole seawater (Fig. 22).

The pico- and nanophytoplankters, whose autofluorescence is routinely analyzed by flow cytometry, are also subjects of immunofluorescence studies. Antibodies to several serotypes of *Synechococcus* (Campbell *et al.* 1983; Campbell and Iturriaga 1988) and several small eukaryotic phytoplankters (Campbell *et al.* 1989) have been developed, and it has been shown that flow cytometry is capable of detecting these fluorescently-labelled probes (Campbell 1990). These polyclonal antibody probes, which have been chosen to reflect specificity at roughly the species or genus level, should make it possible to quantify the contributions of various components of the natural phytoplankton assemblage to biomass and productivity.

Oligonucleotide probes (e.g., to rRNA) offer another approach, in which the taxon specificity of the probe can be tailored to the aims of the study, from kingdom down to species. The sequence of the rRNA for the target organisms must be known, but as this database becomes available for more organisms, probes can be custom designed to be as specific or general as desired (Giovannoni *et al.* 1988).

As in immunofluorescence, fluorescently-tagged oligonucleotide probes are usually detected by epifluorescence microscopy (e.g., DeLong *et al.* 1989), but the potential for flow cytometric detection of individual bacteria labelled with rRNA probes has been recently demonstrated (Amann *et al.* 1990). In this study, a probe specific to sulfate reducers was shown to bind to *Desulfovibrio gigas* but not to *E. coli*, while a kingdom-specific probe to eubacteria labelled both species (Fig. 23). It was also possible to enumerate cells of the sulfate reducing species in mixtures of the two cultures. By hybridizing additional probes to different sites on the rRNA molecule, the signal size could be increased. This is an important point since bacteria in nature typically grow at much slower rates (and thus have lower rRNA contents) than in the laboratory cultures tested.

These results are exciting since antibodies and rRNA probes offer researchers a wide range of specificities. These probes may allow us to study populations of cells that cannot easily be identified in any other way (e.g. heterotrophic flagellates or particular species of phytoplankton), and thus allow field studies that could previously only be done with cultures. However, significant problems remain, including the need to preconcentrate cells for the hybridization procedure and the reduction of background fluorescence levels. When cells can

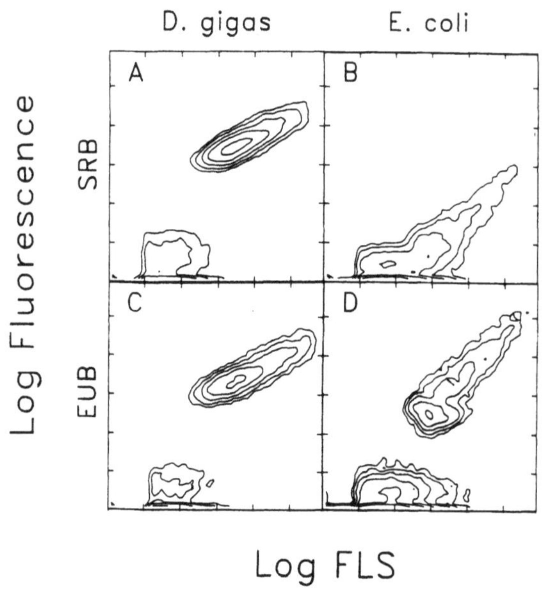

Figure 23. Flow cytometric analysis of the sulfate reducing bacterium *Desulfovibrio gigas* and *E. coli* hybridized with different fluorescently labelled rRNA probes: SRB is genus specific to sulfate reducers (A, B); EUB is kingdom specific to eubacteria (C, D). For each treatment, contour plots of forward light scatter (FLS) vs. fluorescence are plotted on 3-decade log scales. After Amann *et al.* 1990.

be immobilized, as for detection by epifluorescence microscopy, preconcentration can be achieved using filters, and washing steps are relatively simple. Achieving optimum results for flow cytometric samples, which must be suspended, will require methodological development.

Preparative cell sorting: Sorting large numbers of cells for biochemical and other studies remains a tantalizing prospect: obtaining pure samples of chosen subsets of the particles in the ocean would enable us to answer many questions about plankton communities. For example, direct measurements of C:N or C:Chlorophyll ratios in pure samples of phytoplankton uncontaminated by detritus could tell us whether N is limiting growth in the oligotrophic oceans, and stable isotope ratios of selected classes of particles could tell us much about marine food chains. Flow cytometry is a tool with potential for providing these samples, but the sheer numbers of particles required for most analytical techniques is daunting for a sorter that operates on individuals. Consequently, the number of studies in which preparative cell sorting has been utilized is still rather small.

In examples described above, samples were sorted for pigment analyses by fluorescence techniques (Olson *et al.* 1988; Chisholm *et al.* 1988) which can be very sensitive. Similarly, sorting of relatively small numbers of cells can be useful in physiological studies utilizing radioisotopes, though it has been shown that the relatively harsh sorting step must take place at the *end* of the experiment (Rivkin *et al.* 1986). Using this approach, relative uptake rates of ^{14}C-bicarbonate were demonstrated to vary considerably between different constituents of the phytoplankton in a coastal seawater sample (Li 1986). In addition, the first evidence that prochlorophyte picoplankton were actually phototrophic came from ^{14}C sorting experiments (Fig. 24). In another application, preparative cell sorting has been used to "synchronize" phytoplankton cultures growing in continuous conditions by selecting large cells ready to divide (Campbell and Yentsch 1989; Campbell *et al.* 1989; Armbrust *et al.* 1990). Druehl *et al.* (1989) detected differences in the DNA stain fluorescence of male and female brown algal meiospores (previously indistinguishable) by sorting and subsequent cultivation. Subba Rao *et al.* (1991), while investigating the life history and domoic acid production of the diatom *Nitzchia pungens*, went a step further and mixed sorted male and female gametes (distinguished based on light scatter properties) to produce new vegetative cells.

However, for many of the most exciting applications, such as measuring the chemical composition of individual species in the field, the number of cells required is too large to allow sorting to be a practical method, given existing analytical constraints. For example, a typical measurement of particulate N requires a sample of at least 10 μg N (e.g. Lee and

Fuhrman 1987). To analyze a sample of 10-μm diatoms (containing 10^{-5} μg N $cell^{-1}$), then, roughly 10^6 cells are needed. This number of cells can be sorted under optimum conditions from a culture in about an hour using a typical commercial instrument; even a preconcentrated seawater sample would take much longer because the cells of interest are usually only a small

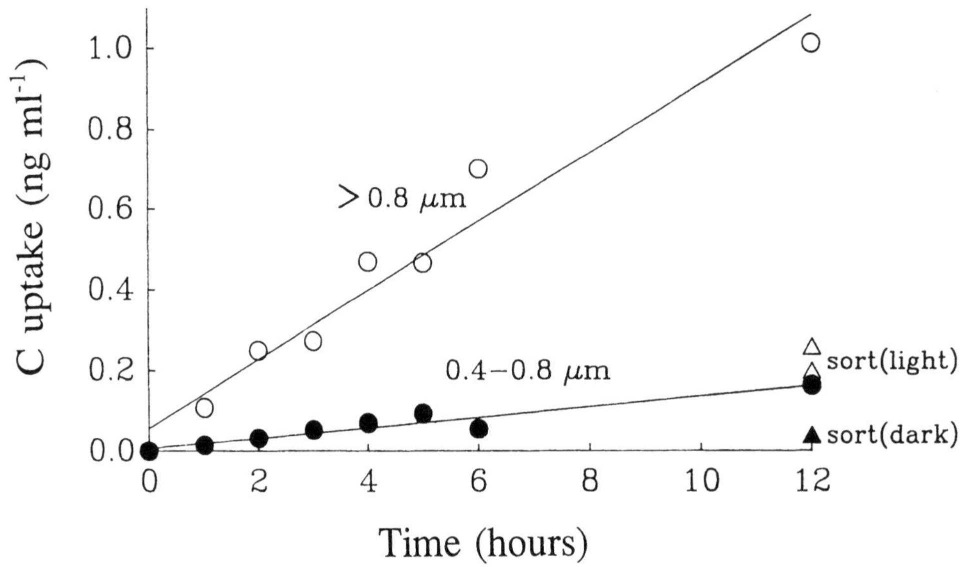

Figure 24. ^{14}C-bicarbonate uptake by filter-fractionated seawater (circles) and sorted prochlorophytes (triangles) from two experiments in the Sargasso Sea (32°45'N 62°01'W) in June 1987. In one experiment, water from the 1 % incident light level was incubated in simulated *in situ* conditions, and fractionated by gravity filtration on Nuclepore membranes (circles). Flow cytometric analysis showed that the phytoplankton in the 0.4-0.8 μm fraction was composed mainly of prochlorophyte cells (95 % of the integrated FLS signal), but that almost half the prochlorophytes were retained on the 0.8 μm filter. The prochlorophytes thus account for a significant fraction of the total production at this depth.

In another experiment at the same station, whole seawater from the same depth was incubated for 12 h, after which the flow cytometer was used to sort pure samples of prochlorophytes from both simulated *in situ* (open triangles) and dark (filled triangle) incubations. The ^{14}C uptake by the sorted prochlorophytes was consistent with that of the filter fractionation experiment. The results suggest an approximate doubling time of 2-3 days for the prochlorophytes under these low-light conditions. Olson *et al.* (unpublished).

fraction of the total particles and the instrument must "look" at all of them during the sorting process. This limits the rate at which a sample can be processed. Sorting enough smaller cells becomes progressively more difficult: 10 μg N of 1-μm *Synechococcus* cells would take hundreds of hours at present sorting rates.

This situation may be changing with improvements in the sensitivity of techniques such as stable isotope analyses. Labelling studies are now being carried out using sorting followed by emission spectroscopy (Lipschultz 1990; P. Wheeler, pers. comm.), and the detection limits for natural abundance mass spectrometry should now be attainable by sorting, at least for some of the larger phytoplankton cells (B. Fry, M. Altabet, pers. comm.).

Data Analysis: Efficient and appropriate analysis of the copious multiparameter data from flow cytometry remains, as always, a challenging area for future research. The amount of data obtained from flow cytometers will continue to increase both in terms of complexity (number of parameters measured) and number of samples. The latter aspect is already being influenced greatly by the introduction of preservation of seawater samples in liquid nitrogen for later analyses of the picoplankton (Vaulot *et al.* 1989).

There is a wide consensus that manual, iterative techniques such as gating and "brushing" (Dean 1990) need to be supplemented with automated techniques for "routine" analyses (e.g., experiments in which the signatures do not change dramatically). Examples of approaches to date include curve-fitting methods (Li 1990) and multivariate analyses such as ordination (Frankel *et al.* 1989) and clustering (Murphy 1985). These techniques are versatile and avoid subjective decisions inherent in interactive analyses such as brushing. Curve-fitting techniques are relatively simple, requiring user-supplied estimates of the number of populations present and the shapes of the distributions in terms of the parameters measured. Ordination techniques use various mathematical methods to derive new axes which can be linear or non-linear combinations of the original variables. The data can be projected onto the new axes, reducing the number of dimensions while preserving the structure of the data. One commonly applied ordination technique, principle component analysis, seeks to derive new, orthogonal (hence uncorrelated) axes which account for most of the variance in the data. Frankel *et al.* (1989) used principal component analysis to analyze seawater run on a flow cytometer and were able to achieve two new principal component axes which separated four of the five particle types in the sample.

Clustering, the other major multivariate analysis technique, seems a logical choice for flow cytometry, since it attempts to divide a set of multivariate observations into similar groups or "clusters." As with ordination techniques, there are many different methods of clustering. Non-hierarchical methods are probably best suited to the large data sets typical of flow cytometry, since they optimize within-group homogeneity and do not require the entire data set to be loaded into computer memory for analysis (Legendre and Legendre 1983).

A promising new technique for flow cytometric analyses is the use of neural net computer systems (Frankel *et al.* 1989, Frankel and Vogt 1990). These computer systems are trained with a known data set, as in discriminant analysis, and can then classify unknown samples quickly and effectively. Neural nets require no mathematical model of the phenomena being studied and make no assumptions about the form or distribution of the data. They can also adapt to non-linear variations such as the increase in phytoplankton cell pigment content with depth, something commonly used multivariate techniques do not handle well. A neural net can be trained to take factors such as depth, season, time of day, and geographical location as well as optical properties into account when classifying particles. Frankel *et al.* (1989) used software versions and achieved near-real-time rates of analysis of marine field samples, so faster hardware implemented versions should be able to process data as it is collected. A similar approach has been used successfully to analyze mixtures of phytoplankton using 6-parameter data from the Optical Plankton Analyzer (Balfoort *et al.* 1990).

CONCLUSION

The study of individual particles in the sea using flow cytometry has changed the way we look at the plankton; this new technology has opened our eyes to organisms and to relationships in the plankton community that were not apparent from traditional bulk analyses. As advances in molecular and cell biology allow us to probe the identity and physiology of individual organisms with increasing specificity, we must develop the capacity to apply these advances to oceanography in meaningful ways. For flow cytometry to play its optimum part in this process, its technical limitations (such as detection of small signals, specificity in the face of a heterogeneous community, and efficient sampling of "rare" particles) must be recognized and addressed.

Even as capabilities for flow cytometric analyses expand, we should recognize that the best way to realize our goals in oceanography entails the combination of complementary approaches, both for single cells and for bulk properties. For example, image analysis (which provides detailed analyses of relatively small numbers of cells) has already been combined with a flow system to study zooplankton (Kachel *et al.* 1990), and the combination confers obvious advantages over either technique alone. An example of the power of coupled bulk and individual cell analyses has already been described, in revealing the "identity" of the prochlorophyte picoplankton using flow cytometry, HPLC and electron microscopy. This story is continuing with analysis of the phylogeny of cultured isolates of this group through molecular biological techniques; strain-specific fluorescent probes from these studies, perhaps coupled with flow cytometric detection, could then allow study of variability among natural populations of related clones of these organisms in the ocean. Individual cell analyses may even play an important role in enhancing our interpretation of satellite images, the ultimate in bulk measurements, for studying the global ocean.

ACKNOWLEDGEMENTS

We thank Ginger Armbrust, Brian Binder, Raffaella Casotti, Michele DuRand, Donald Frankel, Sheila Frankel, Ralf Goericke, Heidi Sosik, Daniel Vaulot, and John Waterbury for valuable discussions and advice. This work was supported in part by NSF grants OCE 88-14332 and OCE 90-12147 (to R.J.O.), OCE 83-16616, OCE 84-21041, OCE 85-08032, OCE 86-14488 (to S.W.C.), by ONR contracts N00014-83-K-0661, 84-C-0278, 87-K-0007 (to R.J.O. and S.W.C.), and by an NSF Graduate Fellowship (to J.A.D.).

REFERENCES

Ackleson SG, Balch WM, Holligan PM (1988) White waters of the Gulf of Maine. Oceanography 1(2):18-22

Ackleson SG, Robins DB (1990) Flow cytometric determinations of North Sea phytoplankton optical properties. Netherlands Journal of Sea Resarch 25:11-18

Ackleson SG, Spinrad RW (1988) Size and refractive index of individual marine particulates: A flow cytometric approach. Applied Optics 27:1270-1277

Amann RI, Binder BJ, Olson RJ, Chisholm SW, Devereux R, Stahl DA (1990) Combination of 16S rRNA-targeted oligonucleotide probes with flow cytometry for analyzing mixed microbial populations. Appl Environ Microbiol 56:1919-1925

Armbrust EV, Chisholm SW, Olson RJ (1990) Role of light and the cell cycle on the induction of spermatogenesis in a centric diatom. J Phycol 26:470-478

Armbrust EV, Bowen JD, Olson RJ, Chisholm SW (1989) Effect of light on the cell cycle of a marine *Synechococcus* strain. Appl Environ Microbiol 55:425-432

Balfoort HW, Smits A, Hofstraat JW, Ringelberg J (1990) Identification of algae: Analyses of optical plankton analyzer data. (Abstract) NATO/ASI on Individual Cell and Particle Analysis in Oceanography, Acquafredda di Maratea, Italy, Oct. 21-30

Binder BJ, Chisholm SW (1990) Relationship between DNA cycle and growth rate in *Synechococcus* sp. strain PC 6301. J Bact 172:2313-2319

Borsheim, KY, Harboe T, Johnsen T, Norland S, Nygaard K (1989) Flow cytometric characterization and enumeration of *Chrysochromulina polylepis* during a bloom along the Norwegian coast. Mar Ecol Progr Ser 54:307-309

Boucher N, Vaulot D, Partensky F (in press) Flow cytometric determination of phytoplankton DNA in cultures and oceanic populations. Mar Ecol Prog Ser

Boyd CM (1978) Perspectives for automated identification and counting. In: Sournia A (ed) Phytoplankton Manual. UNESCO, Paris, pp 216-217

Buonnacorsi G, Cunningham A (1990) Azimuthal inhomogeneity in the forward light scattered from microalgal colonies, and its use as a morphological indicator in flow cytometry. Limnol Oceanogr 35:1170-1175

Burkill PH, Tarran GA, Edwards ES (1990) Flow cytometric analysis of protozoan grazing. (Abstract) NATO/ASI on Individual Cell and Particle Analysis in Oceanography, Acquafredda di Maratea, Italy, Oct. 21-30

Button DK, Robertson BR (1989) Kinetics of bacterial processes in natural aquatic systems based on biomass as determined by high-resolution flow cytometry. Cytometry 10:558-563

Button DK, Robertson BR (1990) Use of flow cytometry in aquatic microbiology: Bacteria. (Abstract) NATO/ASI on Individual Cell and Particle Analysis in Oceanography, Acquafredda di Maratea, Italy, Oct. 21-30

Campbell JW, Yentsch CM (1989) Variance within homogeneous phytoplankton populations, II: Analysis of clonal cultures. Cytometry 10:596-604

Campbell JW, Yentsch CM, Cucci TL (1989) Variance within homogeneous phytoplankton populations, III: Analysis of natural populations. Cytometry 10:605-611

Campbell L, Shapiro LP, Haugen EM, Morris L (1989) Immunochemical approaches to the identification of the ultraplankton: Assets and limitations. In: Cosper EM, Bricelj VM, Carpenter EJ (eds) Novel Phytoplankton Blooms, Lecture Notes on Coastal and Estuarine Studies 35:39-56

Campbell L (1990) Immunochemical approaches to characterization of the picoplankton. (Abstract) NATO/ASI on Individual Cell and Particle Analysis in Oceanography, Acquafredda di Maratea, Italy, Oct. 21-30

Campbell L, Iturriaga R (1988) Identification of *Synechococcus* spp. in the Sargasso Sea by immunofluorescence and fluorescence excitation spectroscopy performed on individual cells. Limnol Oceanogr 35:1196-1201

Campbell L, Carpenter EJ, Iacono VJ (1983) Identification and enumeration of marine chroococcoid cyanobacteria by immunofluorescence. Appl Environ Microbiol 46:553-559
Carpenter EJ, Chang J (1988) Species-specific phytoplankton growth rates via diel DNA synthesis cycles. I. Concept of the method. Mar Ecol Prog Ser 43:105-111
Chang J (1989) Estimating species-specific phytoplankton growth rates by cell cycle analysis. Ph.D. Thesis, State University of New York at Stony Brook, 191 pp
Chang J, Carpenter EJ (1988) Species-specific phytoplankton growth rates via diel DNA synthesis cycles. II. DNA quantification and model verification in the dinoflagellate *Heterocapsa triquetra*. Mar Ecol Progr Ser 44:287-296
Charlson RJ, Lovelock LE, Andrae MO, Warren SG (1987) Oceanic phytoplankton, atmospheric sulfur, cloud albedo and climate. Nature 326:655-661
Chisholm SW, Olson RJ, Zettler ER, Goericke R, Waterbury JB, Welschmeyer NA (1988) A novel free-living prochlorophyte abundant in the oceanic euphotic zone. Nature 334:340-343
Chisholm SW, Armbrust EV, Olson RJ (1986) The individual cell in phytoplankton ecology: Cell cycles and flow cytometry. In: Platt T, Li WKW (eds) Photosynthetic Picoplankton. Can Bull Fish Aquat Sci 214:343-369
Cowles, TJ, Olson RJ, Chisholm SW (1988) Food selection by copepods: Discrimination on the basis of food quality. Mar Biol 100:41-49
Cucci TL, Shumway SE, Brown WS, Newell CR (1989) Using phytoplankton and flow cytometry to analyze grazing by marine organisms. Cytometry 10:659-669
Cucci TL, Shumway SE, Newell RC, Selvin R, Guillard RRL, Yentsch CM (1985) Flow cytometry: A new method for characterization of differential ingestion, digestion and egestion by suspension feeders. Mar Ecol Prog Ser 24:201-204
Cullen JJ, Lewis MR (1988) The kinetics of algal photoadaptation in the context of vertical mixing. J Plankton Res 10:1039-1063
Cullen JJ, Lewis MR, Davis CO, Barber RT (1991) Photosynthetic characteristics and estimated growth rates indicate grazing is the proximate control of primary production in the equatorial Pacific. J Geophys Res
Cullen JJ, Yentsch CM, Cucci TL, MacIntyre HL (1988) Autofluorescence and other optical properties as tools in biological oceanography. Ocean Optics 9, Proc SPIE 925:149-156
Dean PN (1990) Data processing. In: Melamed MR, Lindmo T, Mendelsohn ML (eds) Flow Cytometry and Sorting, 2nd edn. Wiley-Liss, New York, pp 415-444
DeLong EF, Wickham GS, Pace NR (1989) Phylogenetic stains: Ribosomal RNA-based probes for the identification of single cells. Science 243:1360-1363
Demers S, Davis K, Cucci TL (1989) A flow cytometric approach to assessing the environmental and physiological status of phytoplankton. Cytometry 10:644-652
Dorsey J, Yentsch CM, Mayo S, McKenna C (1989) Rapid analytical technique for the assessment of cell metabolic activity in marine microalgae. Cytometry 10:622-628
Druehl LD, Robertson BR, Button DK (1989) Characterizing and sexing laminarialean meiospores by flow cytometry. Marine Biology 101:451-456
Dubelaar GBJ, Groenewegen AC, Stokdijk W, van den Engh GJ, Visser JWM (1989) Optical plankton analyser: A flow cytometer for plankton analysis, II: Specifications. Cytometry 10:529-539

Dusenberry, JA, Chisholm SW, Olson RJ (1990) Characterizing vertical mixing through the use of picophytoplankton as living tracers. (Abstract) NATO/ASI on Individual Cell and Particle Analysis in Oceanography, Acquafredda di Maratea, Italy, Oct. 21-30

Falkowski PG (1983) Light-shade adaptation and vertical mixing of marine phytoplankton: A comparative field study. J Mar Res 41:215-237

Frankel DS, Vogt RF (1990) Neural network FCM analysis of CD4 and CD8 labelled cells. Cytometry Suppl 4:90

Frankel DS, Olson RJ, Frankel SL, Chisholm SW (1989) Use of a neural net computer system for analysis of flow cytometric data of phytoplankton populations. Cytometry 10:540-550

Frankel SL, Binder BJ, Chisholm SW, Shapiro HM (1990) A high-sensitivity flow cytometer for studying picoplankton. Limnol Oceanogr 35:1164-1169

Gerritsen J, Sanders RW, Bradley SW, Porter KG (1987) Individual feeding variability of protozoan and crustacean zooplankton analyzed with flow cytometry. Limnol Oceanogr 32:691-699

Gieskes WW, Kraay GW, Nontji A, Setiapermana D, Sutomo (1988) Monsoonal alternation of a mixed and a layered structure in the phytoplankton of the euphotic zone of the Banda Sea (Indonesia): A mathematical analysis of algal pigment fingerprints. Neth J Sea Res 22:123-137

Gieskes WW, Kraay GW (1986) Floristic and physiological differences between the shallow and the deep nanophytoplankton communities in the euphotic zone of the open tropical Atlantic revealed by HPLC analysis of pigments. Mar Biol 91:567-576

Gieskes WW, Kraay GW (1983) Unknown chlorophyll *a* derivatives in the North Sea and the tropical Atlantic Ocean revealed by HPLC analysis. Limnol Oceanogr 28:757-766

Gieskes WW, Kraay GW, Tijssen SB (1978) Chlorophylls and their degradation products in the deep pigment maximum layer of the tropical North Atlantic. Neth J Sea Res 12:195-204

Giovannoni SJ, DeLong EF, Olsen GJ, Pace NR (1988) Phylogenetic group-specific oligodeoxynucleotide probes for identification of single microbial cells. J Bact 170:720-726

Glover HE, Keller MD, Guillard RRL (1986) Light quality and oceanic ultraphytoplankters. Nature 319:142-143

Glover HE (1985) The physiology and ecology of the marine cyanobacterial genus *Synechococcus*. Adv Aquat Microbiol 3:49-107

Goericke R (1990) Pigments as ecological tracers for the study of the abundance and growth of marine phytoplankton. Ph.D. dissertation, Harvard University, Cambridge, Massachusetts, 418 pp

Goericke R, Repeta (1991) The pigments of *Prochlorococcus minutus*: The presence of chlorophyll a_1 and b_1 in a marine prochlorophyte. Limnology and Oceanography

Haugland RP (1989) Handbook of fluorescent probes and research chemicals. Molecular Probes, Inc., P.O. Box 22010, Eugene OR 97402

Henderson LE (1990) DNA and size distributions in the marine bacterium *Pseudomonas haloduran*. M.S. thesis, MIT, 78 pp

Holligan PM, Viollier M, Harbour DS, Camus P, Champagne-Philippe M (1983) Satellite and ship studies of coccolithophore production along a continental shelf edge. Nature 304:339-342

Iturriaga R, Marra J (1988) Temporal and spatial variability of chroococcoid cyanobacteria *Synechococcus* spp. specific growth rates and their contribution to primary productivity in the Sargasso Sea. Mar Ecol Prog Ser 44:175-181

Johnson PW, Sieburth JM (1979) Chroococcoid cyanobacteria in the sea: A ubiquitous and diverse phototrophic biomass. Limnol Oceanogr 24:928-935

Kachel V, Hueller R, Schmidlechner S (1990) A macro flow planktometer for integrated flow cytometric measurement and imaging of fish larvae and macro plankton. (Abstract) NATO/ASI on Individual Cell and Particle Analysis in Oceanography, Acquafredda di Maratea, Italy, Oct. 21-30

Lavin DP, Fredrickson AG, Srienc F (1990) Flow cytometric measurement of rates of particle uptake from dilute suspensions by a ciliated protozoan. Cytometry 11:875-882

Lee S, Fuhrman JA (1987) Relationships between biovolume and biomass of naturally derived marine bacterioplankton. Appl Environ Microbiol 53:1298-1303

Legendre L, Legendre P (1983) Numerical Ecology. Elsevier Scientific Publishing Company, Amsterdam

Lewin RA (1981) The prochlorophytes. In: Starr MP, Stolp H, Truper HG, Balows A, Schlegel H (eds) The Prokaryotes, vol 1. Springer, Berlin, pp 257-266

Li WKW (1990) Bivariate and trivariate analysis in flow cytometry: Phytoplankton size and fluorescence. Limnol Oceanogr 35:1356-1368

Li WKW (1989) Shipboard analytical flow cytometry of oceanic ultraphytoplankton. Cytometry 10:564-579

Li WKW (1986) Experimental approaches to field measurements: Methods and interpretation. In: Platt T, Li WKW (eds) Photosynthetic Picoplankton, Can Bull Fish Aquat Sci 214:251-286

Li WKW, Wood M (1988) Vertical distribution of North Atlantic ultraphytoplankton: Analysis by flow cytometry and epifluorescence microscopy. Deep-Sea Res 35:1615-1638

Lipschultz F (1990) Scaling of photoadaptations in marine phytoplankton: a species approach. (Abstract) NATO/ASI on Individual Cell and Particle Analysis in Oceanography, Acquafredda di Maratea, Italy, Oct. 21-30

McDuff RE, Chisholm SW (1982) The calculations of *in situ* growth rates of phytoplankton populations from fractions of cells undergoing mitosis: A clarification. Limnol Oceanogr 27:783-788

Melamed MR, Lindmo T, Mendelsohn ML (eds) (1990) Flow cytometry and sorting, 2nd edn. Wiley-Liss, New York

Murphy RF (1985) Automated identification of subpopulations in flow cytometric list mode data using cluster analysis. Cytometry 6:302-309

Neale PJ, Cullen JJ, Yentsch CM (1989) Bio-optical inferences from chlorophyll *a* fluorescence: What kind of fluorescence is measured in flow cytometry? Limnol Oceanogr 34:1739-1748

Neveux JD, Vaulot D, Courties C, Fukai E (1989) Green photosynthetic bacteria associated with the deep chlorophyll maximum of the Sargasso Sea. Comptes Rendus de l'Acadamie des Sciences 308, Serie III:9-14

Newell CR, Shumway SE, Cucci TL, Selvin R (1989) The effects of natural seston particle size and type on feeding rates, feeding selectivity and food resource availability for the mussel *Mytilus edulis* Linnaeus, 1758 at bottom culture sites in Maine. J Shellf Res 8:187-196

Olson RJ, Chisholm SW, Zettler ER, Armbrust EV (1990a) Pigments, size, and distribution of *Synechococcus* in the North Atlantic and Pacific Oceans. Limnol Oceanogr 35:45-58

Olson RJ, Chisholm SW, Zettler ER, Altabet MA, Dusenberry JA (1990b) Spatial and temporal distributions of prochlorophyte picoplankton in the North Atlantic Ocean. Deep-Sea Res 37:1033-1051

Olson RJ, Zettler ER, Anderson OK (1989) Discrimination of eukaryotic phytoplankton cell types from light scatter and autofluorescence properties measured by flow cytometry. Cytometry 10:636-643

Olson RJ, Chisholm SW, Zettler ER, Armbrust EV (1988) Analysis of *Synechococcus* pigment types in the sea using single and dual beam flow cytometry. Deep-Sea Res 35:425-440

Olson RJ, Vaulot D, Chisholm SW (1986) Effects of environmental stresses on the cell cycle of two marine phytoplankton species. Plant Physiol 80:918-925

Olson RJ, Vaulot D, Chisholm SW (1985) Marine phytoplankton distributions measured using shipboard flow cytometry. Deep-Sea Res 32:1273-1280

Olson RJ, Frankel SL, Chisholm SW, Shapiro HM (1983) An inexpensive flow cytometer for the analysis of fluorescence in phytoplankton: Chlorophyll and DNA distributions. J Exp Mar Biol Ecol 68:129-144

Ong LJ, Glazer AN, Waterbury JB (1984) An unusual phycoerythrin from a marine cyanobacterium. Science 224:80-83

Perry MJ, Porter SM (1989) Determination of the cross-section absorption coefficients of individual phytoplankton cells by analytical flow cytometry. Limnol Oceanogr 34:1727-1738

Peters RH (1984) Methods for the study of feeding, grazing and assimilation by zooplankton. In: Downing JA, Rigler FH (eds) A Manual on Methods for the Assessment of Secondary Productivity in Fresh Waters. Blackwell Scientific Publications, Oxford, pp 336-412

Rivkin RB, Phinney DA, Yentsch CM (1986) Effects of flow cytometric analysis and cell sorting on photosynthetic carbon uptake by phytoplankton in cultures and from natural populations. Appl Environ Microbiol 52:935-938

Robertson BR, Button DK (1989) Characterizing aquatic bacteria according to population, cell size, and apparent DNA content by flow cytometry. Cytometry 10:70-76

Sakshaug E, Demers S, Yentsch CM (1987) *Thalassiosira oceanica* and *T. pseudonana*: Two different photoadaptational responses. Mar Ecol Prog Ser 41:275-282

Shapiro HM (1988) Practical flow cytometry, 2nd edn. Alan R. Liss, New York

Sherr BF, Sherr EB, Fallon RD (1987) Use of monodispersed, fluorescently labelled bacteria to estimate *in situ* protozoan bacterivory. Appl Environ Microbiol 53:958-965

Shumway SE, Cucci TL (1987) The effects of the toxic dinoflagellate *Protogonyaulax tamarensis* on the feeding and behaviour of bivalve molluscs. Aquatic Toxicology 10:9-27

Shumway SE, Cucci TL, Newell RC, Yentsch CM (1985) Particle selection, ingestion, and absorption in filter-feeding bivalves. J Exp Mar Biol Ecol 91:77-92

Siegel DA, Dickey TD, Washburn L, Hamilton MK, Mitchell BG (1989) Optical determination of particulate abundance and production variations in the oligotrophic ocean. Deep-Sea Res 36:211-222

Sosik HM, Chisholm SW, Olson RJ (1989) Chlorophyll fluorescence from single cells: Interpretation of flow cytometric signals. Limnol Oceanogr 34:1749-1761

Spinrad RW, Brown JF (1986) Relative real refractive index of marine microorganisms: A technique for flow cytometric estimation. Applied Optics 25:1930-1934

Stoecker DK, Cucci TL, Hulburt EM, Yentsch CM (1986) Selective feeding by *Balanion* sp. (Ciliate : Balanionidae) on phytoplankton that best support its growth. J Exp Mar Biol Ecol 95:113-130

Subba Rao DV, Partensky F, Wohlgeschaffen G, Li WKW (1991) Flow cytometry and microscopy of gametogenesis in *Nitzschia pungens*, a toxic, bloom-forming, marine diatom. J Phycol 27:21-26

Vaulot D, Partensky F, Neveux J, Mantoura RFC, Llewellyn CA (1990) Winter presence of prochlorophytes in surface waters of the northwestern Mediterranean Sea. Limnol Oceanogr 35:1156-1164

Vaulot D, Courties C, Partensky F (1989) A simple method to preserve oceanic phytoplankton for flow cytometric analyses. Cytometry 10:629-635

Vaulot D, Olson RJ, Chisholm SW (1986) Light and dark control of the cell cycle in two marine phytoplankton species. Exp Cell Res 167:38-52

Veldhuis MJW, Kraay GW (1990) Vertical distribution and pigment composition of a picoplanktonic prochlorophyte in the subtropical North Atlantic: A combined study of HPLC-analysis of pigments and flow cytometry. Mar Ecol Prog Ser, in press

Vernet M (1983) Phytoplankton pigments: Methods and ecology. Ph.D. dissertation, University of Washington, Seattle, 159 pp

Ward BB (1990) Immunology in biological oceanography and marine ecology. Oceanography 3(1):30-35

Ward BB, Carlucci AF (1985) Marine ammonia- and nitrite-oxidizing bacteria: Serological diversity determined by immunofluorescence in culture and in the environment. Appl Environ Microbiol 50:194-201

Ward BB, Olson RJ, Perry MJ (1982) Microbial nitrification rates in the primary nitrite maximum off southern California. Deep-Sea Res 29:247-255

Waterbury JB, Watson SW, Valois FW, Franks DG (1986) Biological and ecological characterization of the marine unicellular cyanobacterium *Synechococcus*. In: Platt T, Li WKW (eds) Photosynthetic Picoplankton. Can Bull Fish Aquat Sci 214:71-120

Waterbury JB, Watson SW, Guillard RRL, Brane LE (1979) Widespread occurrence of a unicellular, marine, planktonic, cyanobacterium. Nature 277:293-294

Wood, AM, Horan PK, Muirhead K, Phinney DA, Yentsch CM, Waterbury JB (1985) Discrimination between types of pigments in marine *Synechococcus* spp. by scanning spectroscopy, epifluorescence microscopy, and flow cytometry. Limnol Oceanogr 30:1303-1315

Yentsch CM, Cucci TL, Phinney DA, Topinka JA (1986) Real-time characterization of individual marine particles at sea: Flow cytometry. Lecture notes on Coastal and Estuarine Studies 17:414-448

Yentsch CM, Horan PK, Muirhead K, Dortch Q, Haugen E, Legendre L, Murphy LS, Perry MJ, Phinney DA, Pomponi SA, Spinrad RW, Wood M, Yentsch CS, Zahuranec BJ (1983) Flow cytometry and cell sorting: A technique for analysis and sorting of aquatic particles. Limnol Oceanogr 28:1275-1280

Conclusion

CONCLUDING REMARKS: PROMISES AND LIMITATIONS OF INDIVIDUAL CELL AND PARTICLE ANALYSIS

P. G. Falkowski[1], S. Demers[2], and L. Legendre[3]

[1]Oceanographic and Atmospheric Sciences Division, Brookhaven National Laboratory, Upton, New York 11973, U.S.A.

[2]Maurice Lamontagne Institute, 850 Route de la Mer, Mont-Joli, Québec, Canada, G5H 3Z4

[3]Département de biologie, Université Laval, Québec, Québec, Canada, G1K 7P4

At the present time, there are two philosophical schools of thought addressing the major research questions in biological oceanography. On the one hand, many biological (and chemical) oceanographers are interested in quantifying the major geochemical cycles mediated by biological processes. Up to now, biogeochemists have been attempting to scale up shipboard or *in situ* measurements, using either satellite images and/or large circulation models, in order to assess the importance of specific processes to global geochemical cycles. These researchers have, for the most part, used bulk measurements to reach their objectives. On the other hand, a large number of biologists (and some chemists) are concerned with describing the structure of biological or particle assemblages, and relating these structures to the physical and chemical characteristics of the environment. These scientists are attempting to scale down bulk measurements, and to develop techniques capable of measuring individual cells or particles rapidly and differentially. While biogeochemists and community biologists share some common goals, their approaches differ markedly. We propose to partake, here, our thoughts about the present and potential roles of individual cell and particle analysis in oceanography, and give our opinions about the promises and limitations of this type of analyses in aquatic sciences. Our comments will mainly address biological issues.

The size spectrum of primary producers, as well as the differential rates of primary production and sedimentation, are fundamental attributes of aquatic ecosystems (e.g., Takahashi and

NATO ASI Series, Vol. G 27
Particle Analysis in Oceanography
Edited by S. Demers

Bienfang, 1983; Hopcroft *et al.*, 1990). However, biologists involved in the analysis of individual cells and particles have recently focused their interest on the smaller phytoplankton, as for example cyanobacteria and prochlorophytes (see the review by Olson *et al.*, this volume). Due to technical limitations of modern instruments, the large eukaryotic cells are currently neglected. Yet, as pointed out in a number of studies (e.g., Ryther, 1969; Goldman, 1988; Hopcroft *et al.*, 1990; Legendre, 1990), the larger particles may dominate the export of biogenic carbon into the deep ocean and also toward higher trophic levels. In this perspective, Legendre and Le Fèvre (this volume) developed a conceptual model which stresses the role played by size structure in the biogeochemical cycle of carbon, including the sequestration of carbon in deep waters and sediments, and also in the production of renewable marine resources. Concerning the transfer of primary production to higher trophic levels, Iverson (1990) showed that the production of carnivorous fish is directly proportional to phytoplankton new production, which is often dominated by large cells (e.g., Michaels and Silver, 1988). We therefore believe that a greater effort should be dedicated to studying the large particles, on an individual basis.

A potential catch in the automatic analysis of individual cells and particles is that such techniques are seductive. Even if flow cytometers are technologically sophisticated, they are accessible to many researchers. These instruments are designed to count large numbers of small particles, which is a requisite for many medical applications. In oceanography, researchers are naturally tempted to focus on the large clusters of points corresponding to various types of small cells, since the larger cells are virtually invisible to commercially configured cytometers. However, the small cells may be by far less important, in the carbon and nitrogen economy of the ocean, than the less abundant larger cells. One wonders if the current fascination with picoplankton, which has been a virtual obsession to biological oceanography and limnology over the last decade, is not in great part due to technological biases, resulting in part from the increased use of flow cytometers. Of course, past emphasis on large cells (to the virtual exclusion of anything smaller than ca. 5 μm) was due to biases inherent in optical microscopy, where the larger organisms are identified and counted while the thousands of little green dots in the background remain ignored. Hopefully, a more balanced (i.e., objective) view of the size spectrum of biogenic particles in the ocean will emerge in the next decade, but we feel that this cannot occur unless the limitations of automatic cell analyses are acknowledged.

Phytoplankton and aquatic microbial biologists often spend long, tedious hours looking down a microscope ocular, identifying and counting organisms from natural waters. Devices capable of automating cell counting, and discriminating amongst cell types based on simple pattern recognition signals, potentially provide a means of reducing the tedium, increasing sample throughput, automating and collating the output, and thereby developing a much larger data base than could otherwise be achieved by manual, visual counting. Many of the papers in this volume clearly show that it is technically possible to distinguish, based on cell size, shape, and various other properties, amongst a fairly large number of cell types (e.g. papers in this volume, Ault; DeLong; Olson *et al.*; Salzman *et al.*; Steen). Automatic analysis of individual cells and particles has proven extremely valuable in helping to identify and enumerate new organisms from natural waters (see Olson *et al.*, this volume). However, the flow cytometer is not only a counting tool. The instrument can be used to study the physiological status of organisms as well as various transfer rates, by taking advantage of natural probes such as the autofluorescence of chlorophyll (Owens, this volume) or of various accessory pigments (Gieskes, this volume), or alternatively by addressing fluorochromes, i.e., staining cell compounds such as proteins or DNA with specific fluorescent probes (e.g. DeLong in this volume; see also Table 1 in Legendre and Yentsch, 1989). Thus, automated cell and particle analyses is potentially a powerful tool for understanding biological processes which affect the abundance and fate of phytoplankton and microbes in aquatic ecosystems (Azam and Smith, this volume; Legendre and Yentsch, 1989).

In addition to techniques reviewed in the present volume, some other types of analyses, that have been or are being used in oceanography, are also amenable to single cells or particles (e.g. papers in this volume, Arnd-Jovin; Iturriaga). These concern in particular the elemental composition of particles, which is of interest to geochemists. For example, Biscaye and Olsen (1976) described the use of energy-dispersive X-ray fluorescence in conjunction with scanning electron microscopy (SEM) to analyze metal levels on the surfaces of abiogenic and biogenic particles. This technique is relatively standard with many scanning electron microscopes today. A similar technique, with 5 fm spatial resolution, but much greater sensitivity, has been developed for synchrotron radiation, and this technique has been used to study the spatial distributions of Br, Cu, Fe, and Zn in macroalgae (Keith Jones, personal communication). With increased luminosity, higher spatial scale resolution is achievable. Positron imaging

x-ray emission (PIXE) has been used to examine surface and intracellular concentrations of Ag, Cu, Hg, Zn, Cd, Mg, P, S, Cl, Ar, K, Ca, Fe, Sr, and Pb in *Chara corallina* (Jahnke et al., 1981) *Synechococcus, Oscillatoria, Scenedesmus*, and *Ankistodemus* (MacArthur et al., 1985). Accelerator mass spectrometry (AMS) may be useful for determining long-lived radioisotopic compositions of particles containing as little as a few ng of material (Elmore and Phillips, 1987). These, and other analyses, may prove valuable in helping to elucidate the pathways and fate of sorbed or colloidal materials in the ocean.

In so far as the virtues of individual cell or particle analysis are concerned, the question emerges as to how to abstract useful information from potentially very large data sets. New numerical tools are being developed to specifically deal with flow cytometry data (e.g. Li, 1990; Demers *et al.*, ms). On the other hand, there are attempts, in the present volume, to deal with the problem of scaling from cells or particles up to satellite images (e.g. papers in this volume, Legendre and Le Fèvre; Lewis and Cullen; Holligan and Balch; Morel). This problem is not unique to oceanography, and it plagues terrestrial plant ecologists as well (Osmond, 1989) at such a point that it sometimes completely impedes progress. Satellite images are powerful because they integrate information over large areas. However, attempts to describe the optical properties of the upper ocean as inferred from satellites, using the properties of individual cells or particles, is a bit like resolving whether light is best described as a particle or an electromagnetic wave. Both approaches are valid, but the approach of inferring bulk optical properties from the number and types of individual particles is daunting, perhaps a bit like taking the census of sheep by counting their legs and dividing by four. Instead, it seems that a more useful application of optical or other individual cell properties would be in inferring the effects of the physical or chemical environment from apparent cell responses. Such an approach is exemplified by Dusenbury (pers. comm.), who examined phytoplankton responses to the effects of diurnal heating of the upper mixed layer. This is attractive in that the information is not lost in volumes of data, and also because it would be difficult to obtain the information as cleanly by any other approach. Our suggestion is that analyses of individual cells should be preferably applied to those areas where information about individual cells is important, and not used as sophisticated overkill for bulk property measurements.

It should finally be mentioned that this volume highlights, on the whole, the differences between problem-driven and technology-driven science. A number of papers describe how analytical techniques could be used to resolve problems, while others report research demonstrating the validity of proposed approaches. New technologies always bring new discoveries, which cannot be predicted (e.g., the discovery of prochlorophytes using flow cytometry; see Olson *et al.*, this volume). These are often exciting, precisely because they are unexpected. However, it is problem-driven science which generally appeals to most of the scientific community. In oceanography, however, problem-driven science tends to be staid primarily because oceanographers focus on questions related to "how much" rather than "how." If we, as a scientific community, use the analysis of individual cells and particles as a means of deriving qualitative information about phenomena and processes we know little about, we shall have taken good advantage of the technology. We should not get bogged down in numbers.

ACKNOWLEDGMENT

Paul G. Falkowski is supported by the U.S. Department of Energy, Office of Health and Environmental Research.

REFERENCES

Biscaye PE, Olsen CR (1976) Suspended particulate concentrations and compositions in the New York Bight. In: Middle Atlantic Continental Shelf and the New York Bight. M. G. Gross (ed.) Special Symposia Vol. 2. Amer Soc. Limnol. Oceanogr., 124-137

Demers S, Kim J, Legendre P, Legendre L (ms) Analyzing multivariate flow cytometric data in aquatic sciences

Elmore D, Phillips FM (1987) Accelerator mass spectrometry for measurement of long- lived radioisotopes. Science 236:543-550

Goldman, JC (1988) Spatial and temporal discontinuities of biological processes in pelagic surface waters. In: Rothschild B.J. (ed.) Toward a theory on biological-physical interactions in the world ocean. Kluwer, Dordrecht, 273-296

Hopcroft RR, Roff JC, Berges JA (1990) Size-fractionated sedimentation in a tropical neritic ecosystem near Kingston, Jamaica. Cont Shelf Res 10:795-806

Iverson, RL (1990) Control of marine fish production. Limnol Oceanogr 35:1593-1604

Jahnke A, Shimmen T, Hiroko KI, Toshimitsu Y (1981) PIXE trace element analysis of an aquatic plant as indicator of heavy metal pollution. Chemosphere 10:303-312.

Legendre L (1990) The significance of microalgal blooms for fisheries and for the export of particulate organic carbon in oceans. J Plankton Res 12:681-699

Legendre L, Yentsch CM (1989) Overview of flow cytometry and image analysis in biological oceanography and limnology. Cytometry 10:501-510

Li WKW (1990) Bivariate and trivariate analysis in flow cytometry: Phytoplankton size and fluorescence. Limnol Oceanogr 35:1356-1368

MacArthur JD, Palmer GR, Budd K, Hekman WE, Nichol BE, Casey JR (1985) Element analysis of algae trated with F- or arsenite. Nuclear Inst Meth Phys Res B10/11:653-659

Michaels AF, Silver MW (1988) Primary production, sinking fluxes and the microbial food web. Deep-Sea Res 35:473-490

Osmond CB (1989). Photosynthesis from molecule to the biosphere: A challenge for integration. In W.R. Briggs (ed.) Photyosynthesis, pp. 577. A.R. Liss, New York.

Ryther JH (1969) Photosynthesis and fish production in the sea. Science 166:72-76

Takahashi M, Bienfang PK (1983) Size structure of phytoplankton biomass and photosynthesis in subtropical Hawaiian waters. Mar Biol 7:305-318

List of participants

ORGANIZING COMMITTEE

Demers, Serge (Canada)
Marino, Donato (Italy)
Yentsch, Clarice (U.S.A.)
Lewis, Marlon (U.S.A.)
Chisholm, Sally (U.S.A.)
Burkill, Peter (U.K.)
Vaulot, Daniel (France)
Laflamme, Manon (Canada)

BERMUDA

Lipschultz, Fred

CANADA

Babin, Marcel
Blaise, Christian
Boyd, Carl
Denizeau, Francine
Frenette, Jean-Jacques
Gagnon, Réal
Harrison, Paul J.
Legendre, Louis
Levasseur, Maurice
Li, William
Mazumder, Asit
Partensky, Frederic
Roy, Suzanne
Therriault, Jean-Claude

FINLAND

Leppanen, Juha-Markku

FRANCE

Denis, Michel
Dinet, Marie-Josèphe
Lefevre, Dominique
Maestrini, Serge Y.
Monfort, Patrick
Morel, André
Rassoulzadegan, Fereidoun

GERMANY

Amann Rudolf
Jochem, Frank
Kachel, Volker
Valet, Gunter
Wehrmeyer, Werner

ICELAND

Gudmundsson, Kristinn

ISRAEL

Berman, Thomas
Dubinsky, Zvi
Paran, Nava
Zohary, Tamar

ITALY

Acosta Pomar, Lucila
Benedetti, Alberto
Casotti, Raffaella
Decembrini, Franco
Lazzara, Luigi

Montresor, Marina
Orellana, Monica
Zingone, Adriana

JAPAN

Taguchi, Satoru
Taniguchi, Akira

NORWAY

Gilstad, Mona
Heldal, Mikal
Johnsen, Geir
Sakshaug, Egil
Sandvick, Roar
Steen, Harald

POLAND

Stramsky, Darius

SPAIN

Agusti, Susana
Blanco, Juan
Estrada, Marta
Marrase, Celia
Wyatt, Timothy

THE NETHERLANDS

Balfoort, Hans
Dubelaar, George B.
Gieskes, Winfried
Hofstraat, J.W.
Kraay, Gysbert
Krigsman, Johan
Peeters, Kees
Veldhuis, Marcel

TURKEY

Koray, Tufan

UNITED KINGDOM

Cunningham, Alexander
Happey-Wood, Christine M.
Holligan, Patrick
Medlin, Linda
Taran, Glen

UNITED STATES OF AMERICA

Anderson, Donald
Ault, Kenneth
Azam, Farooq
Bannister, Thomas T.
Binder, Brian
Bruce, Barry D.
Button, D. K.
Campbell, Lisa
Cucci, Terry
De Long, Edward
Dusenberry, Jeffrey
Falkowski, Paul
Frankel, Sheila
Goericke, Ralf
Iturriaga, Rodolfo
Kana, Todd M.
Martin, David
Mazel, Charles
Olson, Robert
Owens, Thomas
Perry, Mary-Jane
Roesler, Collin
Salzman, Gary
Shumway, Sandra E.
Smith, David C.
Sosik, Heidi
Wood, Michelle A.
Yentsch, Charlie

Subject index

NATO ASI Series G

Vol. 1: **Numerical Taxonomy.** Edited by J. Felsenstein. 644 pages. 1983. (out of print)

Vol. 2: **Immunotoxicology.** Edited by P.W. Mullen. 161 pages. 1984.

Vol. 3: **In Vitro Effects of Mineral Dusts.** Edited by E.G. Beck and J. Bignon. 548 pages. 1985.

Vol. 4: **Environmental Impact Assessment, Technology Assessment, and Risk Analysis.** Edited by V.T. Covello, J.L. Mumpower, P.J.M. Stallen, and V.R.R. Uppuluri. 1068 pages.1985.

Vol. 5: **Genetic Differentiation and Dispersal in Plants.** Edited by P. Jacquard, G. Heim, and J. Antonovics. 452 pages. 1985.

Vol. 6: **Chemistry of Multiphase Atmospheric Systems.** Edited by W. Jaeschke. 773 pages. 1986.

Vol. 7: **The Role of Freshwater Outflow in Coastal Marine Ecosystems.** Edited by S. Skreslet. 453 pages. 1986.

Vol. 8: **Stratospheric Ozone Reduction, Solar Ultraviolet Radiation and Plant Life.** Edited by R.C. Worrest and M.M. Caldwell. 374 pages. 1986.

Vol. 9: **Strategies and Advanced Techniques for Marine Pollution Studies: Mediterranean Sea.** Edited by C.S. Giam and H.J.-M. Dou. 475 pages. 1986.

Vol. 10: **Urban Runoff Pollution.** Edited by H.C. Torno, J. Marsalek, and M. Desbordes. 893 pages. 1986.

Vol. 11: **Pest Control: Operations and Systems Analysis in Fruit Fly Management.** Edited by M. Mangel, J.R. Carey, and R.E. Plant. 465 pages. 1986.

Vol. 12: **Mediterranean Marine Avifauna: Population Studies and Conservation.** Edited by MEDMARAVIS and X. Monbailliu. 535 pages. 1986.

Vol. 13: **Taxonomy of Porifera from the N.E. Atlantic and Mediterranean Sea.** Edited by J. Vacelet and N. Boury-Esnault. 332 pages. 1987.

Vol. 14: **Developments in Numerical Ecology.** Edited by P. Legendre and L. Legendre. 585 pages. 1987.

Vol. 15: **Plant Response to Stress. Functional Analysis in Mediterranean Ecosystems.** Edited by J.D. Tenhunen, F.M. Catarino, O.L. Lange, and W.C. Oechel. 668 pages. 1987.

Vol. 16: **Effects of Atmospheric Pollutants on Forests, Wetlands and Agricultural Ecosystems.** Edited by T.C. Hutchinson and K.M. Meema. 652 pages. 1987.

Vol. 17: **Intelligence and Evolutionary Biology.** Edited by H.J. Jerison and I. Jerison. 481 pages. 1988.

Vol. 18: **Safety Assurance for Environmental Introductions of Genetically-Engineered Organisms.** Edited by J. Fiksel and V.T. Covello. 282 pages. 1988.

Vol. 19: **Environmental Stress in Plants. Biochemical and Physiological Mechanisms.** Edited by J.H. Cherry. 369 pages. 1989.

Vol. 20: **Behavioural Mechanisms of Food Selection.** Edited by R.N. Hughes. 886 pages. 1990.

Vol. 21: **Health Related Effects of Phyllosilicates.** Edited by J. Bignon. 462 pages. 1990.

NATO ASI Series G

MIX
Papier aus verantwortungsvollen Quellen
Paper from responsible sources
FSC® C105338

If you have any concerns about our products,
you can contact us on
ProductSafety@springernature.com

In case Publisher is established outside the EU,
the EU authorized representative is:
Springer Nature Customer Service Center GmbH
Europaplatz 3, 69115 Heidelberg, Germany

Printed by Libri Plureos GmbH
in Hamburg, Germany